AF848164

Advances in Wastewater Research

Series Editor

Maulin P. Shah, Department of Research Impact and Outcome, Research and Development Cell, Lovely Professional University, Phagwara, Punjab, India

The Advances in Wastewater Research highlights the prospective future development of wastewater treatment techniques, research and field applications. The series also focuses on the limitations and challenges associated with some generally accepted wastewater strategies and evaluates the possible applications of these corrective strategies to eliminate toxic pollutants from the environment through integrated technologies in wastewater treatment. The series covers the research areas including, but not limited to, toxic pollutants removal from wastewater, methods and procedures, materials, management of wastewater, valorisation possibilities of waste resulting from the removal of toxic pollutants from wastewater, wastewater recycling, environmental impact, and wastewater policies. The series contribute to remove, reduce, or neutralize water contaminants that threaten human health, and ecosystem productivity and integrity. The series includes peer-reviewed monographs, edited volumes, textbooks, and conference proceedings.

Satarupa Dey · Maulin P. Shah
Editors

Microbial Applications in Industrial Wastewater Treatment

Springer

Editors
Satarupa Dey
Department of Botany
Shyampur Siddheswari Mahavidyalaya
Shyampur, West Bengal, India

Maulin P. Shah
Department of Research Impact and Outcome, Research and Development Cell
Lovely Professional University
Phagwara, Punjab, India

ISSN 3004-9725 ISSN 3004-9733 (electronic)
Advances in Wastewater Research
ISBN 978-981-96-9127-2 ISBN 978-981-96-9128-9 (eBook)
https://doi.org/10.1007/978-981-96-9128-9

© The Editor(s) (if applicable) and The Author(s), under exclusive license to Springer Nature Singapore Pte Ltd. 2026, corrected publication 2026

This work is subject to copyright. All rights are solely and exclusively licensed by the Publisher, whether the whole or part of the material is concerned, specifically the rights of translation, reprinting, reuse of illustrations, recitation, broadcasting, reproduction on microfilms or in any other physical way, and transmission or information storage and retrieval, electronic adaptation, computer software, or by similar or dissimilar methodology now known or hereafter developed.
The use of general descriptive names, registered names, trademarks, service marks, etc. in this publication does not imply, even in the absence of a specific statement, that such names are exempt from the relevant protective laws and regulations and therefore free for general use.
The publisher, the authors and the editors are safe to assume that the advice and information in this book are believed to be true and accurate at the date of publication. Neither the publisher nor the authors or the editors give a warranty, expressed or implied, with respect to the material contained herein or for any errors or omissions that may have been made. The publisher remains neutral with regard to jurisdictional claims in published maps and institutional affiliations.

This Springer imprint is published by the registered company Springer Nature Singapore Pte Ltd.
The registered company address is: 152 Beach Road, #21-01/04 Gateway East, Singapore 189721, Singapore

If disposing of this product, please recycle the paper.

This book is dedicated to the passionate researchers, scientists, and environmentalists striving to develop sustainable solutions for wastewater treatment.

We also extend our gratitude to our families and mentors for their unwavering support, encouragement, and guidance throughout this journey.

Preface

With rapid industrialization, both inorganic and organic pollutants have been released in the water bodies. Both these inorganic and organic pollutants can have detrimental effects on public health and biodiversity. This wide plethora of pollutants comprises antibiotic residues, heavy metals, radionuclides, emerging contaminants, and personal care products which are released into surface water, groundwater, and seas through anthropogenic sources.

The efficient removal and degradation of these pollutants are crucial for maintaining our sustainable development goals and prevent environmental and human health hazards. Bioremediation and phytoremediation utilize microorganisms and plants for detoxification, degradation, and immobilization, which offers a promising solution. Additionally, enzyme biotechnology provides a cost-effective, energy-efficient, and environmentally friendly approach. However, bioremediation alone may not be sufficient to eliminate all pollutants. Therefore, integrating physical, chemical, and nanotechnological treatments is often necessary for effective and sustainable pollutant removal. Moreover, the dynamics of microbial communities of this polluted water is unique which may render further insight in wastewater treatment strategies.

This book consists of 16 chapters that explore various aspects of bioremediation of antibiotics, heavy metals, radionuclides, emerging contaminants, personal care products and emergent pollutants, their health hazards, and bioremediation mechanisms. It covers the latest policy aspects and regulation in using wastewater systems. This book serves as a valuable resource and gives an insight of how microalgae and enzymes can be successfully used to treat wastewater. The book serves as a valuable resource for policymakers, researchers, climate change scientists, and students in fields like agriculture, forestry, ecology, and environmental science.

We extend our gratitude to the contributing authors, reviewers, and the Springer Nature team for their support. Special thanks to mentors, teachers, and students for their encouragement, and to colleagues, family, and friends for their unwavering support throughout the preparation of the book.

Shyampur, India Dr. Satarupa Dey
Phagwara, India Dr. Maulin P. Shah

Acknowledgments

We would like to express our heartfelt gratitude and sincere gratitude to all the authors and reviewers for their invaluable contributions to this book. We greatly appreciate their timely submission and insightful revision. We also thank the publishing team at Springer Nature for giving us the opportunity to edit this work. We are deeply grateful to our colleagues for their support and encouragement, which gave us confidence to complete this project. Finally, we extend our profound appreciation to our family members for their unwavering support and continuous encouragement throughout the entire process.

Shyampur, India — Dr. Satarupa Dey
Phagwara, India — Dr. Maulin P. Shah

Contents

Contributors

Lamiaa Ibrahim Ahmed Food Hygiene and Control Department, Faculty of Veterinary Medicine, Cairo University, Giza, Egypt

Shivashis Bikram Banerjee Department of Biotechnology, St Xavier's College (Autonomous), Kolkata, West Bengal, India

Apurba Baruah ICAR-ATARI Zone VI, Guwahati, Kahikuchi, Assam, India

Sayan Bhattacharya School of Ecology and Environment Studies, Nalanda University, Rajgir, Bihar, India

Ashish Bhatt Shri Alpesh N. Patel Post Graduate Institute of Science & Research, Anand, Gujarat, India

Sanjukta Chakraborty Department of Computer Science and Engineering, Seacom Skills University, Santiniketan, Bolpur, Birbhum, West Bengal, India

Siddhartha Das Department of Plant Pathology, M.S. Swaminathan School of Agriculture, Centurion University of Technology and Management, Paralakhemundi, Odisha, India

Udisha Datta PG Department of Botany, Lady Brabourne College, University of Calcutta, Kolkata, India

Satarupa Dey Department of Botany, Shyampur Siddheswari Mahavidyalaya, Ajodhya, Shyampur, Howrah, India

Esha Dutta School of Biotechnology, KIIT Deemed to be University, Bhubaneswar, Odisha, India

Deblina Gangopadhyay School of Forensic Sciences, The West Bengal National University of Juridical Sciences, Kolkata, India

Puspita Ganguly Department of Botany, Bethune College, Kolkata, West Bengal, India

Pranabesh Ghosh School of Life Sciences, Seacom Skills University, Santiniketan, Bolpur, Birbhum, West Bengal, India

Bahadur Singh Gurung School of Ecology and Environment Studies, Nalanda University, Rajgir, Bihar, India

Asif Hossain Department of Zoology, The University of Burdwan, Golapbag, Burdwan, West Bengal, India

Aliza Irfan Department of Biotechnology, Ramaiah University of Applied Sciences, Bangalore, India

Dhruv Kapdi Shri Alpesh N. Patel Post Graduate Institute of Science & Research, Anand, Gujarat, India

Debasish Kar Department of Biotechnology, Ramaiah University of Applied Sciences, Bangalore, India

Jasmine Kaur Department of Applied Biology and Environmental Sciences, Centennial College, Toronto, ON, Canada

Hanan S. Khalefa Veterinary Hygiene and Management Department, Faculty of Veterinary Medicine, Cairo University, Giza, Egypt

Abhishek Konar School of Life Sciences, Seacom Skills University, Santiniketan, Bolpur, Birbhum, West Bengal, India

Chunchu Suchith Kumar Department of Agronomy, School of Agriculture, SR University, Warangal, Telangana, India

Pravin Kumar School of Ecology and Environment Studies, Nalanda University, Rajgir, Bihar, India

Subrat Kumar School of Biotechnology, Kalinga Institute of Industrial Technology (Deemed-to-be-University), Bhubaneswar, Odisha, India

Ghada Abd-Elmonsef Mahmoud Department of Botany and Microbiology, Faculty of Science, Assiut University, Assiut, Egypt

Manisha Mohapatra Department of Biotechnology, Trident Academy of Creative Technology (TACT), Infocity, Chandrasekharpur, Odisha, India

Anjan Mondal Department of Zoology, The University of Burdwan, Golapbag, Burdwan, West Bengal, India

Chhandak Mondal Department of Zoology, The University of Burdwan, Golapbag, Burdwan, West Bengal, India

S. Mounika Department of Agronomy, Professor Jayashankar, Telangana Agricultural University, Hyderabad, Telangana, India

B. B. Nayak School of Sciences, Woxsen University, Telangana, India

Chaitali Patel Shri Alpesh N. Patel Post Graduate Institute of Science & Research, Anand, Gujarat, India

Ritesh Pattnaik School of Biotechnology, Kalinga Institute of Industrial Technology (Deemed-to-be-University), Bhubaneswar, Odisha, India

Gauri Raj Department of Biotechnology, Ramaiah University of Applied Sciences, Bangalore, India

Debleena Roy PG Department of Botany, Lady Brabourne College, University of Calcutta, Kolkata, India

Pranjal Roy School of Biotechnology, KIIT Deemed to be University, Bhubaneswar, Odisha, India

Ushasi Roy PG Department of Botany, Lady Brabourne College, University of Calcutta, Kolkata, India

Pranabesh Sarkar School of Forensic Sciences, The West Bengal National University of Juridical Sciences, Kolkata, India

Sukanya Sasmal School of Forensic Sciences, The West Bengal National University of Juridical Sciences, Kolkata, India

Subhra Satahrada School of Biotechnology, Kalinga Institute of Industrial Technology (Deemed-to-be-University), Bhubaneswar, Odisha, India

Mahi Singh Department of Biotechnology, School of Engineering, IILM University, Greater Noida, UP, India

Meenu Singh Department of Biotechnology, School of Engineering, IILM University, Greater Noida, UP, India

Pankaj Singh Biotechnology Program, Dr. Rammanohar Lohia Avadh University, Ayodhya, Uttar Pradesh, India

Priyanka Singh Department of Biotechnology, School of Engineering, IILM University, Greater Noida, UP, India

Samriddhi Singh Department of Biotechnology, School of Engineering, IILM University, Greater Noida, U.P., India

Sanjana Singh Department of Biotechnology, Trident Academy of Creative Technology (TACT), Infocity, Chandrasekharpur, Odisha, India

Rishit Soni Shri P. M. Patel Institute of Biosciences, Anand, Gujarat, India

R. P. N. Sravya Sree Department of Plant Pathology, M.S. Swaminathan School of Agriculture, Centurion University of Technology and Management, Paralakhemundi, Odisha, India

Bhavya Srivastava School of Forensic Sciences, The West Bengal National University of Juridical Sciences, Kolkata, India

Viral Surti Shri Alpesh N. Patel Post Graduate Institute of Science & Research, Anand, Gujarat, India

Eman M. Taher Food Hygiene and Control Department, Faculty of Veterinary Medicine, Cairo University, Giza, Egypt

Avishek Talukdar Department of Zoology, University of Calcutta, Kolkata, West Bengal, India

Gaurisha Tripathi Department of Biotechnology, Ramaiah University of Applied Sciences, Bangalore, India

Janki Trivedi Shri Alpesh N. Patel Post Graduate Institute of Science & Research, Anand, Gujarat, India

I. A. Zeinab Food Hygiene and Control Department, Faculty of Veterinary Medicine, Cairo University, Giza, Egypt

Chapter 1
Insights on Wastewater Impacts on Animal Health and Production

Ghada Abd-Elmonsef Mahmoud, Eman M. Taher, I. A. Zeinab, Lamiaa Ibrahim Ahmed, and Hanan S. Khalefa

Abstract Water is a critical natural resource for life on earth, as well as one of the most vulnerable ecological niches. Every day, about two million tons of pollutants were released into the water, including sewage, industrial waste, and agricultural waste. This resulted in raising a global concern linked to their potential middle- and long-term impacts on wildlife, aquatic environments, livestock, animals' health, and production. Only about 8% of the least-developed countries carry out wastewater treatment. These practices generate a variety of pollutants, including bacteria, viruses, protozoa, antibiotic-resistant genes, and heavy metals. Animal exposure to wastewater is associated with dangerous animal diseases, which may impede productivity, increase economic losses, and hurt human health. Wastewater pollution diseases are associated with hypertension, proximal renal tubular damage, anemia, cardiac disease, and nervous system damage. Treated wastewater (TWW) was found to have several effects on fish growth rate and survivability. Adult fish exposed to diluted TWW showed markedly higher atretic oocyte production, decreased embryo production, and abnormalities. As a result of these pollutants, aquaculture products may experience convulsions, reduced feeding, slow growth, hypoxia, damaged gills, endocrine and immune system dysfunction, as well as death in serious cases. This chapter provides an overview of the various sources and types of wastewater pollutants, together with the risks they pose to humans and animals and focuses on aquaculture's health and production.

Keywords Wastewater · Environment · Pollutants · Animal disease · Animal production

---

G. A.-E. Mahmoud (✉)
Department of Botany and Microbiology, Faculty of Science, Assiut University, Assiut, Egypt
e-mail: ghadamoukabel@aun.edu.eg

E. M. Taher · I. A. Zeinab · L. I. Ahmed
Food Hygiene and Control Department, Faculty of Veterinary Medicine, Cairo University, Giza, Egypt

H. S. Khalefa
Veterinary Hygiene and Management Department, Faculty of Veterinary Medicine, Cairo University, Giza, Egypt

© The Author(s), under exclusive license to Springer Nature Singapore Pte Ltd. 2026
S. Dey and M. P. Shah (eds.), *Microbial Applications in Industrial Wastewater Treatment*, Advances in Wastewater Research,
https://doi.org/10.1007/978-981-96-9128-9_1

1.1 Introduction

Wastewater is water that has been polluted and contaminated with organic or inorganic substances and microorganisms after being used for any purpose, including industrial, commercial, or agricultural activities, as well as stormwater runoff or "sewer inflow" as defined by Tilley et al. (2014). The 3rd World Water Forum in Kyoto, Japan, revealed that the 2 million tons of pollutants including sewage, industrial waste, and agricultural waste-released into the water every day produce an estimated 1500 km^3 of wastewater every day. This amount is almost equal to the weight of the entire human population on the planet (Geissen et al. 2015; Khater et al. 2024). Globally, 380 billion m^3 of wastewater is produced annually (Qadir et al. 2020). Furthermore, there is a global water shortage, in which millions of people lack access to clean water, and at least 45 countries around the world face moderate-to-severe water scarcity problems. As the population grows, wastewater production increases, which, if unpurified, can contaminate water, soil, and food with pathogens. Treated wastewater is already widely used for agricultural irrigation; roughly 5000 km^2 of land around the globe is irrigated with treated wastewater (Shah 2025; Jiménez and Asano 2008).

Over 2.9 million acres of croplands depend on surface water. They make up 14% of all irrigated crops. These waters likely have a lot of untreated wastewaters. The level is moderate to high (Akhtar et al. 2021). According to Fito and Van Hulle (2021), sewage contributes up to 56% of freshwater withdrawals into streams. Only seventy percent of industrialized nations clean their wastewater and only 8% of people live in the least developed countries. Using untreated wastewater in developing countries has major risks. These risks are to the environment and public health. They occur when suitable regulations are absent. Wastewater contains many organic substances: fats, proteins, acids, carbs, and amino acids. It also has inorganic substances. These are salts and metals, like sulfur, arsenic, sodium, and phosphate. It also has chlorine, bicarbonate, calcium, ammonium, and magnesium salts. Also, organic, and synthetic chemicals and pollutants may move into freshwater via runoff. They can harm aquatic ecosystems (Goswami et al. 2021; Mahmoud 2021; Shah 2024a; Mahmoud et al. 2025).

Crops grown with treated or untreated wastewater Treated wastewater pose severe health risks. They expose humans and livestock to toxic chemicals. Wastewater from slaughterhouses is rich in nutrients like nitrates and phosphates. They cause algae to bloom. These blooms poison fish and other aquatic life by spreading pathogens. Abattoir wastewater must not enter receiving environments. We must reduce pollutants Pollutants. This is critical. It stops eutrophication, soil degradation, and contamination. It also cuts greenhouse gas emissions and disrupts disruptions. Farmers must adopt alternative resources. They need them for sustainable development. They need them to maintain social, economic, and environmental balance. Wastewater reuse is a solution to this challenge. It reduces demand for limited water resources and promotes circular economies. It also has lots of nutrients. This can cut the need for some fertilizer (Akhtar et al. 2021; Bashandy et al. 2022). Treated wastewater Treated wastewater is more suitable than untreated wastewater Treated wastewater. We generally use treated wastewater for non-drinking purposes. These include farming and watering land. They also include refilling the groundwater. They include washing cars, flushing toilets, fighting fires, building, and golf.

1.2 Wastewater Sources

The EPA defines two main types of water pollution. These are point-source pollution and non-point-source pollution. The EPA defines point-source pollution as the direct emission of pollutants. They come from a single and identifiable source. Point pollution comes from farms. They raise many animals, like cows, pigs, and chickens, at high densities (EPA 1996). Also, combined sewer systems treat sewage from streets and storm runoff. They are a major source of pollution. Non-point-source pollution comes from a wide area. There is more than one source for it. This means the pollution does not come from a specific source. Instead, it comes from many sources. It is hard to detect and treat. Farms, cities, and old coal mines are all non-point sources of pollution. During the rainy season, rain collects pollutants from fields and yards. These pollutants include herbicides, insecticides, and fertilizers. They also include oil and animal waste. They also have urban road salt, acids, and noxious elements from abandoned mines. As a result, those pollutants end up in groundwater as well as surface water. Based on their sources and chemicals, we can categorize three types of wastewaters. Researchers have subclassified these types. They are industrial wastewater and household wastewater (sewage/municipal waste). They are also stormwater runoff and agricultural drains (Libutti et al. 2018). For example, trash from cities is a major source of water pollution. Its types include solid waste and liquid waste. One can categorize their chemical composition as organic or inorganic waste. Wastewater has many types of pollutants from different sources. These include chemicals and germs.

1.2.1 Industrial Wastewater

Industrial waste is a waste that has the potential to directly pollute water reservoirs. Mining, electroplating, pulp, and paper, chemical, sugar, distilleries, tanneries, textiles, bleaching and dyeing, food, dairy, and beverage diligences produce a large amount of wastewater worldwide (Ali et al. 2020; Shah 2024b). The nature of wastewater varies according to the type of industry and the materials processed. As an example, wastewater discharged from abattoirs differs from wastewater discharged from wineries; therefore, it is important to understand their chemical characteristics. Overall industrial wastes are rich in organic matter, increasing meat consumption by people makes the meat industry one of the major generators of wastewater. Wastewater from abattoirs contains organic loads derived from a variety of sources. Among other wastewater sources, abattoir effluent has the highest organic load, including COD (8000 mg/L), proteins (70%), and suspended solids (15–30 mg/L) (Kundu et al. 2013). The manufacturing industry contributes more waste than municipal and agricultural operations, especially mining, which produces the most waste around the world. During the last few decades, exponential industrial development has resulted in the discharge of an escalating and diverse assortment of waste products into the environment, which has had a significant impact on water resource quality (Obeng 2010). In the long run, industrial waste can contaminate surrounding water supplies, the air, the soil, and ultimately finish up in the oceans and rivers.

1.2.2 Domestic Wastewater (Urban Activities)

People globally produce municipal or household waste. They come from a range of sectors and human activities. They can be either liquid or solid, which makes it challenging to use them as raw materials. It consists of approximately 99% water and 1% solids. Seventy percent of sewage solids are organic. Thirty percent are inorganic contaminants, like salts, heavy metals, and germs. The germs include viruses, bacteria, fungi, protozoa, parasites, and helminths (Jaramillo and Restrepo 2017; Shah 2024c). Around 1.2 million livestock and poultry farms exist in the USA. Animal waste can move from farmland to nearby lakes, streams, and groundwater. As a result, waste from livestock feedlots goes into water, soil, and the air. This waste may have many negative effects on the environment and human health. Reusing wastewater for crop irrigation spreads diseases. It transmits enteric diseases. Intestinal nematodes and Ascaris/hookworms cause them (Fradette et al. 2022). Also, watering with household wastewater adds nutrients to the soil. These include carbon, macronutrients (nitrogen, phosphorus, and potassium), and inorganic micronutrients. They boost crop yields. However, this disposal method lacks scientific consideration. It can cause major health concerns and land decay over time (Libutti et al. 2018).

1.2.3 Agricultural Wastes

Runoff is the source of agricultural waste. It comes from cultivated fields and animal farms. Farmers have recently used many agrochemicals to grow fruits and vegetables. But this has caused many pollution problems. It can also be toxic to aquatic life. Also, unmanaged stormwater causes many problems. These include pollution with germs. Also, the release of fertilizers, pesticides, and other harmful chemicals into water. The water is near houses and businesses (Lundy et al. 2018). Due to a lack of reliable freshwater, farmers often only use domestic-industrial effluents for irrigation in many places. This results in the introduction of hazardous chemicals and heavy metals into soil and plant systems (Singh 2021). Water from different sources is mixed when transported through conduits, resulting in a wide variety of compositions and

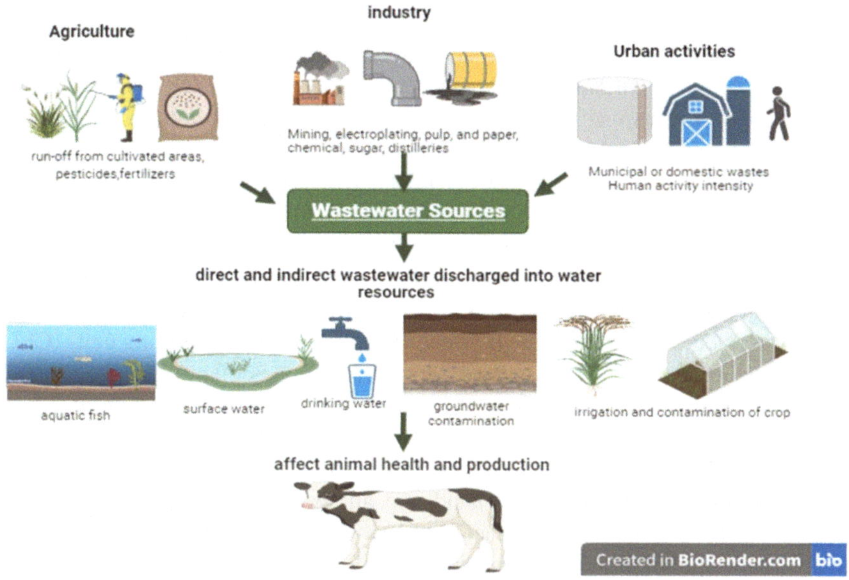

Fig. 1.1 Wastewater sources and its impact on animal health and environment

characteristics. The presence of several chemical contaminants in wastewater poses a serious threat to ecosystems. Heavy metals, emerging organic contaminants (EOCs), nanoparticles, and other chemical toxins such as bleach, salts, and pesticides are among these contaminants (Rai et al. 2019) (Fig. 1.1).

1.3 Wastewater Contaminants

Agricultural practices, industrial emissions, and human activities have a substantial impact on the issue of wastewater contamination. These practices have generated a variety of pollutants, such as biological contaminants like bacteria, viruses, protozoa, and antibiotic resistance genes, as well as non-biological contaminants like heavy metals in wastewater, which have raised global concerns due to their potential medium- and long-term impacts on wildlife, aquatic environments, livestock, animals' health, and production.

1.3.1 Bacteria

Bacteria are the most abundant and diverse pathogens found in wastewater. The most prevalent bacterial species isolated from wastewater treatment plants (WWTPs) are enteric bacteria, which are common indicators of fecal contamination. Many types of bacteria that can make you sick can be found in wastewater. These include coliforms, *Salmonella*, *Shigella*, *Campylobacter*, *Vibrio cholerae*, *Klebsiella* spp., *Yersinia* spp., *Leptospira* spp., *Aeromonas hydrophila*, and *enterococci* (Aghalari et al. 2020; Shah 2024a; Mahmoud et al. 2020). In addition, wastewater is a significant source of *Legionella*, yet only few researches have focused on its incidence and risk in wastewater treatment systems (Caicedo et al. 2016). Moreover, Bonetta et al. (2016) had reported the presence of thermotolerant *Campylobacter*, *Salmonella* spp. and STEC even in treated wastewater.

The natural gut flora of warm-blooded mammals includes genetically diverse bacteria termed *E. coli*. There are between 10^5 and 10^7 colony-forming units (CFU) of this species in 1 L of raw sewage, which is why it is often used to show that feces are contaminated. US-EPA guidelines state that the concentration of *E. coli* in treated wastewater should be less than 100 (MPN; Most Probable Number) per gram (Chahal et al. 2016; Shah 2024b; EPA 1994). Total coliforms, *Salmonella*, and fecal coliforms, on the other hand, are allowed at 1000 MPN/100 mL, not detected at 50 g of the final product, and less than 1000 MPN per gram, respectively. Fluoroquinolones, carbapenems, or extended-spectrum b-lactamases do not kill a wide range of clinically important antimicrobial resistant bacteria (ARB), including methicillin-resistant *Staphylococcus aureus*, vancomycin-resistant *Enterococcus* spp., and Gram-negative bacteria like *Enterobacteria*, *Pseudomonads*, and *Acinetobacter* (Aghalari et al. 2020; ElMokadem et al. 2025). The environmental fate of both ARB and the genetic variants conferring antibiotic resistance is currently of significant interest, particularly since these genes are transferred to and survived in the environmental bacteria, as will be briefly outlined below.

1.3.2 Viruses

Viruses are another type of wastewater contamination. Water that has not been treated may contain up to 10^3–10^7 virus particles per liter. Enteric viruses are the most prominent viral pathogens in wastewater due to their molecular and cellular composition, which renders them resistant to regular water treatment procedures. *Hepatitis A* (HAV), *noroviruses* (NoV), *rotaviruses* (RoV), *adenoviruses* (HAdV), astroviruses, and many other *enteroviruses* (EV) are among them (Chahal et al. 2016). Circoviruses are non-enveloped, single-stranded circular DNA viruses that are heat resistant. *Pico-birnaviridae* are non-enveloped segmented double-stranded RNA viruses that are particularly resistant to UV radiation inactivation. Parvoviruses have single-stranded RNA and can handle high temperatures. Polyomaviruses are heat-stable, non-enveloped double-stranded DNA viruses that are relatively resistant to chlorination. Animal *adenoviruses*, *sapoviruses*, and *hepatitis E* are among the zoonotic viruses that industrial waste from slaughterhouses can spread to wastewater (Omarova et al. 2018). According to US-EPA guidelines, the permissible limits of enteric viruses in treated wastewater should not exceed 1 PFU per 4 g of total dry solids (EPA 1994).

A study by Liu et al. (2021) looked at how common enteric viruses were in untreated wastewater. They found that polyomavirus BK and Nov GII had the highest amounts of DNA and RNA viruses, with more than 5 log10 copies/L. Some enteric viruses, such as RoV, are resistant to UV disinfection. Conversely, encapsulated viruses are more easily disintegrated by wastewater. It is noteworthy that viruses are host-specific, so their abundance in wastewater can be constant or progressively reduced by inactivation in the absence of their hosts. Nonetheless, while WWTPs are effective in reducing viral abundance, the implications related to the reuse of this treated wastewater on various livestock health and production are still limited.

1.3.3 Protozoa

Protozoa are a group of microorganisms that can be up to ten times larger than bacteria and are one of the most important groups of microorganisms in wastewater (Chahal et al. 2016). *Cryptosporidium* spp. and *Giardia duodenalis* are the most common intestinal parasites that are frequently detected in wastewater (Fradette et al. 2022), although their abundance varies by country, environment, and ecosystem. The transmitting stage of Cryptosporidium is an oocyst that does not reproduce and sleeps. *G. duodenalis*, on the other hand, is a flagellated eukaryote with only one cell. Giardia cysts and Cryptosporidium oocysts are very small (1–17 μm) and can get into water treatment systems. They can also survive regular water disinfection (Omarova et al. 2018). Unfortunately, Cryptosporidium and Giardia have no effective medication available that directly kills such parasites in their hosts; hence, such infections can be fatal in immune-compromised cases (Hijjawi et al. 2022; Chahal et al. 2016). *Toxoplasma gondii*, *Entamoeba histolytica*, *Cyclospora cayetanensis*, *Isospora belli*, *Blastocystis hominis*, *Balantidium coli*, *Acanthamoeba* spp., *Sarcocystis* spp., and *Naegleria* spp. are some pathogens that can be spread through water (Omarova et al. 2018). Many studies have been carried out to explore the presence of such protozoa in wastewater, particularly those that compare their abundance in both untreated and treated wastewater. Hijjawi et al. (2022), for example, said that *Cryptosporidium* spp. and *G. duodenalis* assemblages were very common in both wastewater and WWPT in the Middle East and North Africa (MENA) region. Omarova et al. (2018) reported several analogous outcomes. Demonstrating that WWTP could be a major route for such protozoan parasites' transmission, posing an imminent risk to both human and animal health.

1.3.4 Antibiotic Resistance Genes (ARGs)

WWTPs are also important reservoirs and sources of antibiotic-resistant microorganisms and their antibiotic-resistance genes (ARGs) (Hong et al. 2018). ARGs are regarded as emerging wastewater contaminants. Nevertheless, current water reuse legislation and guidelines do not adequately tackle these concerns. According to Hong et al. (2018), ARGs, which enable bacteria to survive in the presence of antibiotics, are categorized into numerous classes based on their mode of action and the antibiotic class they encode resistance for. ARGs are also often found on plasmids, transposons, and integrons, which are types of mobile genetic elements. This has led to the idea that WWTPs are places where horizontal gene transfer (HGT) happens a lot, making it easier for ARGs to spread even more between different types of bacteria (Taher et al. 2020). Schmiege et al. (2021) reported the detection of ESBL-producing *E. coli* in all the samples collected during a wastewater monitoring investigation in Germany. Osińska et al. (2016) investigated the transfer rates of *E. faecalis* resistance plasmids across various strains in activated sludge and indicated that the ARGs located in multidrug-resistance (MDR) plasmids were successfully HGT into an *E. coli* recipient strain, implying that HGT among bacteria in the wastewater environment is even more likely. Emphasizing the urgent need for further scientific understanding to enable appropriate mitigation strategies for ARB and ARGs and the potential risks associated with wastewater reuse in the animal sector.

1.3.5 Heavy Metals

The presence of heavy metals in wastewater has increased in tandem with the growth of industry and human activity. Industrial and municipal effluents are the primary sources of wastewater contamination with heavy metals (Odumbe et al. 2023). Additionally, agricultural activities such as irrigation with wastewater effluent, on which farmers rely mostly due to the lack of reliable freshwater, can also result in heavy metals infiltrating the soil–plant system and groundwater (Singh 2021). The most common heavy metals in wastewater are lead (Pb), zinc (Zn), mercury (Hg), nickel (Ni), cadmium (Cd), copper (Cu), chromium (Cr), arsenic (As), silver (Ag), iron (Fe), manganese (Mn), molybdenum (Mo), boron (B), antimony (Sb), and cobalt (Co) (Singh 2021; Mahmoud 2021). Most of these heavy metals are non-biodegradable and may cause cancer and teratogenicity (Odumbe et al. 2023). Table 1.1 summarizes the WHO and UN acceptable limits for heavy metals in industrial effluent surface water discharges. Despite the detection of certain heavy metals in traces, their presence poses a significant threat to humans, livestock, and ecosystems due to their toxicity effects, bioaccumulation tendencies, and adverse economic impact.

Untreated or inadequately treated heavy metal-contaminated wastewater frequently has adverse health and environmental consequences. Ingested heavy metals tend to accumulate in the vital organs of both humans and animals. Cadmium

Table 1.1 The WHO and UN heavy metals in surface water discharge standards for industrial effluent

Heavy metals	Acceptable limit (mg/L)	
	WHO (2006)	UN (Connor 2017)
Chromium (Cr)	0.05	0.5
Cadmium (Cd)	0.003	–
Copper (Cu)	2.0	–
Nickel (Ni)	0.07	0.5
Zinc (Zn)	1.0	–
Iron (Fe)	2.0	–
Manganese (Mn)	1.0	–
Lead (Pb)	0.01	0.1
Mercury (Hg)	0.006	–

and lead are known to have toxicological effects in even trace amounts; lead and mercury exposure may trigger the emergence of autoimmune disorders. In contrast, while low levels of zinc and copper ions are not very harmful to animals, high doses are extremely toxic. Several studies have detected heavy metal contamination in soil and aquatic ecosystems due to treated wastewater reuse.

1.4 Impact of Wastewater on Animal Health and Production

Exposure to wastewater can result in severe illness in animals. Many water-borne diseases in animals are mild, so farmers do not seek veterinary care. Several other factors, including management (age, education, and farming experience), internal (housing conditions or herd type), and external (climate change, extreme heat, or extreme cold), can also contribute to animal diseases (Sundström et al. 2014). Consequently, rigorous quantitative microbiological studies are required to determine the intensity and specificity of wastewater infections in animals and humans. Animal exposure to wastewater is associated with dangerous animal diseases, which may impede productivity, increase economic losses, and hurt human health. Microbial contamination of surface water may pose a potential health risk when animals are grazing on pastures irrigated with wastewater (Elahi et al. 2017b).

Animals located near wastewater sources or contaminated rivers are more susceptible to pathogenic infections, such as *Cryptosporidium* spp., due to the ease with which they can interact with contaminated water. However, the availability of freshwater and the control of hygiene conditions on well-managed farms may reduce the likelihood of animals contracting diseases. The farm environment, improper management, and direct and indirect access to wastewater may all contribute to the horizontal transmission of pathogens between animals, as in the case of environmental mastitis and FMD (Elahi et al. 2017b). Water quality has a significant impact on animal productivity and efficiency. *E. coli* in wastewater can cause environmental mastitis in dairy animals when directly exposed to water (Elahi et al. 2017a). Sheep, cattle, and young animals are at high risk of Salmonella infection. Humans can easily transmit *Salmonella* to themselves by consuming infected animal products, such as milk from livestock, primarily due to animal waste, through surface water.

Animals can transmit infected meat to other animals by consuming contaminated water, and humans can also contract the infection by consuming contaminated food and water. Inhaling an oocyst accomplishes this. According to Karanis et al. (2007), intensive animal farming areas have significant concentrations of Giardia. Animals may contract this waterborne pathogen, leading to diarrhea. The symptoms of this condition include a loss of appetite and weight, as well as abdominal cramps. In wastewater, 81% of Giardia and 97% of Cryptosporidium are present.

1.4.1 Impact of Wastewater Irrigation on Animal Health on an Indirect Basis

There are several risks associated with wastewater irrigation practices. Depending on the ability of living beings or their interactions with the environment (e.g., pathogenic microorganisms), some contaminants remain in the environment for a short time and exhibit varying levels of toxicity. Due to the level of hygiene and socioeconomic environment in each location, bacterial pathogens and toxic elements may differ. Microorganisms are powerful disease-causing agents for all living creatures, and their effects can last a lifetime. There is evidence that these microorganisms are responsible for significant premature mortality, particularly in developing countries (Junaid et al. 2022). In many developing countries, irrigation plays a pivotal role in ensuring food security. Nevertheless, when things are not controlled, pathogens in wastewater can have several negative effects. These include (a) health problems for people and animals; (b) harmful chemicals in wastewater could affect plants, animals, or people who eat crops; (c) less soil productivity because of some pollutants in wastewater; and (d) the chance that pathogens and organic matter from wastewater could get into aquifers and make them dirty.

In Egypt, treated wastewater (TWW) provides a promising, unconventional water source for irrigation (Abuzaid et al. 2021). Certainly! Wastewater treatment plays a crucial role in maintaining public health and protecting the environment. In Egypt, there are 27 governorates. These governorates are equipped with 415 wastewater treatment plants (WTPs) that are specifically designed to treat municipal wastewater. Currently, these WWTPs collectively handle approximately 4.4 billion cubic meters (BCM) of municipal wastewater annually. However, with the ongoing growth in population and urbanization, we anticipate a significant increase in wastewater generation. The projection indicates that these treatment plants will handle 8.8 BCM of wastewater annually by 2030. There have been several significant concerns regarding the gradual deposition of toxic metals on agricultural land and food crops associated with the use of wastewater for irrigation over a long period, which poses a significant risk to populations dependent on water and pathogenic microorganisms (Zhang and Shen 2019).

Animals are especially susceptible to exposure to these harmful metals through the consumption of contaminated water and vegetation. Although certain elements, such as iron (Fe), copper (Cu), zinc (Zn), and manganese (Mn), are essential for animal growth and development, their excess presence in feed can negatively affect yields as well as development processes in general (Odumbe et al. 2023). Considering that animal-derived products are part of the food chain for humans, zinc toxicity in livestock tissues is ultimately harmful to those who consume them. Abuzaid et al. (2021) say that the Nile Delta in Egypt has been irrigated with wastewater for a long time. Specifically, agricultural drainage water and mixed wastewater have built up six potentially toxic metals (PTMs) in the alluvial soil, berseem plants, and groundwater compared to irrigation with fresh water from the Nile. When cattle consume wastewater-contaminated forage, they may face health- or growth-related complications. However, in certain regions of developing countries where water scarcity is prevalent, cattle not only rely on wastewater-cultivated forage but also have access to it as a drinking source. Although it is not the primary mode of transmission, certain protozoa have the potential to infect animals if they persist in irrigated crops. Individuals who consume meat from cattle grazing on wastewater-irrigated fields or who feed crops from such fields may potentially contract the beef tapeworm (Taenia saginata), according to limited evidence (Stevens et al. 2021). Despite this, there is compelling evidence that cattle grazing on fields recently irrigated with untreated wastewater or drinking water from untreated wastewater canals or ponds in Melbourne, Australia, and Denmark may be at risk of severe infection by Taenia, leading to cysticercosis. No reported issues have been associated with the consumption of wastewater-irrigated fodder. However, irrigating fodder with high levels of nitrogen may result in grass tetany, a condition caused by an imbalance of nitrogen, potassium, and magnesium in pasture grasses. Even at lower levels that do not affect plants, cadmium can be harmful to animals due to its storage in the kidneys and liver, which does not affect meat or milk products (Grzegorczyk et al. 2020). Similarly, copper, although not visibly affecting plants at low concentrations, can be detrimental to ruminant animals (such as cows and sheep) while not affecting monogastric animals (Odumbe et al. 2023). Cattle, particularly ruminants, experience toxicity when consuming crops containing more than five milligrams of molybdenum per kilogram of feed. The toxicity is linked to the consumption of copper and sulfates.

1.4.2 Impact of Treated Wastewater (TWW) on Aquaculture and Human Health

Fish products have gained popularity due to their high protein content and lower cost when compared with meat. Despite this, water shortages continue to be a major limiting factor for aquaculture growth on a global scale. Because of the scarcity of freshwater resources, the search for alternative water resources, such as treated wastewater (TWW), has become an upward strategy. Many countries have adopted the practice of reusing wastewater in aquaculture systems since the first half of the twentieth century. Researchers have experimentally investigated TWW-fed aquaculture in several countries around the world, including Egypt, the USA, Ghana, and Peru. Even so, TWW has been used for aquaculture less frequently in recent years and, in many instances, has ceased to be used. This is reflected in the limited research conducted on the topic. Only a small number of applications have been made due to a limited scientific understanding of the potential health risks and sociocultural barriers associated with TWW reuse in aquaculture (Zaibel et al. 2022).

1.4.3 Growth Rate and Survivability of Fish Cultivated in TWW

In TWW, water quality should meet the recommended criteria for optimal fish health and growth. There has been evidence that fish grown in TWW survive similarly to those in control groups (Zaibel et al. 2020). According to Lan et al. (2007), there were no significant differences in the number of presumptive thermotolerant coliforms between fish farmed in wastewater and non-wastewater-fed fishponds. Furthermore, fish cultured in TWW showed higher final body weights, specific growth rates, and weight gains compared to the controls or were like the controls (Zaibel et al. 2020).

1.4.4 Impacts of Organic Micropollutants (OMPs) on Fish Health

According to Jaramillo and Restrepo (2017), TWW in municipal areas may contain various OMPs that pose potential health and ecological hazards at trace levels. Fish raised in TWW-fed aquaculture are known to accumulate a wide range of OMPs, with the reproductive system bearing the primary health risks associated with their consumption. The study observed an intersex condition where male testicles contained female oocytes. As compared to males upstream of the WWTP discharge point, they also detected higher levels of plasma vitellogenin (VTG), a substance normally detected at low levels in males or immature females. Several studies have evaluated the effects of TWW-derived OMPs on wild fish populations following this pioneering study (Chen et al. 2016).

Furthermore, we found that TWW containing a variety of pharmaceuticals induces stress behavior in fish. A total of 49 OMPs have been evaluated for bioaccumulation in silver perch muscle (*Bidyanus bidyanus*) grown in tertiary-treated wastewater (TTWW). Among the 49 OMPs analyzed, twenty were found at various levels in fish muscle, including carbamazepine (0.1–4.4 g/kg), triclosan (0.1–2.72 g/kg), and bisphenol (0.5–6.22 g/kg). A laboratory study examined the bioaccumulation of pharmaceuticals, hormones, and steroids in the muscle and liver of common carp cultured in TTWW for five months. Zaibel et al. (2020) detected only four out of the 19 compounds tested. Carbamazepine (0.466 g/kg; $n = 12/12$); diclofenac ($n = 9/12$; mean 0.17 g/kg); and diphenhydramine and carbamazepine-epoxide were found at concentrations below the limit of detection (0.05 g/kg), respectively.

Aquaculture systems contain solid particulate matter from feces, uneaten feed, biologics (bacteria or dead bacteria), and fish mucus. Depending on their size, these substances can either precipitate in a solid–liquid separation device or cannot readily precipitate in water, making mechanical filtration an ineffective method for their removal. In aquatic environments, these substances pose an extremely serious threat to the health of the organisms. Aquatic animals can experience a variety of stress responses, which can affect their respiration, reduce their resistance to disease, provide a suitable environment for pathogenic microorganisms to grow, and increase the load on the biological filter by reducing its capacity to nitrify (Schumann and Brinker 2020). Nitrogen pollutants include nitrate, nitrite, and ammonia. Ammonia exists in two forms in water: ionic ammonia (NH_4^+) and non-ionic ammonia (NH_3). Phosphorous pollutants are considered one of the main pollutants in aquaculture systems; they come from feed. When the concentration exceeds the water's absorption capacity, it can indirectly affect aquatic organisms. There is evidence that phosphorus may promote excessive algal reproduction, which can lead to water eutrophication, indirectly affecting animal growth. Raised fish and bacteria produce most of the carbon dioxide (CO_2) in aquaculture water. Stock density increased, resulting in a decrease in the water exchange rate and an increase in the concentration of dissolved CO_2, which will be a limiting factor for aquaculture growth. So, if there is more CO_2 in the water, there would also be more CO_2 in the blood of fish. This would make hemoglobin less likely to bind to oxygen molecules, which would lower the ability of the blood to carry oxygen, lower the ability of fish to feed, lead to excessive protein consumption, and hurt the fish's growth, metabolism, and well-being (Stiller et al. 2015). CO_2 safety levels vary depending on the fish species, stage of fish growth, and water quality. A concentration of 9–30 mg/L of CO_2 is safe for trout; tilapia and striped bass can tolerate a concentration of up to 60 mg/L.

1.4.5 Bioaccumulation of Heavy Metals in Fish and Their Public Health Effects

The presence of heavy metals (HM) in TWW is one of the most reported challenges for aquaculture production. As they are non-degradable, they accumulate inside fish tissues, causing ecological damage. Fish tissues can accumulate HM through three main mechanisms: the gills, the body surface, and the digestive tract. The gills in fish directly absorb metals from the water, while the body surface absorbs other elements at a lower concentration. While heavy metals are essential for maintaining optimal health, high concentrations can be toxic (Li et al. 2022). Depending on the concentration of heavy metals in the water and the exposure period, heavy metals accumulate in fish tissues. Thus, continuous exposure to such toxic metals leads to physiological (respiration, immunity, behavior, metabolism, bone growth, embryonic development, and genetic abnormalities) as well as morphological and behavioral abnormalities in fish (Bristy et al. 2021). Table 1.2 shows the heavy metal concentrations in fish cultivated in TWW. The muscles of fish grown in TWW were assessed for a variety of heavy metals according to EU and FAO MRLs (maximum residue levels). Before the establishment of the first international standard in 2003, there was no consideration of heavy metals in the muscles of wastewater-reared fish. Researchers discovered that the concentration of detected metals in fish samples exceeded the concentration in water samples for all treatments, except for Hg, whose values exceeded the WHO's permissible limits. In addition, El-Degwy et al. (2023) found that Fe > Zn > Mn > Pb > Cu > Ni > Cd levels in water were Fe > Zn > Mn > Pb > Cu > Ni > Cd. As indicated by the pollution index, a measure of the effects of each metal on aquatic life, the presence of Cd and Pb in water results in serious pollution effects. However, copper, nickel, iron, and zinc did not exhibit significant pollution effects throughout the lake (Khalefa et al. 2022).

1.4.6 Effect of Microbial Contamination of TWW on the Cultured Fish and Public Health

Water-borne diseases are considered one of the major global health problems in aquatic environments. Although the complex microbial community found in TWW was significantly reduced, it may harbor pathogens that threaten the health of humans and aquatic life (Popović et al. 2019). Bacterial contamination of fish is a serious threat to aquaculture systems, triggering severe damage, high mortality rates, and large economic losses. Among the bacterial pathogens that contaminate fish and subsequently infect humans are *Campylobacter jejuni*, *Salmonella* spp., *Vibrio cholerae*, *E. coli*, *Shigella* spp., and *Enterococcus* spp. (WHO 2006). The presence of different strains of *E. coli* is an indicator of sewage pollution. *S. aureus* and *Salmonella* are within the pathogenic bacterial group of food poisoning. Pseudomonas spp. is one of the most potent bacteria and spoilage organisms for human food (Fahim et al. 2021). Furthermore, researchers have reported Enterobacteriaceae, a common water-borne bacteria, as opportunistic pathogens in fish and as an indicator of sewage pollution in the tissues of seemingly healthy fish. When their concentration in the water was about 10^3/100 mL, Rajasekaran (2008) found no coliforms in muscle, or the majority of the fish organs tested.

The fish samples tested did not contain Salmonella, sulfur-reducing clostridia, Listeria spp., or S. aureus. The total bacterial count, *coliforms*, *Enterobacteriaceae*, and *E. coli* were all abundant in the digestive system, followed by the skin, and

Table 1.2 Heavy metal concentrations in fish cultivated in treated wastewater (TWW)

Heavy metals	MRLs (mg/kg)	Concentration in fish muscle (mg/kg body weight)	Refs.	Effect on human health The toxic effect includes	Effect on fish health	Refs.
Lead (Pb)	0.2[a], 0.3[b], 0.1[c]	<0.01	Ahamdi et al. (2014)	Hypertension, proximal renal tubular damage, anemia, and cardiac disease Damage to the nervous system. Acute lead poisoning can cause symptoms such as headaches, abdominal pains, irritability, as well as neurological manifestations such as encephalopathy, neurotoxicity (sleeplessness and restlessness), and hemorrhagic sunstroke	Large concentrations of lead prevent fish from maturing after hatching	Kinuthia et al. (2020) Briffa et al. (2020)
		<0.46	Cuevas-Uribe and Mims (2015)			
		0.04	Darko et al. (2016)			
		ND	Zaibel et al. (2020)			
Cadmium (Cd)	0.05[a,b], 0.1[c]	0.02–0.1	Ahamdi et al. (2014)	Itai-itai disease, rheumatoid arthritis, complete skeletal deformities, depressed growth hypertension and fetal deformities Cadmium is a group 1 human carcinogen that has been linked to osteoporosis, prostate cancer, and renal cell carcinoma, according to the International Agency Research on Cancer (IARC)	Various catalysts involved in oxidative metabolism are degraded. These include citrate synthase, succinate dehydrogenase (SDH), dehydrogenase (G6PDH), glucose-6-phosphate, and lactate dehydrogenase	Kinuthia et al. (2020) Briffa et al. (2020)
		<0.18	Cuevas-Uribe and Mims (2015)			
		0.016	Darko et al. (2016)			
		<0.025	Zaibel et al. (2020)			

(continued)

Table 1.2 (continued)

Heavy metals	MRLs (mg/kg)	Concentration in fish muscle (mg/kg body weight)	Refs.	Effect on human health The toxic effect includes	Effect on fish health	Refs.
Mercury (Hg)	0.5[a,b], 0.1[c]	0.08–1.26	Ahamdi et al. (2014)	Anxiety, sleeping problems, speech and speech disorders, cerebral palsy, kidney damage, cerebral palsy, down syndrome, and mental retardation in the fetus are among the conditions associated with down syndrome There are defects, complications of vision, and deformities of the fetus	The following disorders are demonstrated in embryos and cardiovascular problems: skeletal abnormalities; cranial distortions; neurological damage (brain damage in fish); and gill damage There has been fetal deformity of the fingerlings, permanent disability, etc. Damage to the immune system due to fatty liver	Kinuthia et al. (2020) Briffa et al. (2020)
		0.03	Cuevas-Uribe and Mims (2015)			
		0.257	Darko et al. (2016)			
		<0.1	Zaibel et al. (2020)			
Arsenic (As)	<0.01 (WHO 2006)	<0.01	Ahamdi et al. (2014)	The skin, bladder, kidney, lung, and skin manifestations of cancer gastrointestinal disorders, bone marrow depression, hemolysis, neurological effects, and hormonal effects Disruption and infertility, peripheral arterial disease, which can cause gangrene (black foot disease in Taiwan)	In the case of fish with severe kidney and nervous system damage Gill damage (severe pathological changes) A structural lesion, for example	Kinuthia et al. (2020); Kumar et al. (2020)
		ND	Cuevas-Uribe and Mims (2015)			
		ND	Darko et al. (2016)			
		0.246	Zaibel et al. (2020)			

MRLs according to the [a] EU (2015), the [b] FAO (2003) and [c] WHO (2006)
UN's Food and Agriculture Organization (FAO) in 2003 followed by standards set by the EU in 2006
ND: Not detected

negligible in the fillets. Furthermore, the study (Lan et al. 2007) determined that fish grown in urban wastewater-fertilized fishing ponds had extremely low levels of thermotolerant coliforms in their muscle tissue, but substantial amounts were present on their skin and digestive tract. The percentages of *Pseudomonas* spp. were found to be approximately 22.5% in fish samples and 25% in TWW samples. In 15% of fish samples and 35% of TWW samples, *E. coli* was present. *Proteus* spp. were present in 14.2% of fish samples, but not in TWW samples. *Klebsiella* spp. were found in 11.7% of fish samples and 25% of TWW samples. *Citrobacter* spp. were present in 10% of fish samples and 15% of TWW samples. Hassan et al. (2016) mentioned a relatively low incidence of Salmonella spp., with rates of 1.6 and 0%. However, Hassan et al. (2016) identified aeromonads as the most prevalent bacteria in both water and fish samples cultured in TWW (treated wastewater). Among the water samples, *A. hydrophile* was the most prevalent species, followed by *A. veronii*, *Rheinheimera* soli, and Ochrobactrum anthropi. In the fish tissues, *V. cholerae*, *Enterobacter cloacae*, and *E. sakazakii* were the most predominant bacteria during the studies applied by Rajasekaran (2008). Therefore, the most significant control point for ensuring the safety of fish flesh from wastewater-fed aquariums is the application of good hygienic practices during the degutting, processing, and cooking of fish, with the prevention of fecal cross-contamination.

The main sources of fish contamination with fungi were soil, water, sewage, workers, and equipment. The presence of mold in fish or fish products makes them unsuitable for human consumption, leading to significant financial losses. Additionally, mold contamination poses a public health concern due to its ability to produce harmful substances called mycotoxins, including aflatoxins, ochratoxins, zearalenone, patulin, and zearalenone. These mycotoxins have acute toxicity when consumed in large quantities and have been proven to be carcinogenic, teratogenic, and mutagenic in chronic instances (Hassan et al. 2016). Also, he has estimated Aspergillus spp. in 90 and 40% of the tested fish and water samples. Whereas the most frequently isolated *Aspergillus* spp. were *A. flavus* (40 and 10%), *A. niger* (20 and 1%), and *A. ochraceous* (20 and 5%). Aflatoxin was present in 25% of fish samples, with a mean value of 13 $\pm$ 2.0 ppb. However, water samples did not contain any detectable levels of aflatoxin. Additionally, the examined samples revealed the presence of *Candida* spp. (70 and 5%) and *Rhodotorula* spp. (70 and 10%).

Viruses are among the most abundant microorganisms that are present within the microbial community of aquatic environments' microbial communities. Enteric viruses are a diverse group of pathogens responsible for a wide range of human disorders, including respiratory and eye infections, hepatitis, gastroenteritis, aseptic meningitis, and myocarditis (Gonçalves et al. 2018). Norovirus (NoV), rotavirus (RoV), adenovirus (AdV), and astrovirus (AsV) are among the most commonly occurring viruses that contaminate TWW and cause the main viral gastroenteritis (Cioffi et al. 2020). Aquatic water describes these viruses by their long persistence and stability. Mostly, untreated and treated wastes discharge enteric viruses into the aquaculture environment (Cioffi et al. 2020). Shellfish, recognized for many years as the primary reservoir for human intestinal pathogens, can bioaccumulate high titers of viruses in a dynamic and fast process.

In addition, one of the food safety risks associated with the utilization of wastewater in fish farming is the presence of zoonotic parasites, specifically those found in fish. The improper disposal of sewage, the illegal release of wastewater, and the runoff of feces from people, pets, and both wild and domesticated animals can all bring in a lot of cysts and oocysts of protozoan parasites. These parasites have sturdy walls that make them highly resistant to a variety of environmental conditions and strategies aimed at their inactivation. Zoonotic protozoa, such as Toxoplasma gondii, Giardia, and Cryptosporidium, are significant pathogenic parasites that can infect marine organisms (Tedde et al. 2019).

Many studies have reported marine bivalve mollusks as the main disseminators of protozoan parasites for animals and humans (Tedde et al. 2019). Recently, researchers have found that cetaceans and fish are carriers of Toxoplasma gondii due to their ability to harbor cysts and oocysts of various protozoa in their tissues. Pig manure has the potential to act as a reservoir for trematodes and their eggs, which contaminated pig waste can introduce into fishponds. In Serbia, available studies indicate that trematodes do not pose a health problem for cultivated carp. Fish reared in wastewater-fed fishponds are certainly at risk of infection with trematode parasites (Table 1.3).

1.5 Influence of Antimicrobial Resistance (AMR) on Cultured Fish and the Public Health

Aquatic environments have become one of the key reservoirs and transmission routes for the spread of antimicrobial and antibiotic resistance (AMR/AR). The proliferation of antibiotic-resistant strains has led to treatment failures, resulting in escalated healthcare expenses. It is projected that by 2050, this trend could contribute to an additional ten million deaths per year (Amarasiri et al. 2020). The dissemination of antibiotic resistance genes (ARGs) among antibiotic-resistant bacteria (ARB) results in increased tolerance in certain bacterial communities, leading to multiple resistances and the emergence of super-bacteria. This poses a significant challenge to disease treatment. Numerous studies have documented the presence of antibiotic resistance in treated wastewater and wastewater systems (TWW). Florfenicol was reported to accumulate in fish tissues and TWW, with a negative influence on fish growth and physiological activities, causing an oxidative stress response. Tetracycline, chloramphenicol, macrolides, sulfonamides, and quinolones are the main antibiotics that are no longer effective. Bacteria and protozoa in biofilters could be potential hosts. Furthermore, the ESKAPE pathogens (*Enterococcus faecium*, *S. aureus*, *Klebsiella pneumoniae*, *Acinetobacter baumannii*, *Pseudomonas aeruginosa*, and *Enterobacter* species) are also very dangerous to human health because they have a "priority status" These germs have learned to resist many different types of drugs, such as oxazolidinones, lipopeptides, macrolides, fluoroquinolones, tetracyclines, β-lactams, β-lactam–β-lactamase inhibitor combinations, and even carbapenems and

Table 1.3 Microbiological parameters (CFU/cm^2 or g) estimated in fish grown in treated wastewater (TWW)

Refs.	Studied microorganisms	Reported results on the affected skin (cfu/cm^2), muscle (cfu/g)/TWW		
		Skin (cfu/cm^2)	Muscle (cfu/g)	Water
Easa et al. (1995)	TBC	8 × 10^2–10^3 cfu/cm^2 skin, while it could not be detected in fish muscle		
	E. coli	Recovered from skin and not detected in fish muscle		
	Faecal coliform	3–10 cfu/cm^2 skin and not detected in fish muscle		
	Salmonella (cfu/25 g)	Not detected in fish muscle and skin		
	S. aureus	Not detected in fish muscle and skin		
Lan et al. (2007)	TBC	2.7 × 10^5 cfu/cm^2 skin and 1.1 × 10^3 cfu/g muscle		
Harnisz and Tucholski (2010)	TBC	8.2 × 10^3–1.1 × 10^6 cfu/cm^2 skin and 20–30 cfu/g muscle		
	Fecal coliform	Not detected in fish muscle and skin		
	Salmonella (cfu/25 g)	Not detected in fish muscle and skin		
Ahamdi et al. (2014)	*E. coli*	2–400 MPN/g in TWW		
	Salmonella (cfu/25 g)	Not detected in TWW		
	S. aureus	<1.0 × 10^2–10^3 cfu/100 mL water		
Zhang and Shen (2019)	*E. coli*	1–2 cfu/cm^2 skin and 0–1 cfu/g muscle		
Popović et al. (2019)	*E. coli*	Not detected in fish muscle and skin while it was 5.5 × 10^5 cfu/100 mL TWW		
	Fecal coliform	Not detected in fish skin and 6.76 × 10^6 cfu/100 mL in TWW		
	Faecal streptococci	1.12 × 10^7 cfu/100 mL TWW		
	Pseudomonas aeruginosa	Not detected in fish skin and 3.5 × 10^3 cfu/100 mL water		
	Sulfite-reducing clostridia	Not detected in fish skin and it was 3.5 × 10^3 cfu/100 mL water		

glycopeptides, which are thought to be the last line of defense. WHO established the One Health Approach (OHA) (a six-point strategy) in response to the growing threat of AMR. The OHA is envisioned as involving animal health, human health, and environmental health and is focused on infectious disease-related challenges, such as AMR, that pose a threat to public health (Mazimba et al. 2021). Furthermore, the USA and others have banned the use of antibiotics in aquaculture.

1.6 Conclusion

Long-term use of untreated wastewater for irrigation in the pharmaceutical, medical, industrial, and residential sectors can expose these resources to hazardous compounds, bacteria, viruses, and helminths, which, if in contact with soil, roots, and underground water, can cause harm to natural resources. The drawbacks of irrigation with wastewater include metal toxicity combined with other pollutants, as well as the possibility of bioaccumulation in food. When these toxic heavy metals are present in concentrations that exceed permissible levels, humans and animals may be at risk for severe health problems. Consuming crops cultivated with either partially treated or untreated wastewater exposes these populations to several toxic contaminants and induces severe health risks in both humans and livestock. Farmers must embrace alternative resources for sustainable development to maintain social, economic, and environmental equilibrium. As a solution to this challenge, wastewater reuse provides an opportunity to reduce the demand on limited water resources, promoting the growth of circular economies when used in limited and under-health conditions.

References

Abuzaid AS, Jahin HS, Asaad AA, Fadl ME, AbdelRahman MA, Scopa A (2021) Accumulation of potentially toxic metals in Egyptian alluvial soils, berseem clover (Trifolium alexandrinum L.), and groundwater after long-term wastewater irrigation Agriculture 11(8):713

Aghalari Z, Dahms HU, Sillanpää M, Sosa-Hernandez JE, Parra-Saldívar R (2020) Effectiveness of wastewater treatment systems in removing microbial agents: a systematic review. Glob Health 16:1–11

Ahamdi HM, Jasper I, Karoli NN (2014) Effects of wastewater characteristics on fish quality from integrated wastewater treatment system and fish farming in urban areas, Tanzania. Agric, For Fish 3(4):292–298. https://doi.org/10.11648/j.aff.20140304.21

Akhtar N, Syakir Ishak MI, Bhawani SA, Umar K (2021) Various natural and anthropogenic factors responsible for water quality degradation: a review. Water 13(19):2660

Ali S, Abbas Z, Rizwan M, Zaheer IE, Yavaş İ, Ünay A et al (2020) Application of floating aquatic plants in phytoremediation of heavy metals polluted water: a review. Sustainability 12(5):1927

Amarasiri M, Sano D, Suzuki S (2020) Understanding human health risks caused by antibiotic resistant bacteria (ARB) and antibiotic resistance genes (ARG) in water environments: current knowledge and questions to be answered. Crit Rev Environ Sci Technol 50(19):2016–2059

Bashandy SR, Hemida Abd-Alla M, Mahmoud, GA-E (2022) Using fermentation waste of ethanol-producing yeast for bacterial riboflavin production and recycling of spent bacterial mass for enhancing the growth of oily plants. J Appl Microbiol 132:2020–2033

Bonetta S, Pignata C, Lorenzi E, De Ceglia M, Meucci L, Bonetta S et al (2016) Detection of pathogenic Campylobacter, E. coli O157: H7 and Salmonella spp. in wastewater by PCR assay. Environ Sci Pollut Res 23:15302–15309

Briffa J, Sinagra E, Blundell R (2020) Heavy metal pollution in the environment and their toxicological effects on humans. Heliyon 6(9)

Bristy MS, Sarker KK, Baki MA, Quraishi SB, Hossain MM, Islam A, Khan MF (2021) Health risk estimation of metals bioaccumulated in commercial fish from coastal areas and rivers in Bangladesh. Environ Toxicol Pharmacol 86:103666

Caicedo C, Beutel S, Scheper T, Rosenwinkel KH, Nogueira R (2016) Occurrence of Legionella in wastewater treatment plants linked to wastewater characteristics. Environ Sci Pollut Res 23:16873–16881

Chahal C, Van Den Akker B, Young F, Franco C, Blackbeard J, Monis P (2016) Pathogen and particle associations in wastewater: significance and implications for treatment and disinfection processes. Adv Appl Microbiol 97:63–119

Chen TH, Chou SM, Tang CH, Chen CY, Meng PJ, Ko FC, Cheng JO (2016) Endocrine disrupting effects of domestic wastewater on reproduction, sexual behavior, and gene expression in the brackish medaka Oryzias melastigma. Chemosphere 150:566–575

Cioffi B, Monini M, Salamone M, Pellicanò R, Di Bartolo I, Guida M et al (2020) Environmental surveillance of human enteric viruses in wastewaters, groundwater, surface water and sediments of Campania region. Reg Stud Mar Sci 38:101368

Connor, R (2017) The United Nations world water development report 2017. Wastewater: the untapped resource

Cuevas-Uribe R, Mims SD (2015) Contaminants in muscle tissue from paddlefish (Polyodon spathula) and hybrid striped bass (Morone chrysops x M. saxatilis) after being raised in reclaimed effluent water. J Appl Ichthyol 31:71–74

Darko G, Azanu D, Logo NK (2016) Accumulation of toxic metals in fish raised from sewage-fed aquaculture and estimated health risks associated with their consumption. Cogent Environ Sci 2(1):1190116

Easa MES, Shereif MM, Shaaban AI, Mancy KH (1995) Public health implications of waste water reuse for fish production. Water Sci Technol 32(11):145–152

Elahi E, Zhang L, Abid M, Javed MT, Xinru H (2017a) Direct and indirect effects of wastewater use and herd environment on the occurrence of animal diseases and animal health in Pakistan. Environ Sci Pollut Res 24:6819–6832

Elahi E, Abid M, Zhang L, Alugongo GM (2017b) The use of wastewater in livestock production and its socioeconomic and welfare implications. Environ Sci Pollut Res 24:17255–17266

El-Degwy AA, Negm NA, El-Tabl AS, Goher ME (2023) Assessment of heavy metal pollution in water and its effect on Nile tilapia (Oreochromis niloticus) in Mediterranean lakes: a case study at Mariout Lake. Appl Water Sci 13(2):50

ElMokadem MM, Mahmoud GA-E, Hifney AF (2025) Inhibitory effects of pomegranate peel extract against staphylococcus aureus causing animals mastitis in Sohag Governorate. J Appl Molecular Biol 3:17–31

Environmental Protection Agency (EPA), Office of Wastewater Management, and United States, Environmental Protection Agency, Office of Wastewater Management, Municipal Technology Branch (1994) A plain English guide to the EPA part 503 biosolids rule. US Environmental Protection Agency, Office of Wastewater Management

EPA (1996) Point and non-point sources of water pollution. United States Environmental Protection Agency, Washington, DC, USA

EU (2015) 495/2015, Commission Implementing Decision (EU) 2015/495 of 20 March 2015 establishing a watch list of substances for union-wide monitoring in the field of water policy pursuant to Directive 2008/105/EC of the European Parliament and of the Council. Off J Eur Union 78:40–42

Fahim KM, Badr AN, Shehata MG, Hassanen EI, Ahmed LI (2021) Innovative application of postbiotics, parabiotics and encapsulated Lactobacillus plantarum RM1 and Lactobacillus paracasei KC39 for detoxification of aflatoxin M1 in milk powder. J Dairy Res 88(4):429–435

FAO (Food and Agriculture Organization) (2003) Heavy metals regulations (Legal notice no. 66/2003)

Fito J, Van Hulle SW (2021) Wastewater reclamation and reuse potentials in agriculture: towards environmental sustainability. Environ Dev Sustain 23:2949–2972

Fradette MS, Culley AI, Charette SJ (2022) Detection of Cryptosporidium spp. and Giardia spp. in environmental water samples: a journey into the past and new perspectives. Microorganisms 10(6):1175

Geissen V, Mol H, Klumpp E, Umlauf G, Nadal M, van der Ploeg M et al (2015) Emerging pollutants in the environment: a challenge for water resource management. Int Soil Water Conserv Res 3(1):57–65

Gonçalves J, Gutiérrez-Aguirre I, Balasubramanian MN, Zagorščak M, Ravnikar M, Turk V (2018) Surveillance of human enteric viruses in coastal waters using concentration with methacrylate monolithic supports prior to detection by RT-qPCR. Mar Pollut Bull 128:307–317

Goswami RK, Mehariya S, Verma P, Lavecchia R, Zuorro A (2021) Microalgae-based biorefineries for sustainable resource recovery from wastewater. J Water Process Eng 40:101747

Grzegorczyk S, Grabowski K, Olszewska M (2020) Content of potassium, calcium, magnesium, phosphorus and sodium in meadow sward irrigated with wastewater. J Elementol 25(1)

Harnisz M, Tucholski S (2010) Microbial quality of common carp and pikeperch fingerlings cultured in a pond fed with treated wastewater. Ecol Eng 36(4):466–470

Hassan AA, Howayda ME, Mogda KM, Samira AMS, Khairy FA (2016) Effect of microbiological contamination and pollution of water on the health status of fish. Eur J Acad Essays 3(5):178–192

Hijjawi N, Zahedi A, Al-Falah M, Ryan U (2022) A review of the molecular epidemiology of Cryptosporidium spp. and Giardia duodenalis in the Middle East and North Africa (MENA) region. Infect, Genet Evol 98:105212

Hong PY, Julian TR, Pype ML, Jiang SC, Nelson KL, Graham D et al (2018) Reusing treated wastewater: consideration of the safety aspects associated with antibiotic-resistant bacteria and antibiotic resistance genes. Water 10(3):244

Jaramillo MF, Restrepo I (2017) Wastewater reuse in agriculture: a review about its limitations and benefits. Sustainability 9(10):1734

Jiménez B, Asano T (2008) Water reuse: an international survey of current practice, issues and needs

Junaid M, Liu S, Liao H, Liu X, Wu Y, Wang J (2022) Wastewater plastisphere enhances antibiotic resistant elements, bacterial pathogens, and toxicological impacts in the environment. Sci Total Environ 841:156805

Karanis P, Kourenti C, Smith H (2007) Waterborne transmission of protozoan parasites: a worldwide review of outbreaks and lessons learnt. J Water Health 5(1):1–38

Khalefa HS, Attia MM, Abdelsalam M, Mahmoud MA, Zaki Ewiss MA (2022) Immunological status of some edible fishes exposed to parasitic infections in relation to heavy metals pollution. J Parasit Dis 46(3):653–663

Khater DZ, Amin RS, Fetohi AE, El-Khatib KM, Mahmoud GA-E (2024). Microbial fuel cells: bio-battery for environmental bioremediation and bioelectricity generation. In: Prasad (ed) Environmental materials and waste (second edition): Circular economy and pollution abatement. Elsevier, pp 813–833

Kinuthia GK, Ngure V, Beti D, Lugalia R, Wangila A, Kamau L (2020) Levels of heavy metals in wastewater and soil samples from open drainage channels in Nairobi, Kenya: community health implication. Sci Rep 10(1):8434

Kumar A, Sharma G, Naushad M, Ala'a H, García-Peñas A, Mola GT et al (2020) Bio-inspired and biomaterials-based hybrid photocatalysts for environmental detoxification: a review. Chem Eng J 382:122937

Kundu P, Debsarkar A, Mukherjee S (2013) Treatment of slaughter house wastewater in a sequencing batch reactor: performance evaluation and biodegradation kinetics. BioMed Res Int 2013

Lan TPN, Dalsgaard A, Cam PD, Mara D (2007) Microbiological quality of fish grown in wastewater-fed and non-wastewater-fed fishponds in Hanoi, Vietnam: influence of hygiene practices in local retail markets. J Water Health 5(2):209–218

Li Z, Junaid M, Chen G, Wang J (2022) Interactions and associated resistance development mechanisms between microplastics, antibiotics and heavy metals in the aquaculture environment. Rev Aquac 14(2):1028–1045

Libutti A, Gatta G, Gagliardi A, Vergine P, Pollice A, Beneduce L et al (2018) Agro-industrial wastewater reuse for irrigation of a vegetable crop succession under Mediterranean conditions. Agric Water Manag 196:1–14

Liu P, Li Z, Che Z, Hu X, Ying M, Ren H, Zhang XX (2021) Prevalence of common enteric viruses in municipal wastewater treatment plants and their health risks arising from wastewater reuse. Blue-Green Syst 3(1):95–107

Lundy L, Revitt M, Ellis B (2018) An impact assessment for urban stormwater use. Environ Sci Pollut Res 25:19259–19270

Mahmoud, GA-E (2021) Microbial scavenging of heavy metals using bioremediation strategies. In: Kumar V, Prasad R, Kumar M (eds), Rhizobiont in bioremediation of hazardous waste. Springer, Singapore, pp 265–289

Mahmoud GA-E, Abd-El Salam AB, Ahmed Li (2025) Insecticides: current status, properties, and the biological degradation strategies. In: Shah M, Bharali P (eds), Development in waste water treatment research and processes, Applied technologies for clean-up of environmental contaminants. Elsevier, pp 45–70

Mahmoud GA-E, Osman YA, Abdel-Hakeem SS (2020) Hydrolytic bacteria associated with natural helminth infection in the midgut of red sea marbled spinefoot rabbit fish Siganus rivulatus. Microb Pathog 147:104404

Mazimba A, Rupasinghe N, Gomez ACC, Zhao F, Pate M (2021) Landscape analysis of tools to address antimicrobial resistance. World Bank, Washington, DC. www.worldbank.org

Obeng AMD (2010) The impact of harbour and associated activities on the water quality of the Durban Bay and their effects on indwelling ecosystems. Doctoral dissertation

Odumbe E, Murunga S, Ndiiri J (2023) Heavy metals in wastewater effluent: causes, effects and removal technologies.

Omarova A, Tussupova K, Berndtsson R, Kalishev M, Sharapatova K (2018) Protozoan parasites in drinking water: a system approach for improved water, sanitation and hygiene in developing countries. Int J Environ Res Public Health 15(3):495

Osińska A, Harnisz M, Korzeniewska E (2016) Prevalence of plasmid-mediated multidrug resistance determinants in fluoroquinolone-resistant bacteria isolated from sewage and surface water. Environ Sci Pollut Res 23:10818–10831

Popović NT, Kazazić SP, Barišić J, Strunjak-Perović I, Babić S, Bujak M et al (2019) Aquatic bacterial contamination associated with sugarplant sewage outfalls as a microbial hazard for fish. Chemosphere 224:1–8

Qadir M, Drechsel P, Jiménez Cisneros B, Kim Y, Pramanik A, Mehta P, Olaniyan O (2020, February) Global and regional potential of wastewater as a water, nutrient and energy source. In: Natural resources forum, vol 44, No 1. Blackwell Publishing Ltd., Oxford, UK, pp 40–51

Rai PK, Lee SS, Zhang M, Tsang YF, Kim KH (2019) Heavy metals in food crops: health risks, fate, mechanisms, and management. Environ Int 125:365–385

Rajasekaran P (2008) Enterobacteriaceae group of organisms in sewage-fed fishes. Adv Biotech 8:12–14

Schmiege D, Zacharias N, Sib E, Falkenberg T, Moebus S, Evers M, Kistemann T (2021) Prevalence of multidrug-resistant and extended-spectrum beta-lactamase-producing Escherichia coli in urban community wastewater. Sci Total Environ 785:147269

Schumann M, Brinker A (2020) Understanding and managing suspended solids in intensive salmonid aquaculture: a review. Rev Aquac 12(4):2109–2139

Shah MP (2024a) Microbial remediation of hazardous chemicals from water & industrial wastewater treatment plant. Springer Nature

Shah MP (2024b) Development in waste water treatment research and processes: applied technologies for clean up of environmental contaminants. Elsevier

Shah MP (2024c) Aerobic and anaerobic microbial treatment of industrial wastewater. CRC Press

Shah MP (2025) Extraction of precious metals from industrial waste using microbial technology. Springer Nature

Singh A (2021) A review of wastewater irrigation: environmental implications. Resour Conserv Recycl 168:105454

Stevens D, Surapaneni A, Deere D, O'Connor N, Crosbie N, Keegan A et al (2021) The probability of cysticercus bovis detection in livestock from exposure to recycled water in non-endemic countries. Microb Risk Anal 18:100164

Stiller KT, Vanselow KH, Moran D, Bojens G, Voigt W, Meyer S, Schulz C (2015) The effect of carbon dioxide on growth and metabolism in juvenile turbot Scophthalmus maximus L. Aquaculture 444:143–150

Sundström JF, Albihn A, Boqvist S, Ljungvall K, Marstorp H, Martiin C et al (2014) Future threats to agricultural food production posed by environmental degradation, climate change, and animal and plant diseases—a risk analysis in three economic and climate settings. Food Secur 6:201–215

Taher EM, Hemmatzadeh F, Aly SA, Elesswy HA, Petrovski KR (2020) Molecular characterization of antimicrobial resistance genes on farms and in commercial milk with emphasis on the effect of currently practiced heat treatments on viable but nonculturable formation. J Dairy Sci 103(11):9936–9945

Tedde T, Marangi M, Papini R, Salza S, Normanno G, Virgilio S, Giangaspero A (2019) Toxoplasma gondii and other zoonotic protozoans in Mediterranean mussel (Mytilus galloprovincialis) and blue mussel (Mytilus edulis): a food safety concern? J Food Prot 82(3):535–542

Tilley E, Ulrich L, Lüthi C et al (2014) Compendium of sanitation systems and technologies. Swiss Federal Institute of Aquatic Science and Technology (Eawag), Dübendorf

WHO-World Health Organization (2006) Guidelines for the safe use of wastewater, excreta and greywater. In: Wastewater use in agriculture, 2nd edn

Zaibel I, Dagan G, Arnon S, Schwartsburd F, Britzi M, Snyder SA, Zilberg D (2020) Tertiary-treated wastewater as a potential water source for sustainable aquaculture: a laboratory-scale experiment with Cyprinus carpio. Aquaculture 522:735161

Zaibel I, Arnon S, Zilberg D (2022) Treated municipal wastewater as a water source for sustainable aquaculture: a review. Rev Aquac 14(1):362–377

Zhang Y, Shen Y (2019) Wastewater irrigation: past, present, and future. Wiley Interdiscip Rev Water 6(3):e1234

Chapter 2
Dynamics of Microbial Communities in Industrial Wastewater: Monitoring and Role in Treatment Process

Sukanya Sasmal, Deblina Gangopadhyay, Bhavya Srivastava, and Pranabesh Sarkar

Abstract The dynamics of microbial communities in industrial wastewater is a complex and critical area of study. These microbial communities play essential roles in biodegradation, pollutant removal, and overall wastewater treatment processes. The microbial community's composition, structure, and functions are influenced by multiple factors including the type of industry, chemical composition of the wastewater, treatment methods, and environmental conditions. This chapter focuses on types of microorganisms with special emphasis on factors, such as wastewater composition, pH, and temperature, affecting dynamics of microorganisms in industrial wastewater. Apart from these, in this chapter authors delve into the process of microbial monitoring, bioremediation, and microbial role in wastewater treatment process as well as challenges in wastewater treatment process and future perspectives. Understanding and utilizing these microbial communities is essential for optimizing industrial wastewater treatment and addressing challenges related to pollutant removal, resistance, and environmental safety.

Keywords Contaminants · Industrial wastewater · Microbial dynamics · Monitoring techniques · Bioremediation

S. Sasmal · D. Gangopadhyay · B. Srivastava · P. Sarkar (✉)
School of Forensic Sciences, The West Bengal National University of Juridical Sciences, Kolkata, India
e-mail: pranabeshsarkar@nujs.edu

S. Sasmal
e-mail: sukanyaphd2024@nujs.edu

D. Gangopadhyay
e-mail: deblinagangopadhyay@nujs.edu

B. Srivastava
e-mail: bhavyasrivastava@nujs.edu

© The Author(s), under exclusive license to Springer Nature Singapore Pte Ltd. 2026

S. Dey and M. P. Shah (eds.), *Microbial Applications in Industrial Wastewater Treatment*, Advances in Wastewater Research,
https://doi.org/10.1007/978-981-96-9128-9_2

2.1 Introduction

Industrial wastewater refers to the water that has been used in industrial processes and contains various pollutants, chemicals, and other waste materials. Recent industrialization has made environmental degradation a key concern for many nations (Ahmed et al. 2021). Development drives modernization of agriculture, urbanization, and industry, all of which tend to discharge different kinds of pollutants into the environment. Natural resource quality and services are seriously threatened by the unsafe release of industrial chemicals into the environment (Ahada and Suthar 2018).

Industrial wastewater can come from various industries, such as chemical manufacturing, oil refineries, metal finishing, and textile production releasing heavy metals which cause apoptosis in animals that reside close to disposal sites as well as in soil deposits along wastewater streams. People are frequently exposed to contaminated wastewater in their daily lives, especially in crowded cities and places where wastewater is utilized for farming (Sharma et al. 2021). Aquatic life adversely affects a variety of manufacturing processes, including tanneries, pulp and paper mills and batteries. A large number of waterborne outbreaks are thought to be caused by microbial infections. Numerous microbial infections found in wastewater can result in long-term, expensive chronic illnesses, such as stomach ulcers and degenerative heart disease. Industrial wastewater is treated before it is put back into the environment or utilized again (Shah 2024a; Akpor et al. 2014).

The microbial communities found in industrial wastewater are mainly made up of a variety of bacteria, including species from the phyla Proteobacteria, Bacteroidetes, and Chloroflexi. Based on the types of industrial waste, these bacteria play a crucial role in the breakdown of organic pollutants. Important genera like Pseudomonas, Comamonas, and Thauera frequently dominate because of their capacity to break down complex compounds like heavy metals and hydrocarbons. The precise composition of these communities varies depending on the wastewater's characteristics and the type of industry. Additionally, some communities contain fungi, protozoa, and viruses that aid in the treatment process (Selvarajan et al. 2018).

2.1.1 Common Contaminants in Industrial Wastewater

The contaminants found in industrial wastewaters include suspended particles, pathogenic microbes, synthetic dyes and chemicals, heavy metals, and a large amount of organic debris, discussed in detail in Table 2.1.

Table 2.1 Contaminants and factors influencing microbial dynamics in wastewater

Type of contaminants	Examples	Sources	Factor	References
Heavy metals	Mercury (Hg), cadmium (Cd), arsenic (As), chromium (Cr), thallium (Tl), zinc (Zn), nickel (Ni), copper (Cu), and lead (Pb)	Urban runoff, mining wastes, industrial wastewaters, and municipal wastewaters	Chemical	Palani et al. (2021)
Toxic chemicals	Solvents, pesticides, dyes	Industrial waste from chemical processing plants (releasing lead, mercury, chromium), agricultural runoff containing pesticides (chlordane, dieldrin, heptachlor), improper disposal of industrial wastewater, accidental spills from industrial facilities, and surface runoff from suburban lawns treated with pesticides	Chemical	Palani et al. (2021)
Organic pollutants	Oils, grease, phenols	Petrochemical sector, edible oil production unit, petrol stations, metal finishing houses, engineering vehicle (automobile) workshops or garages, and kitchens in homes and restaurants	Chemical	Pintor et al. (2016)
Suspended solids	Dirt, debris	Industrial emissions, automobiles, and building and demolition operations	Physical	Ali et al. (2019)
Nutrients	Nitrogen and phosphorus	Combustion of fossil fuels, agriculture, wastewater, stormwater	Chemical	Nazari-Sharabian et al. (2018)
Pathogens	*Salmonella*, *Escherichia*, *Shigella*, *Yersinia*, *Leptospira*, *Vibrio cholerae*, *Aeromonas hydrophila*, *Legionella pneumophila*, *Mycobacterium*, and *Pseudomonas*	Human, animal, and pharmaceutical waste	Biological	Varela and Manaia (2013) and Chowdhary et al. (2018)
Thermal pollutants	Heated water discharged from power plants, industrial cooling systems, factories, and even urban runoff	Industrial effluent from power plants (especially coal-fired and nuclear), manufacturing facilities using water as a coolant, urban runoff from paved surfaces, and certain agricultural practices	Physical	Raman Vinnå et al. (2017)
Micro-organisms	*Salmonella*, *Shigella*, *Staphylococcus*, *Streptococcus*, *Pseudomonas*, *Clostridium* and *Enterococcus*, as well as viruses, fungi, protozoa, and parasites	Municipal sewage, agriculture pollution and storm water runoff	Biological	Selvarajan et al. (2018)

2.1.1.1 Heavy Metals

Heavy metals are metals that are extremely toxic, even at low concentrations, and have a very high density. 'Mercury (Hg), cadmium (Cd), arsenic (As), chromium (Cr), thallium (Tl), zinc (Zn), nickel (Ni), copper (Cu), and lead (Pb)' are some of these heavy metals. The major causes of pollution are urban runoff, mining wastes, industrial wastewaters, and municipal wastewaters. Heavy metals are usually found in effluent from industries. Numerous metals are present in effluents from the electroplating business. Wastewater contains a significant amount of metal from the wood processing industry. Water bodies are also exposed to nickel-containing waste products from the paint and enameling industries. The probability of an adverse health effect resulting from exposure to dangerous and hazardous chemicals found in drinking water is commonly used to describe the health risk associated with these compounds (Shah 2024b; Palani et al. 2021).

2.1.1.2 Toxic Chemicals

Industries often release dyes and pesticides into the water, low concentrations, which urges the need to development of sophisticated innovations, such as the separation or filtering of chemical combinations in conjunction with multi-technique detection methods. Stream courses reveal the most intricate aquatic systems, where microelements interact and transfer heavy metals. Fertilizers, pesticides, poultry processing factory waste, livestock farm drainage, etc. are the main sources of water pollution in the agricultural area. Pesticide-induced pollution from fertilizer heavy metals and nitrate contamination is extremely hazardous (Shah 2024c; Palani et al. 2021).

2.1.1.3 Organic Pollutants

In the present decade, petroleum discharge is a major environmental contamination issue. Oily wastewater pollution from activities associated with the petrochemical sector, edible oil production unit, petrol stations, metal finishing houses, engineering vehicle (automobile) workshops or garages, and kitchens in homes and restaurants, among other places are dangerous due to the nature of certain of its ingredients which are 'hydrocarbons, fatty acids, soaps, lipids and waxes' (Sanghamitra et al. 2021). The biosphere may be impacted when the majority of the materials in this class are released into the environment through wastewater because of their extremely poor biodegradability. Aquatic life will be impacted by even the smallest coating of oil since it will reduce light penetration and oxygen transport between air and water (Pintor et al. 2016).

2.1.1.4 *Suspended Solids*

The main man-made causes of pollution in metropolitan areas are industrial emissions, automobiles, and building and demolition operations. Fine particles linked to potentially harmful substances can enter the respiratory system deeply and cause a growing number of harmful health consequences even at lower quantities (Shah 2025; Ali et al 2019).

2.1.1.5 Nutrients

Eutrophication of water, which happens when the water has excessive levels of nutrients like phosphorus and nitrogen. Combustion of fossil fuels, agriculture, wastewater, stormwater are the sources of these nutrients. Algal blooms may disturb the regular functioning of ecosystems by causing a variety of issues, including water toxicity and deoxygenation (Nazari-Sharabian et al. 2018).

2.1.1.6 *Pathogens*

Numerous microorganisms that are hazardous to human health can be found in wastewater streams as wastewater can harbor pathogens from a variety of sources. The microbial communities in industrial wastewater are highly diverse and include bacteria, archaea, fungi, protozoa, and algae. The most dominant groups are bacteria and archaea, which are involved in various biochemical cycles. '*Salmonella, Escherichia, Shigella, Yersinia, Leptospira, Vibrio cholerae, Aeromonas hydrophila, Legionella pneumophila, Mycobacterium,* and *Pseudomonas*' are among the main pathogens that humans might get from wastewater (Chahal et al. 2016). Enteric pathogens can enter wastewater through animal and human feces or through fecal-contaminated water used for other domestic purposes like washing or bathing. Pathogens including '*Salmonella, Escherichia, Shigella, Yersinia, and V. cholerae*', commonly induce infections in the gastrointestinal tract, including 'diarrhea, dysentery, and gastroenteritis' (Varela and Manaia 2013). The dark-colored wastewater produced from distillery businesses contains 'Melanoidins, di-n-octyl phthalate, dibutyl phthalate, benzene propanoic acid, 2-hydroxysocaproic acid, etc.' which are highly genotoxic, carcinogenic, mutagenic and endocrine disrupting. It reduces solar penetration power, photosynthetic activity, and dissolved oxygen concentration in aquatic resources, causing major environmental issues (Chowdhary et al. 2018).

2.1.1.7 *Thermal Pollutants*

Water that has been heated or other materials that alter a waterbody's temperature are considered thermal pollutants. This can occur when warm water is released into natural waterbodies by human activity like power plants and industrial production. Thermal pollution can be caused by both natural and human-caused sources. However, cooling for power plants and industrial machinery is the main source of thermal pollution. Thermal pollution comes from a variety of sources, including power plants that release large amounts of hot water into nearby waterbodies through the burning of fossil fuels like coal or natural gas. Warm water runoff from urbanized concrete and asphalt surfaces that absorb and retain heat, and excessive nutrient runoff from agricultural operations promotes algal blooms that deteriorate in water and release heat (Råman Vinnå et al. 2017).

2.1.1.8 Microorganisms

Various microorganisms, such as *Salmonella, Shigella, Staphylococcus,Streptococcus, Pseudomonas, Clostridium and Enterococcus*, as well as viruses, fungi, protozoa, and parasites, can be found in wastewater produced from different industries. If people are exposed to certain microorganisms found in industrial wastewater, they may become hazardous and cause diseases (Selvarajan et al. 2018).

2.1.2 Environmental Impact of Industrial Wastewater

Industrial effluent discharge into the ecosystem is an imminent threat that has sparked several environmental, public health, and safety concerns.

2.1.2.1 Aquatic Ecosystem

Untreated or poorly treated industrial wastewater often contains toxic substances that can harm aquatic ecosystems. The wastewater is a serious environmental problem because it harms the aquatic ecosystem by obstructing light flow through water bodies and significantly lowering oxygen concentrations owing to the presence of hydrosulfides (Khan and Malik 2014). Heavy metals, chemicals, and excessive nutrients lead to the death of aquatic organisms, reduce biodiversity, and disrupt food chains.

Eutrophication is a key element influencing the quality of coastal and inland waterways. Toxic cyanobacteria blooms in lakes and streams and the growth of green macro algae leading to oxygen depletion, which suffocates fish and other aquatic species due to the presence of nitrogen and phosphorus in wastewater along the coast, are its most well-known indicators (Le Moal et al. 2019). Other notable impacts include the hypoxia of bottom water, the alteration of biogeochemical processes like sediment anoxia, the accumulation of toxic hydrogen sulfide, and nutrient cycling and the release of CO_2 associated with the decomposition of intensively produced organic matter which contributes to water acidification. The varying nutrient requirements of various phytoplankton groups and the ratios between the various nutrients in these waters have led to the frequent reports of phytoplankton community shift in many eutrophic coastal areas (Dorgham 2014).

2.1.2.2 Soil Contamination

When industrial wastewater is improperly disposed of on land, it can leach into the soil and contaminate it with harmful chemicals and heavy metals. This affects plant growth and can make soil toxic for agricultural use. However, when released without proper treatment, it lowers the alkalinity and manganese availability of the soil, which inhibits seed germination and depletes vegetation on agricultural land (Chowdhary et al. 2018). Heavy metal-containing food is hazardous to the health of individuals. Heavy metal poisoning and exposure from eating contaminated fruits and vegetables are serious concerns. The mineral content of vegetables is influenced by several variables, such as fertilizer doses, the concentrations of trace elements in mineral fertilizers, and their natural concentration in the environment. Bedrock is a naturally occurring source of these metals in the soil. Along with organic and mineral fertilizers, certain amounts of metals are added to soils used for agricultural reasons (Zwolak et al. 2019) Additionally, pesticides for plants are another source of the metals. The dynamics of heavy metals in the soil, their bioavailability, and plant uptake are all significantly influenced by soil properties (Antisari et al. 2015).

2.1.2.3 Groundwater Contamination

Industrial pollutants can infiltrate into groundwater reserves, making the water unsafe for drinking and agricultural use. Groundwater contamination is often more challenging to detect and clean up than surface water pollution. Consuming water with a high NO_3 content can be extremely harmful to one's health, particularly for young children (less than 5 years old). Some of the sources of NO_3 in groundwater include excretions from animals, industrial effluents, N-based fertilizers, surface leaching from wastewater and waste disposal sites, etc. (Ahada and Suthar 2018). However, it is regarded as a major source of potable and drinking water for both urban and rural people in developing nations. However, the quality and quantity of groundwater supplies have been seriously threatened in recent decades by ongoing human activity and the industrial revolution (Niu et al. 2014).

2.1.2.4 Air Pollution

Considerable consideration should be given to its released volatile organic compounds (VOCs) into the atmosphere, contributing to air pollution and global warming as well as its possible negative impacts on the health of humans. Numerous VOCs, such as benzene, have teratogenic, mutagenic, and carcinogenic qualities, which has led to serious worries regarding the health of employees and the local populations around textile dyeing factories (Liang et al. 2020).

2.1.2.5 Human Health Risks

Contaminated water sources can pose direct health risks to communities living near industrial sites. Ingesting, using, or encountering polluted water causes serious health issues like cancer, neurological damage, reproductive issues, and respiratory problems. Many health diseases and environmental deterioration are caused by wastewater from textile factories. Approximately 40% of colorants used worldwide include organically bound chlorine, which is known to cause cancer. The disruption of normal cell function brought on by this chemical pollution may affect a human's physiology and biochemical processes, impairing vital processes including breathing, osmoregulation, reproduction leading to death. Heavy metals in wastewater from the textile industry accumulate in the major organs of the body resulting in a variety of symptoms of different diseases since they are not biodegradable (Khan and Malik 2014).

2.1.2.6 Climate Change

Water resources are critical for practically every ecosystem, not just humans. Water is a necessity for energy generation, several industries, and agriculture. Water supplies are thus under increased strain due to a changing climate and its possible adverse consequences. Climate change leads to a decrease in water supply and increase in water demand due to evaporation and rising temperatures (Nazari-Sharabian et al. 2018). Treatment of industrial discharges often requires significant energy, especially in processes like distillation or chemical treatment. Energy-intensive wastewater treatment systems contribute to emissions of greenhouse gases that are released directly from biological activities (Campos et al. 2016).

2.1.2.7 Thermal Pollution

Deterioration of the quality of water by any method that alters water temperature or 'an accumulation of unusable heat from human activities that disrupts ecosystems in the natural environment' are two causes of thermal pollution. In addition to being a significant physical stressor on aquatic ecosystems, thermal pollution may accelerate the effects of chemical pollution. In the form of condenser cooling water, companies generate a large amount of heat. 'Petroleum refineries, steel mills, chemical plants, paper and pulp mills, thermal power plants, nuclear power plants, and the electricity generation industry' are the main users of water as a coolant. Warm water treatments can prevent oxygen from leaking into deeper bodies of water, which may be beneficial for bacteria but harmful to aquatic life. Aquatic plants and animals may be at risk from algal blooms brought on by the reduced oxygen levels. Perhaps the most prevalent and well-known consequence of thermal pollution is this algal bloom phenomenon (Råman Vinnå et al. 2017).

2.2 Types of Microorganisms in Industrial Wastewater

Numerous bacteria have developed the ability to use or unintentionally break down a wide range of pollutants and byproducts, and the full breakdown of pollutants into safe byproducts depends on the interactions and collective actions of distinct microbial assemblages within different communities. Investigating the connections between microbial functional guilds and functions is essential (Sharma et al. 2021).

These microorganisms can be categorized into different functional groups based on their role in wastewater treatment:

- Decomposers (heterotrophs): The primary decomposers in industrial effluent are fungi and bacteria. Bacteria are the most prevalent and are frequently used in wastewater treatment facilities to break down organic materials like fats, proteins, and carbohydrates, while fungi are especially good at breaking down complex compounds like lignin and can also facilitate heavy metal biosorption (Hajam et al. 2023).
- Nitrifiers: Nitrifying bacteria convert ammonia to nitrate during wastewater treatment when aerobic conditions are present (Johnston et al. 2019).
- Denitrifiers: Denitrifying bacteria can then convert nitrate to gaseous forms of nitrogen (such N_2 and N_2O) in anaerobic environments, hence reducing the nitrogen concentrations in dissolved effluent (Johnston et al. 2019).
- Sulfate-reducing bacteria (SRB): These microbes are useful for treating sulfate-containing wastewater. In addition to eliminating heavy metals and other contaminants, SRB may decompose organic materials in wastewater by utilizing sulfate as a terminal electron acceptor, producing hydrogen sulfide (Yun et al. 2019).
- Methanogens: In the process of treating wastewater anaerobically, methanogenic archaea are very much essential. They produce methane under anaerobic conditions, important in anaerobic digestion. '*Methanosarcina* sp. and *Methanotrix* sp.' are the most prevalent acetoclastic methanogens in anaerobic reactors used to treat industrial effluent (Vítězová et al. 2020).

2.2.1 *Factors Affecting/Influencing Microbial Dynamics*

Depending on their individual requirements and tolerances, environmental factors can either stimulate or inhibit microbial development and change the composition of microbial communities the description of which has been provided in Fig. 2.1.

2.2.1.1 Physical Factors

Temperature is the key physical factor affecting the composition of hydrocarbons and the viability of microorganisms. The slow degradation of natural oil in freezing climates, such as the Arctic, increases the need for microbes to clean up petroleum spills. Most oleophilic bacteria become metabolically inactive due to the sub-zero water temperature which may freeze the whole cytoplasm or block the transport channels inside the microbial cells (Macaulay 2015). An optimal temperature is necessary for the biological enzymes that are engaged in the breakdown process because it influences their metabolic turnover. In addition, many compounds require a particular temperature to decompose. The physiological characteristics of microorganisms are greatly influenced by temperature, which may either accelerate or slow down the bioremediation process. Temperature causes microbial activity to rise, peaking at the ideal temperature. The temperature instantly starts to drop and eventually stops rising or dropping once it reaches a particular level (Yang et al. 2009).

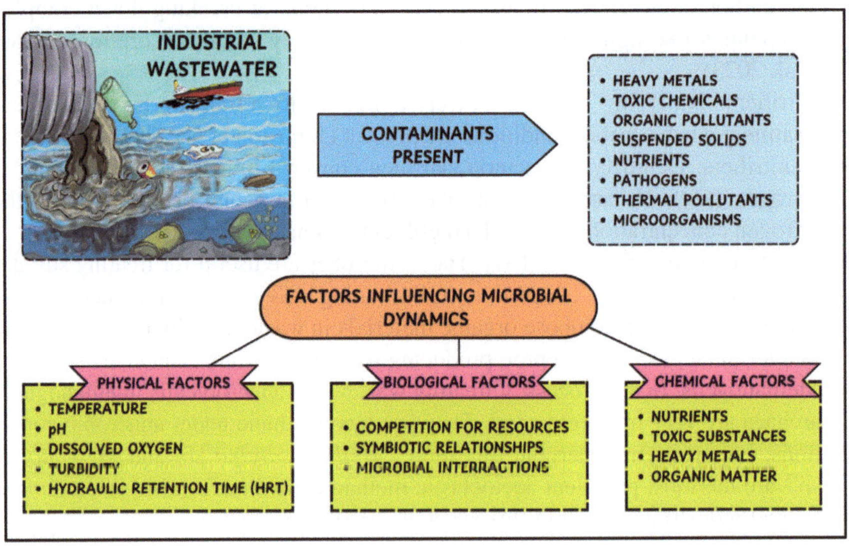

Fig. 2.1 Elements affecting the dynamics of microorganisms in industrial wastewater

The pH can speed up or slow down the elimination process and affects the metabolic activity of bacteria. Measuring the pH of soil ascertains whether microbial growth can happen there or not (Asira 2013).

Certain microorganisms need oxygen which even helps to improve the pace of biodegradation. The majority of living organisms need oxygen to survive; therefore, both aerobic and anaerobic environments can cause biological degradation. Oxygen may generally enhance the metabolism of hydrocarbons (Macaulay 2015).

The turbidity of wastewater is a measurement of the capacity of water to transmit light which indicates the quantity of colloidal and suspended solid matter present. Light cannot be transmitted because colloidal stuff will scatter or absorb it. Consequently, wastewater turbidity will increase with colloidal matter concentration. Turbidity and the amount of suspended particles in raw wastewater have no correlation with one another. However, there is a probable correlation between the suspended particles present in the settled secondary effluent and turbidity (Farrell et al. 2018).

Hydraulic retention time (HRT) is the average duration in which a material stays in a reactor or tank. This is a crucial component for digestion systems and wastewater treatment as it influences process efficiency. A longer HRT can improve treatment performance, but it can also increase costs. Whereas the amount of time that microorganisms have to develop and turn substrates into biogas is also impacted by HRT. The kind of substrate and the parameters of the procedure determine the appropriate HRT (Trebuch et al. 2020).

2.2.1.2 Chemical Factors

Microorganisms require several nutrients such as 'carbon, nitrogen, and phosphorus', to survive and perform their microbiological activities. The distribution of essential nutrients like N and P will boost the biodegradation efficiency that can improve the bacterial 'C:N:P ratio' by nutrition balancing (Couto et al. 2014). Hydrocarbon decomposition also reaches a limit at low concentrations. The addition of an appropriate quantity of nutrients is a successful way to increase the metabolic activity of microorganisms and the pace of biodegradation in cold conditions (Thavasi et al. 2011).

Certain pollutants can slow down decontamination and have harmful effects on microorganisms when present in high quantities. Certain toxicants, their concentrations, and the microorganisms exposed all affect the degree and mechanisms of toxicity. Certain substances, both inorganic and organic, are poisonous to specific organisms (Naik and Duraphe 2012).

Pollution from toxic metals is growing progressively more significant for the environment, human health, and economy. When it comes to biological wastewater treatment, a variety of helpful microorganisms including 'denitrifiers, nitrifiers, heterotrophs, and polyphosphate accumulating organisms (PAOs)', are essential for removing nitrogen, phosphorus and biomass from wastewater. Heavy metals appear to bioaccumulate in the food chain and remain in nature due to their non-biodegradable properties which can create major health and environmental problems

(Sharma et al. 2021). Fungi and bacteria require trace levels of metals but substantial levels of metals prevent cells from metabolizing. The rate of deterioration is influenced by metal compounds both directly and indirectly (Abatenh et al. 2017).

Factors such as 'temperature, level of oxygen, and the composition' of the existing population of the microbes affect the extent to which microbial bioremediation works for organic contaminants. Because of the unexpected abundance of hydrocarbons as a food source, oil-degrading bacteria that were previously relatively rare become prominent in an impacted area's microbial ecology, which is altered by organic contaminants. In the presence of an organic pollutant, some microorganisms, referred to as hydrocarbonoclastic bacteria, may use oil hydrocarbons as food, which increases their population (Dashti et al. 2015).

2.2.1.3 Biological Factors

Microbes are usually surrounded by several species and strains that compete with them for limited space and limited nutrition due to which they exhibit a variety of competing phenotypes. For example, exploitative phenotypes include metabolic alterations that boost growth rates or the release of molecules to gather resources, whereas interference competition takes place through contact-dependent death or antimicrobial secretions (Ghoul and Mitri 2016).

Applications of microalgae provide ingenious and environmentally friendly ways to remediate various wastewater kinds. The fungal–microalgal association is a cheap and efficient way to concentrate microalgal cells and maximize the additive contribution of their contents to the synthesis of biofuels. Fungal–microalgal consortiums have cumulative and synergistic impacts on wastewater treatment efficiency, since both biological components are known to be engaged in the absorption of important nutrients and microelements from growing environments. Similarly, reducing pollutants and carbon emissions is one of the two benefits of the microalgal–bacterial symbiosis (MABS) system which is essential for advancing wastewater treatment sustainability (Tang et al. 2024).

Mechanisms like competition for nutrients, synergistic degradation of complex compounds, and communication via quorum sensing where different microbial species cooperate or antagonize each other depending on their needs and the prevailing environmental conditions within the wastewater system microbial interactions in wastewater have a significant impact on microbial dynamics by influencing the population structure, metabolic activity, and pollutant removal efficiency (Ju and Zhang 2015).

2.3 Microbial Community Monitoring

Identifying microorganisms that are essential to detoxification and profiling microbial populations can improve bioremediation techniques. Advances in molecular biology have revolutionized the study of microbial communities in wastewater. Techniques like 16S rRNA sequencing, metagenomics, and meta-transcriptomics allow detailed profiling of the microbial communities, helping in understanding shifts in response to changing conditions or operational challenges as shown in Fig. 2.2.

2.3.1 Monitoring Techniques

Monitoring community composition of microbes can inform treatment performance and detect the presence of harmful or resistant microbes.

2.3.1.1 Culture-Based Method

Effective culturable techniques are employed to demonstrate complex microbial groupings and their metabolic capabilities, offering fascinating insights into how microbial communities are organized and operating in a range of situations, including areas contaminated by oil. Researchers can identify and measure the relative abundance of culturable microbes within a community by using a culture-based method for monitoring microbial communities which involves isolating and growing specific microorganisms from a sample using selective nutrient media. Agar plate counting,

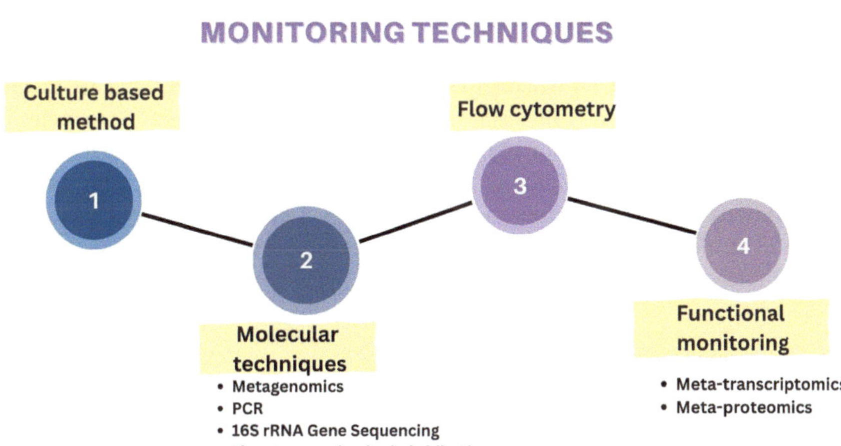

Fig. 2.2 Techniques currently used for monitoring the microbial community

broth dilution assays, and selective enrichment procedures are frequently used to target particular microbial groups (Ferone et al. 2020).

2.3.1.2 Molecular Techniques

In order to monitor microbial communities, molecular techniques such as '16S rRNA gene sequencing, Fluorescence in situ hybridization (FISH), Polymerase Chain Reaction (PCR), Quantitative PCR (qPCR), Amplified Ribosomal DNA Restriction Analysis (ARDRA), Next-Generation Sequencing (NGS), and Ribosomal Intergenic Spacer Analysis (RISA)' are frequently employed. Such techniques are useful for studying complex microbial communities because they enable the identification and quantification of microbial species within a sample without the need for their culture.

- DNA sequencing (meta genomics): The process of applying shotgun sequencing techniques to millions of randomly selected genomic fragments taken from a microbial population is known as metagenomics. Metagenomic profiling identifies functional genes and metabolic pathways by sequencing all of the DNA in a sample, eliminating the need to cultivate separate organisms (Ghosh and Das 2020).
- PCR: The polymerase chain reaction may create millions of copies of a selected gene, whole gene, or gene clusters in 3–4 h. It was the most used method for amplifying 16S rRNA or its gene prior to fingerprinting research. PCR-based methods have also been used to quantify and identify microorganisms in soil and water (Malik et al. 2008).
- 16S rRNA sequencing: This method allows the identification of many bacterial species based on sequence alterations by targeting a conserved area of 16S rRNA gene. It is regarded as the gold standard for microbial community analysis (Numberger et al. 2019).
- Fluorescence in situ hybridization (FISH): This method uses probes with fluorescent tags that attach to distinct DNA sequences to make it possible to see and count certain microbial cells inside a sample. Therefore, it is possible to quantify the presence and distribution of populations of microbes in a sample that contains various communities. Microbial cells on a glass slide are treated with fixative, hybridized with certain probes (often probes that are fluorescently tagged with 15–25 bp oligonucleotides) and then seen using either epifluorescence or confocal laser microscopy (Malik et al. 2008).

2.3.1.3 Flow Cytometry

Flow cytometry is a valuable tool for studying diverse microbial ecosystems because it enables researchers to quickly analyze and quantify the various types of microorganisms present in a sample, including their size, relative abundance, and even specific physiological states without requiring extensive culturing. The use of numerous fluorescent dyes allows flow cytometry to evaluate multiple cell characteristics at once, giving precise information on the components of microbial community, including the existence of certain bacterial species, their viability and possible metabolic activity. This makes it a powerful technique for monitoring microbial communities (Fiedler et al. 2018).

2.3.1.4 Functional Monitoring

The term "functional monitoring of microbial communities" describes the process of evaluating a microbial community's metabolic activities and potential functions, going beyond merely identifying the species that are currently present. This is done by examining the genes and enzymes that are present in the community to determine what ecological roles they are most likely playing, rather than just their taxonomic diversity. While conventional microbial community analysis employs 16S rRNA sequencing to identify the various species present, functional monitoring focuses at the metabolic pathways and functional genes of those species to determine their capability (Franzosa et al. 2015).

It includes:

- Meta-transcriptomics: The process of sequencing microbial community RNA to reveal information about genes that are actively expressed and ongoing metabolic processes (Franzosa et al. 2015).
- Meta-proteomics: Meta-proteomics is the process of identifying and quantifying proteins from microbial communities on a massive scale. It provides an initial information of the molecular characteristics of microbes (Kleiner 2019).

2.4 Bioremediation

Bioremediation is the process of breaking down, neutralizing, or eliminating hazardous chemicals from a polluted environment, such as soil, water, or air, by employing microorganisms like fungi, bacteria, or plants. In order to break down or detoxify pollutants that are detrimental to the environment and biodiversity, bioremediation aims to support their metabolic processes by giving them the ideal quantities of nutrients and other substances. The microbes degrade pollutants by co-metabolizing them with an energy source or by utilizing them as an energy source. To be more precise, bioremediation is the process by which cells of microbes produce energy through a redox reaction. Examples of such reactions include respiration and other biological functions necessary for cell survival and reproduction. Cleaning up pollution from oil spills, industrial operations, agricultural runoff, and other environmental pollutants may be done sustainably and effectively via bioremediation (Abatenh et al. 2017).

Bioremediation methods encompass a variety of treatment systems or approaches. Biostimulation, attenuation, augmentation, and venting are the fundamental bioremediation techniques, discussed in detail in Table 2.2.

2.4.1 Types of Bioremediations

In situ or ex situ bioremediation might be implemented.

2.4.1.1 In-Situ Bioremediation

Pollution is addressed at its source by this kind of bioremediation. It can be artificial, which adds bacteria or nutrients, or intrinsic, which employs endogenous microbes. This strategy could be beneficial since it might lower the expenses associated with handling materials and certain environmental effects. However, the capacity to regulate or alter the chemical and physical environment during bioremediation may place restrictions on in situ procedures (Azubuike et al. 2016).

2.4.1.2 Ex Situ Bioremediation

Pollution is treated off-site by this kind of bioremediation. Contaminated material must be removed and treated using techniques, such as slurry phase bioremediation or land cropping. Ex situ bioremediation procedures are often evaluated in light of the following factors: the geology and geographic location of the contaminated site, the kind and degree of pollution, the depth of contamination, and the cost of treatment (Azubuike et al. 2016).

2.4.2 Types of Organisms Used in Bioremediation

2.4.2.1 *Phytoremediation*

In this type of bioremediation, plants are utilized to restore soil and water in order to eliminate contaminants. The kinds of phytoremediation include:

- Phytoextraction: Plant roots extract pollutants from soil or water which is then transferred and stored in the biomass or shoots present on the surface which are subsequently harvested. This process is called phytoextraction (Muthusaravaran et al. 2018).
- Phytostabilization: The process of using plants that can reduce the mobility of pollutants or bioavailability by adsorption by roots or the production of insoluble

Table 2.2 Bioremediation techniques

Process	Type	Description	References
Bio-stimulation	In situ bioremediation	The activity of local microorganisms is promoted by this type of strategy which operates by injecting certain nutrients into the soil or ground water at the location. It focuses on promoting the native or naturally existing fungal and bacterial communities	Adams et al. (2015)
Bio-attenuation	In situ bioremediation	Natural attenuation or bio-attenuation is the process of removing contaminant concentrations from the environment. It may involve physical phenomena like 'advection, dispersion, dilution, diffusion, volatilization, sorption/desorption', chemical reactions such as 'ion exchange, complexation, abiotic transformation', and biological processes including 'aerobic and anaerobic biodegradation, plant and animal uptake'	Adams et al. (2015)
Bio-augmentation	In situ bioremediation	Bioaugmentation is the technique of introducing natural, exotic, or manufactured pollutant-degrading microorganisms to increase biodegradative potential of existing microbial populations in polluted regions	Niu et al. (2009)
Bioventing	In situ bioremediation	Bioventing is providing oxygen to pre-existing soil microorganisms through the soil to encourage the development of imported or naturally occurring bacteria and fungi. It even functions with materials that decompose aerobically. Bioventing uses low air flow rates to provide just enough oxygen to sustain microbial activity	Agarry and Latinwo (2015)
Phytoremediation	In situ bioremediation	In this type of bioremediation, plants are utilized to restore soil and water in order to eliminate contaminants	Azubuike et al. (2016)
Bioslurping	In situ bioremediation	This method uses bioventing, soil vapor extraction, and vacuum-enhanced pumping to remediate soil and groundwater by indirectly supplying oxygen and promoting the biodegradation of contaminants	Azubuike et al. (2016)
Biosparging	In situ bioremediation	This method is slightly similar to bioventing, it involves injecting air beneath the soil to encourage microbial activity and aid in the removal of pollutants from contaminated areas. Air is pumped at the saturated zone, which might induce volatile organic compounds to travel higher to the unsaturated zone to enhance biodegradation, in contrast to bioventing. Soil permeability, which controls the bioavailability of pollutants to microorganisms, and pollutant biodegradability are the two main determinants of biosparging efficacy	Azubuike et al. (2016)
Biopiles	Ex situ bioremediation	In biopile-mediated bioremediation, excavated toxic soil is piled above ground, followed by nutrient enrichment and occasionally aeration to improve bioremediation by essentially boosting microbial activity. Aeration, irrigation, leachate and nutrient collection systems, and a treatment bed are the elements that constitute this method. The application of this specific ex situ approach is becoming more and more popular because of its advantageous aspects, such as its affordability and ability to facilitate efficient biodegradation provided that temperature, aeration, and nutrients are properly managed	Whelan et al. (2015)

(continued)

Table 2.2 (continued)

Process	Type	Description	References
Landfarming	Ex situ bioremediation	Land farming is amongst the simplest bioremediation techniques owing to its low cost and less equipment requirement for operation. Polluted soils are often dug and/or tilled in land farming, however the sort of bioremediation seems to depend on the treatment location	Azubuike et al. (2016)
Bioreactors	Ex situ bioremediation	A bioreactor is a vessel that uses a sequence of biological processes to transform raw materials into one or more specialized products, as the name suggests. A bioreactor can operate in a variety of ways, including batch, fed-batch, sequencing batch, continuous, and multistage. Compared to conventional ex situ bioremediation methods, using a bioreactor to clean polluted soil has a number of benefits. Polluted samples can be fed into the bioreactor as slurry or dry materials	Zangi-Kotler et al. (2015)
Windrows	Ex situ bioremediation	Windrows, a type of ex situ bioremediation technology, use heaped contaminated soil that is periodically turned to improve bioremediation by boosting the degrading activities of native and/or transitory hydrocarbonoclastic bacteria. The rate of bioremediation, which can be achieved through assimilation, biotransformation, and mineralization, is accelerated by the regular turning of polluted soil and the addition of water, which increases aeration and uniformly distributes pollutants, nutrients, and microbial degradative activities	Azubuike et al. (2016)

compounds in the root zone to stop it from leaching into groundwater or entering the food chain is known as phytostabilization (Khan et al. 2019).

- Phytofiltration: This is also known as rhizofiltration, which is the process by which contaminants from solutions will either adsorb or precipitate onto plant roots or be absorbed into the roots that surround the root zone (Khan et al. 2019).
- Phytovolatilization: This is another phytoremediation technique that uses the absorption of pollutants by plants which converts them into volatile chemicals. Because of the plants' metabolic and transpiration pull, these compounds are ultimately released into the atmosphere in their original or modified form (Khan et al. 2019).
- Phytodegradation: The process by which pollutants and nutrients are drawn from water, sediment, or soil and then chemically altered as a direct result of plant metabolism is known as phytodegradation or phytotransformation. This process frequently results in the inactivation, degradation, or immobilization of contaminants in plant roots as well as shoots (Conceição Gomes et al. 2016).

2.4.2.2 Microorganisms

The two primary processes of microbial remediation of polluted media are pollutant immobilization and bioavailability decrease. Microorganisms may vary their form due to their distinct chemical and physical properties, yet they are unable to degrade inorganic pollutants such as heavy metals. The primary microbial remediation processes include 'precipitation, oxidation–reduction reactions, intracellular accumulation, and extracellular complexation'. Among these processes, bioaccumulation and biosorption are crucial because they allow microorganisms, or biomass, to attach to and concentrate environmental pollutants (Wang et al. 2021).

2.4.3 Mechanisms of Bioremediation

2.4.3.1 Biodegradation

Biodegradation is the process by which bacteria or other biological agents break down organic compounds chemically. One significant pathway for the elimination of these substances is the microbial breakdown of chemicals in the environment. The components include plastics, organic chemicals (including trace chemicals like endocrine disruptors and industrial chemicals used in significant numbers), and organometallics like methylmercury. These substances frequently undergo a complicated sequence of biochemical events throughout their biodegradation, which typically differs depending on the microorganisms involved (Wu et al. 2012).

2.4.3.2 *Bioaccumulation*

The rise in pollutant concentrations in aquatic species after absorption from the ambient environmental medium is known as bioaccumulation. The bioaccumulation of contaminants can be attributed to many exposure sources. The dissolved absorption rate constant, efflux, and food assimilation efficiency are examples of toxicokinetic characteristics that are crucial to the bioaccumulation processes (Wang 2016). The presence of toxic metal ions in their surroundings causes bioaccumulation species to manufacture special thiol-group-rich proteins termed phytochelatins (PCs) and metallothioneins (MTs), which have an internal binding mechanism. It is possible to exclude the complex containing such contaminants from regular metabolic activities (Chojnacka 2010).

2.4.3.3 *Biotransformation*

The process of biotransformation involves changing the shape of organic substances in order to lessen their toxicity and persistence. Enzymes, fungus, and bacteria are among the many microorganisms and their byproducts that help in this process. Compounds or materials can also be synthesized using biotransformations if synthetic methods prove difficult. The process of natural transformation is less efficient, slow, and nonspecific. Microbial biotransformations, also known as microbial biotechnology, are becoming more popular and widely used to produce metabolites in greater quantities and with more selectivity (Smitha et al. 2017).

2.4.3.4 *Biosorption*

Metal ions attach themselves to the surface of a biosorbent through a physiochemical process called biosorption. Potential metal sorbents include biopolymers, bacteria, yeasts, fungi, algae, and industrial and agricultural waste. Biosorption is regarded as an important method for both recovering heavy metals and removing them from even lower quantities. To ensure that all of the sorbate is extracted from the biomass using a little volume of solution and that the biosorbent is not damaged and maintains its sorptive qualities so that it may be utilized again in the next biosorption cycle, it is crucial to carefully select the desorbing agent (Shrestha et al. 2021).

2.5 Role of Microbes in Wastewater Treatment Process

Microbial dynamics are key to several wastewater treatment processes.

2.5.1 Activated Sludge

Microbial populations create bioflocs in aerobic treatment systems such as activated sludge, which break down organic contaminants. Microorganisms in the aeration tank of an activated sludge system may break down the organic materials in wastewater. The chemical environment and unique characteristics of the organisms that make constitute the biological mass determine the sorts of cells. A clarifier or sedimentation tank, utilizes gravity to extract the suspended particles from the wastewater after it has been treated. Following their recycling, the concentrated biological solids are introduced back into the aeration tank to continue treating the wastewater by sustaining a concentrated population of microorganisms. The balance between microbial growth and sludge settling efficiency is critical (Akpor et al. 2014).

It involves the following processes:

- Growth of microorganisms: In an aeration tank, the microorganisms, such as fungi, bacteria, and algae, are cultivated.

- Decomposition of organic matter: The microbes break down organic materials and dissolved oxygen in wastewater into water, carbon dioxide, and other inorganic substances.
- Formation of flocs: The oxidized material is contained in flocs which are basically the biological sludges produced by the microorganisms.
- Separation: In a settling tank, the sludge and treated wastewater are separated.
- Recycling: The sludge is recycled back into the aeration tank in portions.

2.5.2 *Anaerobic Digesters*

Anaerobic digesters are airtight vessels that utilize "anaerobic" bacteria like fermentative bacteria and methanogens to break down organic, solid waste slurry that is based on carbon, such food waste or cow dung, into smaller molecules with reduced odor. As the anaerobic bacteria break down the waste, they produce about equal volumes of carbon dioxide (CO_2) and methane (CH_4, often known as natural gas). This biogas is collected in contemporary anaerobic digesters and utilized to recover energy usually in an internal combustion engine connected to an electric generator. The carbon dioxide produced by the following burning of the methane releases energy that may be used to power the engine or generate heat for other purposes (Ali Shah et al. 2014).

2.5.3 *Biofilm Reactors*

A biofilm reactor is a kind of bioreactor in which the biocatalyst is fixed to a surface, like the reactor wall or inert material particles, enabling the development of microorganisms that are joined to one another as biogranules. The transfer of substrate into the biofilm limits the substrate conversion rate in this kind of reactor. The five main compartments of a biofilm reactor are as follows:

(a) a distribution system for influent wastewater
(b) a containment structure
(c) a biofilm carrier
(d) a system for collecting effluent water
(e) a system for aeration or mixing.

Microorganisms form biofilms on surfaces, which can enhance their stability and resistance to stress. Biofilm reactors are used in treating difficult-to-degrade pollutants (Lewandowski and Boltz 2011).

2.5.4 Membrane Bioreactors (MBR)

Membrane bioreactors (MBRs) are biological wastewater treatment systems that utilize membranes to separate the solid–liquid components of sludge suspension instead of the gravity settling used in the traditional activated sludge process. The technique enables all suspended particles and bacterial flocs to be physically retained inside the bioreactor (Akpor et al. 2014). Microbes are essential to a membrane bioreactor because they actively break down organic pollutants in wastewater through biological processes. They are essentially the "cleaning agents" that break down waste materials into simpler compounds, while the membrane component filters out the treated water to produce high-quality effluent. By keeping microorganisms inside the reactor, the membrane enables the biomass to be maintained at a steady, high concentration. However, fouling, a major problem in MBR operation, can also be caused by microbial growth on the membrane surface (Wu et al. 2012).

2.5.5 Metabolic Pathways

Depending on aerobic or anaerobic conditions the metabolic pathways of degradation in wastewater mainly involve the breakdown of complex organic matter, such as proteins, carbohydrates, and lipids by microorganisms through processes like hydrolysis, acidogenesis, acetogenesis, and methanogenesis. Under anaerobic conditions, the production of volatile fatty acids (VFAs) like acetate and propionate occurs, while the Krebs cycle, glycolysis, and electron transport chain are important pathways for aerobic degradation (Wang et al. 2022).

2.5.6 Enzymatic Degradation

Enzymatic degradation is a method of degrading both natural and manmade polymers, which is one way to treat or recover many kinds of contaminants. Enzymatic degradation utilizes naturally occurring enzymes made by microorganisms to "digest" the microbes in the sewage, breaking down complex organic matter and facilitating its purification by breaking down large molecules into smaller, more easily biodegradable components. Microbes in the wastewater release a variety of enzymes that help break down organic debris, such as amylases, cellulases, lipases, and proteases. The enzymatic degradation process may be carried out in a variety of enzymatic reactor designs to treat industrial wastewater (Al-Mutwalli et al. 2021).

2.6 Limitation and Challenges

2.6.1 Resistance to Pollutants

Antibiotics or specific industrial pollutants can cause some microorganisms to become resistant, which can result in ineffective treatment or the spread of resistant strains. Antibiotic-resistant bacteria and antibiotics are frequently found in wastewater, which causes Antibiotic-resistant genes (ARGs) to proliferate among treatment plant microorganisms. These ARGs can then transmit to other bacteria, further spreading resistance. Numerous processes such as the synthesis of enzymes to break down contaminants, changes in membrane permeability, or the acquisition of resistant genes through horizontal gene transfer might help the microorganisms to become resistant (Li et al. 2022).

2.6.2 Microbial Community Shifts

Sudden changes in wastewater composition can lead to microbial community shifts, impacting treatment efficiency. It leads to modifications in the relative abundance of various microorganisms within the treatment process which are mainly caused by changes in parameters like hydraulic retention time (HRT), temperature, dissolved oxygen levels, and nutrient availability. These changes in microbial populations can have an effect on the effectiveness of wastewater treatment processes (Tsai et al. 2022).

2.6.3 Emerging Contaminants

Synthetic or naturally occurring chemicals or microbes that are not often observed in the environment but have the potential to do so and have hazardous ecological or human health impacts that are known or suspected are known as emerging contaminants (ECs). Industrial wastewaters increasingly contain emerging contaminants like microplastics, pharmaceuticals, personal care products (PPCPs), plasticizers, surfactants, fire retardants, nanomaterials and pesticides which can alter microbial dynamics and complicate treatment. The detrimental effects of certain ECs on endocrine systems have led to their classification as endocrine disrupting chemicals (EDCs) (Kumar et al. 2022).

2.7 Future Perspectives

The use of sophisticated molecular techniques, such as next-generation sequencing (NGS) to identify diverse microbial communities, real-time monitoring systems for quick pathogen detection, and data analytics integration to optimize treatment processes based on microbial community dynamics, are some of the future perspectives for monitoring microbes in wastewater treatment. This will enable more efficient and targeted removal of pollutants while enhancing the overall quality of wastewater.

2.8 Conclusion

Microbial communities in industrial wastewater are dynamic and highly adaptable. The study of microbial communities in industrial wastewater underscores their crucial role in biodegradation and pollutant removal. The dynamic nature of these communities, shaped by physical, chemical, and biological factors, is pivotal in optimizing wastewater treatment systems. Wastewater treatment is primarily done to protect water sources from contamination and stop the spread of diseases.

Biological treatment is being used to prevent the unpleasant conditions in natural water resources because of several shortcomings with chemical treatment over the years. Many studies have been conducted on the prevalence of biological nitrogen and phosphorus in wastewater treatment systems. The majority of research on nutrient removal relies on the presence of bacteria, and their roles in removing nitrogen and phosphorus are widely recognized. The eutrophication of water sources is caused by a large number of microorganisms that actively contribute to nutrient removal. Thus, additional nutrient-related research monitoring units are required to accomplish uncontaminated wastewater discharge into water bodies.

Advances in molecular and functional monitoring have enhanced our understanding of microbial dynamics, allowing for more efficient and sustainable bioremediation strategies. Despite the progress, challenges such as resistance to pollutants, emerging contaminants, and shifts in microbial populations require continuous research and innovation. Biological treatment methods, complemented by modern techniques like metagenomics and bioaugmentation, hold promise for addressing these challenges while minimizing environmental and human health risks.

Future efforts should focus on integrating microbial community insights with advanced technologies to refine treatment processes and mitigate the impact of industrial wastewater on ecosystems. A holistic approach to monitoring, managing, and leveraging microbial diversity is vital for achieving cleaner water resources and a more sustainable environment.

References

Abatenh E, Gizaw B, Tsegaye Z, Wassie M (2017) The role of microorganisms in bioremediation—a review. Open J Environ Biol 2(1):038–046

Adams GO, Fufeyin PT, Okoro SE, Ehinomen I (2015) Bioremediation, biostimulation and bioaugmention: a review. Int J Environ Bioremed Biodegrad 3(1):28–39

Agarry S, Latinwo GK (2015) Biodegradation of diesel oil in soil and its enhancement by application of bioventing and amendment with brewery waste effluents as biostimulation-bioaugmentation agents. J Ecol Eng 16(2)

Ahada CP, Suthar S (2018) Groundwater nitrate contamination and associated human health risk assessment in southern districts of Punjab, India. Environ Sci Pollut Res 25:25336–25347

Ahmed J, Thakur A, Goyal A (2021) Industrial wastewater and its toxic effects

Akpor OB, Ogundeji MD, Olaolu DT, Aderiye BI (2014) Microbial roles and dynamics in wastewater treatment systems: an overview. Int J Pure Appl Biosci 2(1):156–168

Ali MU, Liu G, Yousaf B, Ullah H, Abbas Q, Munir MAM (2019) A systematic review on global pollution status of particulate matter-associated potential toxic elements and health perspectives in urban environment. Environ Geochem Health 41:1131–1162

Ali Shah F, Mahmood Q, Maroof Shah M, Pervez A, Ahmad Asad S (2014) [Retracted] Microbial ecology of anaerobic digesters: the key players of anaerobiosis. Sci World J 2014(1):183752

Al-Mutwalli SA, Korkut S, Kilic MS, Imer DY (2021) Enzymatic degradation of industrial wastewater pollutants. In: Removal of emerging contaminants through microbial processes, pp 373–398

Antisari LV, Orsini F, Marchetti L, Vianello G, Gianquinto G (2015) Heavy metal accumulation in vegetables grown in urban gardens. Agron Sustain Dev 35:1139–1147. https://doi.org/10.1007/s13593-015-0308-z

Asira EE (2013) Factors that determine bioremediation of organic compounds in the soil. Acad J Interdiscip Stud 2(13):125

Azubuike CC, Chikere CB, Okpokwasili GC (2016) Bioremediation techniques—classification based on site of application: principles, advantages, limitations and prospects. World J Microbiol Biotechnol 32:1–18

Campos JL, Valenzuela-Heredia D, Pedrouso A, Val del Río A, Belmonte M, Mosquera Corral A (2016) Greenhouse gases emissions from wastewater treatment plants: minimization, treatment, and prevention. J Chem 2016(1):3796352

Chahal C, Van Den Akker B, Young F, Franco C, Blackbeard J, Monis P (2016) Pathogen and particle associations in wastewater: significance and implications for treatment and disinfection processes. Adv Appl Microbiol 97:63–119

Chojnacka K (2010) Biosorption and bioaccumulation—the prospects for practical applications. Environ Int 36(3):299–307

Chowdhary P, Raj A, Bharagava RN (2018) Environmental pollution and health hazards from distillery wastewater and treatment approaches to combat the environmental threats: a review. Chemosphere 194:229–246

Couto N, Fritt-Rasmussen J, Jensen PE, Højrup M, Rodrigo AP, Ribeiro AB (2014) Suitability of oil bioremediation in an Artic soil using surplus heating from an incineration facility. Environ Sci Pollut Res 21:6221–6227

da Conceição Gomes MA, Hauser-Davis RA, de Souza AN, Vitória AP (2016) Metal phytoremediation: general strategies, genetically modified plants and applications in metal nanoparticle contamination. Ecotoxicol Environ Saf 134:133–147

Dashti N, Ali N, Eliyas M, Khanafer M, Sorkhoh NA, Radwan SS (2015) Most hydrocarbonoclastic bacteria in the total environment are diazotrophic, which highlights their value in the bioremediation of hydrocarbon contaminants. Microbes Environ 30(1):70–75

Dorgham MM (2014) Effects of eutrophication. In: Eutrophication: causes, consequences and control, vol 2, pp 29–44

Farrell C, Hassard F, Jefferson B, Leziart T, Nocker A, Jarvis P (2018) Turbidity composition and the relationship with microbial attachment and UV inactivation efficacy. Sci Total Environ 624:638–647

Ferone M, Gowen A, Fanning S, Scannell AG (2020) Microbial detection and identification methods: bench top assays to omics approaches. Compr Rev Food Sci Food Saf 19(6):3106–3129

Fiedler CJ, Schönher C, Proksch P, Kerschbaumer DJ, Mayr E, Zunabovic-Pichler M et al (2018) Assessment of microbial community dynamics in river bank filtrate using high-throughput sequencing and flow cytometry. Front Microbiol 9:2887

Franzosa EA, Hsu T, Sirota-Madi A, Shafquat A, Abu-Ali G, Morgan XC, Huttenhower C (2015) Sequencing and beyond: integrating molecular' omics' for microbial community profiling. Nat Rev Microbiol 13(6):360–372

Ghosh S, Das AP (2020) Microbial metagenomics: current advances in investigating microbial ecology and population dynamics. Front Soil Environ Microbiol 247–254

Ghoul M, Mitri S (2016) The ecology and evolution of microbial competition. Trends Microbiol 24(10):833–845

Hajam YA, Kumar R, Kumar A (2023) Environmental waste management strategies and vermi transformation for sustainable development. Environ Chall 100747

Johnston J, LaPara T, Behrens S (2019) Composition and dynamics of the activated sludge microbiome during seasonal nitrification failure. Sci Rep 9(1):4565

Ju F, Zhang T (2015) Bacterial assembly and temporal dynamics in activated sludge of a full-scale municipal wastewater treatment plant. ISME J 9(3):683–695

Khan I, Iqbal M, Shafiq F (2019) Phytomanagement of lead-contaminated soils: critical review of new trends and future prospects. Int J Environ Sci Technol 16:6473–6488

Khan S, Malik A (2014) Environmental and health effects of textile industry wastewater. In: Environmental deterioration and human health: natural and anthropogenic determinants, pp 55–71

Kleiner M (2019) Metaproteomics: much more than measuring gene expression in microbial communities. Msystems 4(3):10–1128

Kumar R, Qureshi M, Vishwakarma DK, Al-Ansari N, Kuriqi A, Elbeltagi A, Saraswat A (2022) A review on emerging water contaminants and the application of sustainable removal technologies. Case Stud Chem Environ Eng 6:100219

Le Moal M, Gascuel-Odoux C, Ménesguen A, Souchon Y, Étrillard C, Levain A et al (2019) Eutrophication: a new wine in an old bottle? Sci Total Environ 651:1–11

Lewandowski Z, Boltz JP (2011) Biofilms in water and wastewater treatment

Li S, Ondon BS, Ho SH, Jiang J, Li F (2022) Antibiotic resistant bacteria and genes in wastewater treatment plants: from occurrence to treatment strategies. Sci Total Environ 838:156544

Liang Z, Wang J, Zhang Y, Han C, Ma S, Chen J, Li G, An T (2020) Removal of volatile organic compounds (VOCs) emitted from a textile dyeing wastewater treatment plant and the attenuation of respiratory health risks using a pilot-scale biofilter. J Clea Prod 253:120019

Macaulay BM (2015) Understanding the behaviour of oil-degrading micro-organisms to enhance the microbial remediation of spilled petroleum. Appl Ecol Environ Res 13(1):247–262

Malik S, Beer M, Megharaj M, Naidu R (2008) The use of molecular techniques to characterize the microbial communities in contaminated soil and water. Environ Int 34(2):265–276

Muthusaravanan S, Sivarajasekar N, Vivek JS, Paramasivan T, Naushad M, Prakashmaran J et al (2018) Phytoremediation of heavy metals: mechanisms, methods and enhancements. Environ Chem Lett 16:1339–1359

Naik MG, Duraphe MD (2012) Review paper on—Parameters affecting bioremediation

Nazari-Sharabian M, Ahmad S, Karakouzian M (2018) Climate change and eutrophication: a short review. Eng, Technol Appl Sci Res 8(6):3668

Niu GL, Zhang JJ, Zhao S, Liu H, Boon N, Zhou NY (2009) Bioaugmentation of a 4-chloronitrobenzene contaminated soil with Pseudomonas putida ZWL73. Environ Pollut 157(3):763–771

Niu B, Loáiciga HA, Wang Z, Zhan FB, Hong S (2014) Twenty years of global groundwater research: a science citation index expandedbased bibliometric survey (1993–2012). J Hydrol 519:966–975

Numberger D, Ganzert L, Zoccarato L, Mühldorfer K, Sauer S, Grossart HP, Greenwood AD (2019) Characterization of bacterial communities in wastewater with enhanced taxonomic resolution by full-length 16S rRNA sequencing. Sci Rep 9(1):9673

Palani G, Arputhalatha A, Kannan K, Lakkaboyana SK, Hanafiah MM, Kumar V, Marella RK (2021) Current trends in the application of nanomaterials for the removal of pollutants from industrial wastewater treatment—a review. Molecules 26(9):2799

Pintor AM, Vilar VJ, Botelho CM, Boaventura RA (2016) Oil and grease removal from wastewaters: sorption treatment as an alternative to state-of-the-art technologies. A critical review. Chem Eng J 297:229–255

Råman Vinnå L, Wüest A, Bouffard D (2017) Physical effects of thermal pollution in lakes. Water Resour Res 53(5):3968–3987

Sanghamitra P, Mazumder D, Mukherjee S (2021) Treatment of wastewater containing oil and grease by biological method—a review. J Environ Sci Health Part A 56(4):394–412

Selvarajan R, Sibanda T, Venkatachalam S, Kamika I, Nel WA (2018) Industrial wastewaters harbor a unique diversity of bacterial communities revealed by high-throughput amplicon analysis. Ann Microbiol 68(7):445–458

Shah MP (2024a) Microbial remediation of hazardous chemicals from water & industrial wastewater treatment plant. Springer Nature

Shah MP (2024b) Development in waste water treatment research and processes: applied technologies for clean up of environmental contaminants. Elsevier

Shah MP (2024c) Aerobic and anaerobic microbial treatment of industrial wastewater. CRC Press

Shah MP (2025) Extraction of precious metals from industrial waste using microbial technology. Springer Nature

Sharma P, Pandey AK, Kim SH, Singh SP, Chaturvedi P, Varjani S (2021) Critical review on microbial community during in-situ bioremediation of heavy metals from industrial wastewater. Environ Technol Innov 24:101826

Shrestha R, Ban S, Devkota S, Sharma S, Joshi R, Tiwari AP et al (2021) Technological trends in heavy metals removal from industrial wastewater: a review. J Environ Chem Eng 9(4):105688

Smitha MS, Singh S, Singh R (2017) Microbial biotransformation: a process for chemical alterations. J Bacteriol Mycol Open Access 4(2):85

Tang CC, Hu YR, He ZW, Li ZH, Tian Y, Wang XC (2024) Promoting symbiotic relationship between microalgae and bacteria in wastewater treatment processes: technic comparison, microbial analysis, and future perspectives. Chem Eng J 155703

Thavasi R, Jayalakshmi S, Banat IM (2011) Application of biosurfactant produced from peanut oil cake by Lactobacillus delbrueckii in biodegradation of crude oil. Biores Technol 102(3):3366–3372

Trebuch LM, Oyserman BO, Janssen M, Wijffels RH, Vet LE, Fernandes TV (2020) Impact of hydraulic retention time on community assembly and function of photogranules for wastewater treatment. Water Res 173:115506

Tsai SW, Schwinghammer L, Lee CH, Lin CF, Hou CH (2022) Shifts of microbial community structure along substrate concentration gradients in immobilized biomass for nitrogen removal. NPJ Clean Water 5(1):21

Varela AR, Manaia CM (2013) Human health implications of clinically relevant bacteria in wastewater habitats. Environ Sci Pollut Res 20:3550–3569

Vítězová M, Kohoutová A, Vítěz T, Hanišáková N, Kushkevych I (2020) Methanogenic microorganisms in industrial wastewater anaerobic treatment. Processes 8(12):1546

Wang S, Ping Q, Li Y (2022) Comprehensively understanding metabolic pathways of protein during the anaerobic digestion of waste activated sludge. Chemosphere 297:134117

Wang M, Chen S, Jia X, Chen L (2021) Concept and types of bioremediation. In: Handbook of bioremediation. Academic Press, pp 3–8

Wang WX (2016) Bioaccumulation and biomonitoring. In: Marine ecotoxicology. Academic Press, pp 99–119

Whelan MJ, Coulon F, Hince G, Rayner J, McWatters R, Spedding T, Snape I (2015) Fate and transport of petroleum hydrocarbons in engineered biopiles in polar regions. Chemosphere 131:232–240

Wu Y, Li T, Yang L (2012) Mechanisms of removing pollutants from aqueous solutions by microorganisms and their aggregates: a review. Biores Technol 107:10–18

Yang Si-Zhong, Hui-Jun JIN, Zhi WEI, Rui-Xia HE, Yan-Jun JI, Xiu-Mei LI, Shao-Peng YU (2009) Bioremediation of oil spills in cold environments: a review. Pedosphere 19(3):371–381

Yun YM, Lee E, Kim K, Han JI (2019) Sulfate reducing bacteria-based wastewater treatment system integrated with sulfide fuel cell for simultaneous wastewater treatment and electricity generation. Chemosphere 233:570–578

Zangi-Kotler M, Ben-Dov E, Tiehm A, Kushmaro A (2015) Microbial community structure and dynamics in a membrane bioreactor supplemented with the flame retardant dibromoneopentyl glycol. Environ Sci Pollut Res 22:17615–17624

Zwolak A, Sarzyńska M, Szpyrka E, Stawarczyk K (2019) Sources of soil pollution by heavy metals and their accumulation in vegetables: a review. Water Air Soil Pollut 230:1–9

Chapter 3
Utilization of Microalgae for Wastewater Treatment and Sustainable Biofuel Production: Challenges and Environmental Perspective

Pranjal Roy, Esha Dutta, Subhra Satahrada, Subrat Kumar, and Ritesh Pattnaik

Abstract The global population and demand for alternative energy sources have proliferated extensively due to extensive depletion of fossil fuels and non-renewable energy sources. Biofuels offers a promising alternative resource which are renewable and possess the ability to mitigate greenhouse gas emissions. Organic matter rich in nutrients including C, N, and P are optimized by wastewater treatment plants. Interestingly, wastewater treatment facilities can be transformed to generate biofuels through incorporation of technologies. Biofuel extraction from wastewater containing mixed pollutants includes a number of cutting-edge methods, microbial fermentation, anaerobic digestion, and algae cultivation. In the absence of oxygen, anaerobic digestion breaks down organic waste via microbial activities, yielding biogas as a byproduct. Besides biogas production, organic substrates can be converted by microbial fermentation into ethanol and biodiesel as an alternative energy and transportation applications. Approaches toward next generation biofuel solutions promote the cultivation of microalgae coupled with wastewater treatment to remove lipids from mixed micropollutants to obtain potential extracts of biofuel which are further treated by biological aerated filters (BAF). Wastewater purification provides optimal conditions for microalgal growth, which are extensively effective at converting light and nutrients into biomass. Algal biomass can be hydrothermally liquefied to produce biocrude directly, or by bio decomposition to remove lipids for synthesis and upgradation to biodiesel. The absorption of more nutrients and carbon dioxide (CO_2) allows improvement of overall treatment efficiency toward carbon footprinting and sustainability. More importantly, wastewater treatment strategy encourages resource conservation, lowers environmental risk, and supports circular bioeconomy. The present chapter provides insights into the utilization of microalgae for wastewater treatment and sustainable biofuel production.

P. Roy · E. Dutta · S. Satahrada · S. Kumar · R. Pattnaik (✉)
School of Biotechnology, KIIT Deemed to be University, Bhubaneswar, Odisha, India
e-mail: rpattnaik@kiitbiotech.ac.in

© The Author(s), under exclusive license to Springer Nature Singapore Pte Ltd. 2026
S. Dey and M. P. Shah (eds.), *Microbial Applications in Industrial Wastewater Treatment*, Advances in Wastewater Research,
https://doi.org/10.1007/978-981-96-9128-9_3

Keywords Biofuels · Wastewater treatment · Renewable energy · Sustainability · Microalgal treatment

3.1 Introduction

The production of biofuels has become imperative in today's world because of their ability to reduce greenhouse gas emissions and dependence on fossil-fuel energy. Research investments, development, and implementation of biofuel production have the potential to encounter new economic prospects, stimulate environmental development, and enhance energy security. Aquatic organisms known as microalgae combined with wastewater treatment provide a respite to major global concerns, water pollution and energy scarcity with no negative impact on the environment. Microalgae have the ability to photosynthesize which is cost-efficient and avoids use of renewable energy resources. A cleaner aquatic environment can be achieved through wastewater management, and the algal biomass generated by this technique can be further processed to have the potential to manufacture biofuels and other value-added products. Biofuels processed from microalgae cultivation are a practical green energy alternative to fossil fuels. Lipids that are produced by microorganisms can be utilized to produce biodiesel. Furthermore, by using a fermentation process, microalgae have a great deal of potential for producing bioethanol and biogas (Priya et al. 2022). Maximizing the growth parameters—light, temperature, and nutrient availability—is crucial for growing microalgae on an industrial scale. The issue of efficiently and economically extracting algal biomass also needs to be addressed. It's also critical to examine the energy input required for the entire process to determine whether producing biofuels actually results in a net energy gain. By lowering our reliance on fossil fuels, we can use microalgae to generate electricity which in turn contributes to the reduction of greenhouse gas emissions. Therefore, considering microalgae as a sustainable solution, a comprehensive life cycle model is required in light of environmental factors (Shah 2025; Muller et al. 2014). Biofuel production through microalgae cultivation is being constantly examined and tested for effects on biodiversity, water use, and land use. To further reduce adverse ecotoxic consequences, the byproducts produced during the manufacture of biofuels are also being recycled or reused to reduce carbon emission and pave a way toward sustainable environment.

3.1.1 The Global Energy Crisis and Importance of Sustainable Alternatives

The world is currently amid an unprecedented rise in the global energy crisis, which has prompted the urgent need for viable long-term solutions to ensure environmental sustainability. The depletion of finite resources and increasing demand for energy are at the center of crisis, contributing to environmental degradation and climate change. It has been explicitly evident that relying solely on non-renewable sources of energy expedites has immediate implications on the energy demand market and economies. The major reasons for the global energy deficit are the effects of unequal population growth, degradation in infrastructure owing to energy waste, and widespread overconsumption, which leads to greater energy exploitation. The negative impact of fossil-fuel depletion has led to a growing concern for future energy production and consumption and requires an immediate reorientation of energy policies revolving around fossil-fuel infrastructure. It is crucial to exploring and implementing sustainable alternatives through technological advancements and innovation without causing harmful effects on the environment. Investments in zero-emission energy systems are a sustainable alternative approach that can meet the increasing energy demand of a growing population and mitigate the potential risks associated with the global energy crisis.

3.1.2 Biofuels: Potential Alternative as Renewable Energy

Biofuels are a promising alternative to fossil fuels produced from organic biomass materials known as feedstocks. The production of biofuels involves the fermentation, distillation, and transesterification of biomass into reusable fuel sources, mainly ethanol, biodiesel, and biogas. These derivatives of fuel from organic materials such as plant, algae, and animal waste can be integrated into existing infrastructure with minute modifications, making them suitable alternatives to traditional non-renewable sources of energy. Wood and straw were the most common sources of natural fuels in the early days and were eventually replaced by fossil fuels. In modern times, advancing technologies have come forward that foster the reuse of biotic sources by converting them into solid, liquid, and gaseous fuels known as biofuels (Shah 2024a; Quéméneur and Marty 1994). The utilization of biofuels offers a significant reduction in the release of carbon dioxide during combustion into the atmosphere, and the promotion of the growth of biofuel-producing crops can stimulate agricultural economies through diversification of the energy mix and reduction of reliance on imported finite energy resources (Subhash 2022).

3.1.3 Production of Biofuel from Wastewater

The constant increase in the activities of industrialization, urbanization, and overpopulation has driven a worldwide increase in wastewater production. The process of extracting biofuel from wastewater involves a series of techniques where lipid molecules are removed from mixed microbial pollutants and processed in biological wastewater treatment plants to obtain potential feedstocks for biofuel processing

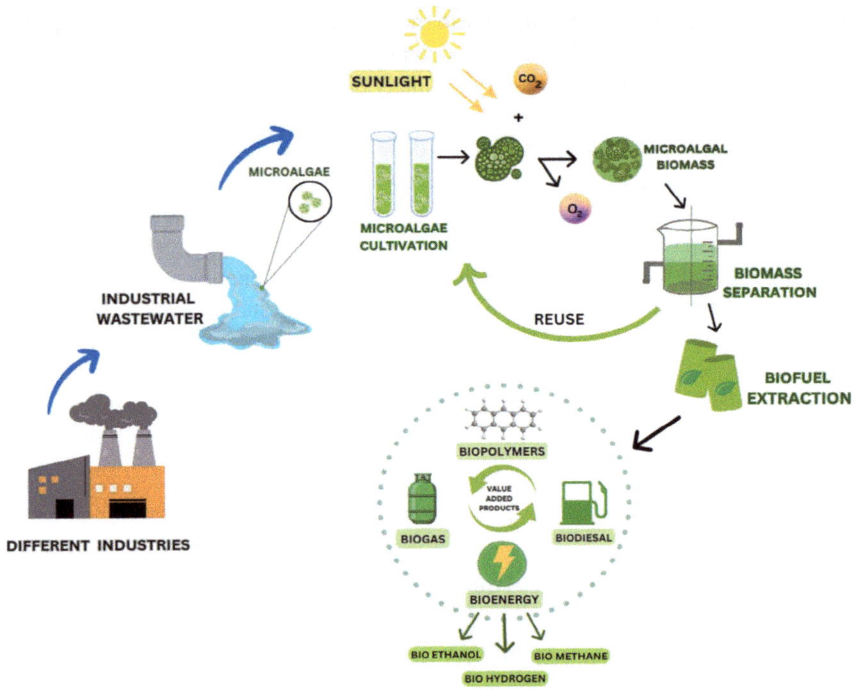

Fig. 3.1 An overview of biofuel production from microalgae

(Fig. 3.1). The wastewaters contain heavy amounts of lipids, which represent huge portions of the organic fraction, with large majorities of triacylglycerols (TAGs) and minor parts of free long fatty acid chains (Shah 2024b; Cambero et al. 2016). These lipid-rich biomasses are assimilated from the wastewater with the help of specialized microalgae present in biological wastewater plants. They undergo microbial oxidation followed by primary and secondary sludge treatment by chemical transesterification to produce fatty acid methyl esters (FAMEs), a form of biodiesel. Biofuel production facilitates the mitigation of environmental impacts in wastewater disposal and avenues for progress toward achieving a sustainable energy future (Arampantzi and Minis 2017).

3.2 Wastewater as a Resource for Biofuel

The organic matter of wastewater can be transformed into biogas and biodiesel, which can be used as energy sources for the production of biofuel. Biofuel produced from wastewater are a sustainable, resource- and environmentally-friendly alternative to fossil fuels. The process of anaerobic digestion (AD) is one of the most promising ways to turn wastewater into biofuels. In this method, bacteria that function in the absence of oxygen break down organic materials to produce biogas, which is mostly composed of carbon dioxide and methane (Jayaseelan et al. 2021). Biogas can be used directly for heating or electricity generation, or it can be processed further to produce biomethane, which can be used as fuel for automobiles (Shah 2024c; Kothari et al. 2014). Another possible approach is to grow microalgae in sewage ponds. Lipids with relatively high lipid productivity, such as microalgae, have been harvested for biodiesel production. In addition, they absorb nutrients such as nitrogen and phosphorus, which pollute untreated water bodies. More importantly, biofuel production can be combined with water treatment to increase overall efficiency, which also results in lower operating costs. In addition, these plants recycle waste to produce energy that can be used for other purposes, while minimizing environmental pollution from sewage sludge (Bhatt et al. 2014). Overall, using wastewater as a resource to produce biofuels is a promising path toward the development of sustainable energy. Along with addressing waste management issues, it helps move the energy industry in the direction of greater environmental friendliness.

3.2.1 Wastewater Specification for Biofuel

Significantly, the composition of effluents can vary greatly depending on the source or treatment they undergo. This pertains to the successful conversion of biofuels like biogas or biodiesel and involves the specifications and features of wastewater for biofuel production.

(a) **Organic Content**: Factors like Chemical Oxygen Demand (COD) and Biochemical Oxygen Demand (BOD) affect how renewable the wastewater is. Furthermore, a high organic content could indicate a promising future for carbons that can be fermented by microorganisms in an anaerobic method to produce biogas. The results showed that increased wastewater COD/BOD increases the amount of biogas mobilized (Rawat et al. 2013).
(b) **Nutrient Level**: Phosphorus (P) and Nitrogen (N) are important because they help microorganisms flourish, like bacteria in anaerobic digestion or algae in algae farming. Certain substances, like phosphates, are harmful to aquatic life but also promote the growth of algae, which is necessary for the generation of biodiesel (Ayala-Parra et al. 2017). Negative handling of high nutrient concentrations that end up in the water, however, can also result in eutrophication (Mo et al. 2018).
(c) **Alkalinity and pH**: The biological makeup of the microorganisms now living in the wastewater line streams is influenced by the pH of the water (Hadhoum et al. 2021). The ideal pH ranges for enzymatic digestion are thus generally comparable, but they vary depending on the microorganisms used and are often in the range of neutral to slightly alkaline 6.5–8.5. Alkalinity is important because it provides a supply of acidity or base to regulate pH fluctuations in a body of water and maintain stability for microbiological processes.

(d) **Toxicity**: Toxic substances that degrade water quality, such as organics, heavy metal ions, and inhibitors, may be present in industrial effluent. These have the potential to slow down microbial growth, which in turn may affect how quickly metabolic events necessary to produce fuel occur. Because of previous treatment or dilution to safe levels that won't put consumers at risk of contamination, the harmful levels frequently need to be increased.
(e) **Solids Content**: Because suspended solids add particle matter into the process, the amount of sludge or suspended solids in wastewaters can have an impact on some phases of the synthesis of biofuels. Elevated solid content can obstruct the microbial conversion process or algal formation, making it difficult for the organism to consume or develop. The feed must frequently be filtered or sedimented before being added to the biological process.
(f) **Temperature**: Differences in wastewater's temperature have an impact on the rates of microbial growth and the processes that occur. Many of the biological processes that produce biofuels are most effective in specific temperature ranges (e.g., mesophilic or thermophilic bioconversion for anaerobic digestion). Seasonality's effect on the temperature of wastes collected for treatment periodically causes possible conditions to shift.
(g) **Fatty Acid Composition**: On the other hand, the fatty acid composition of the algal biomass is crucial for using microalgae produced from wastewater to produce biodiesel. Higher lipid contents and specific fatty acid properties, such as high linoleic or oleic acid values, are advantageous for the generation of biodiesel.
(h) **Water Quality Regulations**: In order to ensure that treated water effluent complies with environmental legal reforms, it is critical to adhere to local water quality requirements. According to these standards, the concentrations of pollutants in the effluent must be kept to a minimum. These levels may include BOD, TSS, and nutrient values.

3.2.2 Potential Benefits of Wastewater as Feedstocks

Wastewater offers significant potential as a feedstock for various processes aimed at biofuel production, presenting several benefits across environmental, economic, and resource utilization perspectives. Wastewater also has high potential as a feedstock in the processes intended for biofuel generation, proving advantages on multiple levels under environmental, economic, and resource utilization considerations.

(a) **Resource Utilization**: This is accurate since wastewater is regarded as a plentiful resource that can be replaced because it is a steady stream. Rather than being limited to a disposal problem, the organic contents can now be utilized to create biofuels. Wastewater is essentially treated, disposed of, and reused. Because this method repurposes garbage into goods that are valuable, it can also be said to align with the concepts of the circular economy.

(b) **Energy Recovery**: Wastewater contains biological waste that can be effectively digested to create energy-dense gases and fuels, such as biogas (methane). They can be converted into biomethane for use in automobiles or utilized directly for heating and electricity production. This procedure also contributes to the development of alternative energy sources and lessens dependency on fossil fuels.

(c) **Reduction of Greenhouse Gas Emissions**: If methane is extracted from wastewater by the anaerobic digestion process, biofuel production can lower greenhouse gas emissions. Since methane, a potent greenhouse gas, is the byproduct of biogas production, it is better for the environment to use methane generated from waste products for energy generation.

(d) **Water Quality Improvement**: To produce biofuel, wastewater must be filtered, which basically means that organic molecules and nutrients like nitrogen and phosphorus must be removed. In addition to improving the quality of the water, this procedure lessens the impact of releasing wastewater into other natural water sources. Being sources of wastewater purification, they also help remove nutrients from the solution.

(e) **Biodiesel Production**: Wastewater has the ability to support microalgae and contains lipids that can be converted into biodiesel. Algae farming in wastewater ponds not only produces sustainable feedstock but also improves wastewater treatment efficiency because of the algae's capacity to absorb fertilizers and carbon dioxide, which has positive effects on the environment and economy (Plácido and Capareda 2016).

(f) **Promotion of Sustainable Practices**: Wastewater management is an essential component of the biofuel production process because it promotes the most environmentally friendly ways to manage industrial and human waste while producing electricity. Thus, it benefits the water and energy industries by advancing the development of novel approaches and techniques in wastewater treatment together with bioprocess engineering (Kothari et al. 2012; Abdo et al. 2016).

Consequently, there are many benefits to using wastewater as feedstock for biofuel production, including resource management, energy recovery, reduced greenhouse gas emissions, better water quality, financial opportunities for producing decentralized energy, and the effective application of sustainable practices (Steven et al. 1943). This innovative concept of using wastewater for biofuels addresses issues related to wastewater management and simultaneously facilitates the deployment of renewable energy sources and the circular economy in resource management (Abdo et al. 2016).

3.3 Microalgae Production Technology

Algae are a diverse group of species found in the water columns and sediment. Microalgae are also a type of algae, microscopic and single cellular in nature, that inhabit a variety of aquatic environments. Microalgal cells have high growth rate, high biomass productivity with high extent of bioactivities and metabolic secretion making them ideal feedstocks for biofuel production. The microalgae biomass production processes consist of three important steps: cultivation followed by harvesting or collection and final processing.

3.3.1 Microalgae Cultivation Growth and Requirements

Microalgae are autotropic in nature and are grown in the presence of carbon dioxide and inorganic compounds like phosphorous and nitrogen. Microalgae production technology involves cultivation of microalgae under optimal conditions for their proper growth, development, and reproduction. The optimal environmental conditions required for cultivation of microalgae include abundant exposure to light and maintenance of stable temperature ranging from 20 to 30 °C. Nutrient availability serves as a critical factor in microalgae cultivation as the growth of microalgae requires a balanced supply of nutrients and trace elements to ensure microalgal growth and metabolism (Khan et al. 2021). A proper monitoring of all the necessary parameters influencing the microalgal cultivation is to be ensured. The harvesting and processing methods involve various techniques such as flotation, centrifugation, filtration, and drying. To obtain large-scale growth of microalgae, industrial growth cultivation systems are used where paddle wheels are used to circulate the microalgal cells, nutrients, and water. These are open systems which are constantly being exposed to the environment. In order to increase productivity and lower the contamination risks, closed cultivation systems such as photobioreactors and biorefineries are preferred (Huang et al. 2010).

3.3.2 Harvesting Techniques and Bioprocess

Microalgae harvesting techniques are a series of complicated processes which includes both mechanical and chemical approaches. The cell size, cell density, and the volume of the culture affect the overall harvesting techniques for microalgae cultivation. The main focus of the harvest procedures is to ensure maximum extraction of microalgal biomass from the growth medium (Deepa et al. 2023). Bioprocess techniques encompass the use of specially designed vessels called bioreactors which provide a resilient and controlled environment for microalgae cultivation. Genetic modifications in the microalgae coupled with wastewater treatment methods increase the production of lipids and proteins in microalgal biomass. The extraction procedures focus on cell wall disruption for release of intracellular components, cell separation from the growth culture medium and other contaminants, and concentration methods to increase the density of microalgae biomass to facilitate processing (Sivaramakrishnan et al. 2022; Mehrgan and Shekarabi 2022).

The various harvesting techniques approached for obtaining microalgal biomass are highlighted in Table 3.1.

3.3.3 *Biofuel Extraction Through Biorefineries*

Biofuel extraction through biorefineries is defined as the process where multiple extracts of biofuel are obtained from the microalgal biomass. Biorefineries utilizes feedstocks and conversion procedures to produce more sustainable biofuels and reduce carbon emissions. The microalgal biorefinery processing technique follows two main stages known as upstream and downstream processing.

The upstream processing involves the microalgae strains, nutrient sources for growth and cultivation, and proper light source. The downstream processing involves harvesting of microalgae cells, extraction, and purification of the targeted microalgae cells. In general. the various types of biofuels extracted from biorefineries are biodiesel, bioethanol, biogas, biohydrogen, bio methanol, pyrolysis oil, and aviation biofuels (Kumari et al. 2022).

3.4 Biofuel Generation Through Anaerobic Digestion of Wastewater Effluents

Anaerobic digestion (AD) is a biological process that produces biogas through the utilization of microorganisms to break down organic materials in the absence of oxygen. Methane (CH_4) and carbon dioxide (CO_2) make up the majority of this biogas, which is used as a renewable energy source and input in the manufacturing of biofuels (Skaggs et al. 2018).

The process of microbially digesting down organic feedstocks in a regulated setting, such as an anaerobic digester, is known as anaerobic digestion. Crop leftovers, post-harvest food leftovers, animal manure, sewage, and other organic waste sources high in carbon can all be used as substitute feedstocks. There are various phases in the digestive process:

1. **Hydrolysis**: In this context, end products of hydrolysis include carbohydrates, proteins, and lipids produced by fermentative bacteria and enzymes. Soluble organic compounds that are further broken down by acidogenic bacteria are released during this phase.

Table 3.1 The various harvesting techniques approached for obtaining microalgal

Sl. No.	Various harvesting techniques	Principle	Types
1	Centrifugation	This particular method applies centrifugal force field at high speeds which allows the denser algal cells to separate from the liquid culture medium. Centrifugation allows efficient isolation of microalgal cells from the growth media	The different kinds of centrifugation techniques are continuous flow centrifugation, batch centrifugation, disc-stack centrifuges, tubular bowl centrifugation
2	Filtration	It is a form of physical method which separates the solids from liquids as the liquid passes through the various sized pores and is membrane bounded process	The different kinds of filtration techniques suitable for large-scale separating operations are microfiltration, nanofiltration, ultrafiltration, vacuum filtration, and pressure filtration
3	Flocculation	A special type of separation technique where microalgal cells are accumulated from surface. It involves addition of various kinds of chemicals and organic compounds known as flocculants and bio-flocculants into the culture making the microalgae cells to clump together as flocs and settle down for harvesting	The different kinds of flocculation techniques are chemical flocculation, bio-flocculation, pH induced, electrolyte addition, magnetic flocculation
4	Flotation	Focuses on the natural buoyancy of the lipid-rich microalgae cells by addition of various flotation agents to the microalgae culture medium which attaches to the cells and allows them to float to the surface layers	The different kinds of flotation techniques are dissolved air flotation, induced air floatation, electroflotation, bio-flotation
5	Sedimentation	It is one of the most traditional and simple techniques which exerts gravitational force on the culture medium and allows the microalgae cells to settle down over time	The different kinds of sedimentation techniques are plain sedimentation, settling tanks or ponds, inclined plate settlers, lamella clarifiers
6	Electro-chemical harvesting	It is a new emerging technique which uses electric currents for coagulation, destabilization, and aggregation of microalgae cells which separates them from the culture medium	The different kinds of electro-chemical harvesting techniques are electro-coagulation, electro-floatation, electro-dewatering

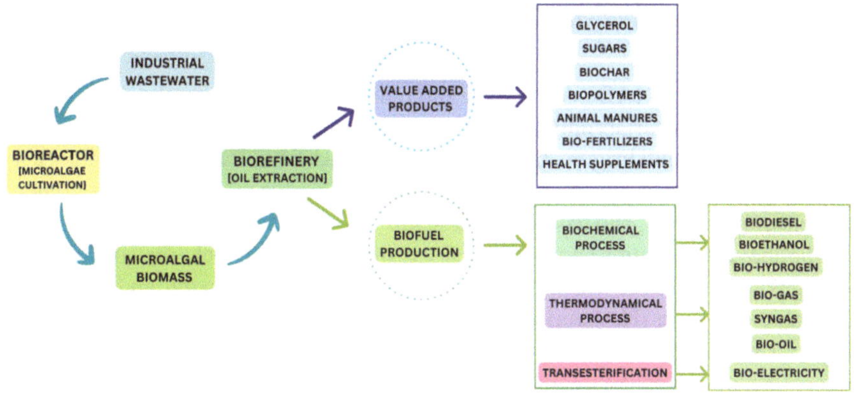

Fig. 3.2 Schematic representation of biofuels and value-added products

2. **Acidogenesis**: From the soluble components of the hydrolysis process, acid-generating biomass is transformed by acidogenic bacteria into volatile fatty acids, alcohols, and other organic acids, possibly including acetic acid (Kumari et al. 2022). Intermediate products produced by this fermentation process serve as food for methanogenic bacteria.
3. **Methanogenesis**: Utilizing biochemical processes, methanogenic archaea subsequently turn the intermediate products—such as acetic acid, hydrogen, and carbon dioxide—into methane and carbon dioxide (CO_2). Methane produced spontaneously by the biological process is collected as biogas, which can be used for a variety of energy-related purposes (Kumari et al. 2022; Skaggs et al. 2018). The production of biofuel via anaerobic digestion has several important and advantageous aspects. The production of biofuel via anaerobic digestion has several important and advantageous aspects.

    1. **Energy Production from Renewable Sources**: Anaerobic digestion processes produce biogas, which can be burned directly for heating, power generation, and vehicle fuel. Alternatively, it can be processed for natural gas pipeline injection or sufficiently refined to be used as vehicle fuel.
    2. **Mitigation of Greenhouse Gas**: Anaerobic digestion captures methane, a more potent greenhouse gas than CO_2 with a far higher potential for global warming, while also controlling greenhouse gas emissions (Ghosh et al. 2020; Banks and Heaven 2013).
    3. **Management of Waste Products**: Anaerobic digestion (AD) in Relation to Waste Management by offering a carbon-neutral method of handling their organic waste streams.
    4. **Recovery of Nutrients**: The digestate left over after anaerobic digestion can be applied to the land to improve soil health and productivity. This closed-loop method replenishes the soil with nutrients from organic waste, promoting sustainable agriculture.

3.4.1 Types of Feedstocks Suitable for Anaerobic Digestion

Anaerobic digestion (AD) is a diverse biological process that breaks down a wide range of organic materials into digestate, a nutrient-rich product, and biogas (Maneein et al. 2018; Guiot and Frigon 2012). In general, a feedstock (Table 3.2) suitability for anaerobic digestion is determined by its amount of organic matter, moisture content, nutrient composition, and ease of breakdown.

The above information shows that a variety of organic feedstocks may be converted to biogas and a digestate with high nutritional content through flexible and sustainable anaerobic digestion (Hashemi et al. 2021). Because anaerobic digestion (AD) can handle a wide range of feedstocks, different industries (such as agriculture, food processing, municipalities, and industry) may simply manage organic waste ethically and offset their carbon liability while creating high-quality fertilizer and renewable energy (Elsayed et al. 2020; Torres et al. 2013). The efficiency and economics of generating these feedstocks for the production of bioenergy would also be improved by additional study and technology advancements on AD-related procedures.

3.5 Technical Challenges and Limitations of Using Microalgae

Making use of microalgae for wastewater treatment as well as sustainable biofuel production has a massive potential but at the same time comes up with several difficulties and restraints that need to be solved so that it can be practically implemented and can be made environmentally sustainable.

(a) **Nutrient Requirements**: The growth of microalgae depends on phosphorus and nitrogen levels. While excessive removal of nutrients could cause eutrophication, efficient nutrient removal in wastewater treatment applications necessitates careful consideration of the balance of nutrients (Mir et al. 2016).
(b) **Light and Temperature**: Subsequently microalgae need light to thrive; having an adequate supply of light at the right temperature is also crucial for promoting the most effective for growth in microalgae (Mir et al. 2016). It's frequently required to have highly regulated settings or employ alternative cultivating techniques because variations in natural light and temperature have an impact on the rate of growth (and overall productivity).
(c) **Strain Selection**: Choosing algae strains with high lipid content and wastewater tolerance is important for producing biofuels. Variations in the lipid content of strains as a result of their varied uptake of nutrients and growth affect the suitability of microalgae for dual-purpose applications.
(d) **Harvesting and Dewatering**: Harvesting and dewatering the microalgae biomass effectively are the most difficult technological difficulty in the production of microalgae biomass. Several procedures are used, such as centrifugation,

Table 3.2 Feedstocks suitable for anaerobic digestion to produce biofuels

Sl. No.	Types of feedstocks	Sub-types	Examples
1	Agricultural waste	i. Crop residue	Crops residues: wheat, rice, maize, crop leftovers like stalks, leaves, husks, and straw
		ii. Animal manure	Animal manure: waste from animal including cattle, poultry, pigs, and dairy operation
		iii. Poultry litter	Poultry litter: animal waste and bedding components from chicken farms are combined to create poultry litter
		iv. Haylage and silage	Haylage and silage: hay and silage are frequently utilized in livestock feeding operations
2	Agro-industrial and food processing waste	i. Food waste	Food waste: leftover food from homes, grocery, and food processing facilities
		ii. Distillery/brewery waste	Distillery/brewery waste: remainders and wasted grains from the brewing and distilling processes, basically yeast slurry
		iii. Waste from dairy industry	Waste from dairy industry: waste products from dairy processing—dairy processing residues, lactose
		iv. Starches and sugar industry residue	Starches and sugar industry residues: products left over from the processing of sugar beets, bagasse from sugarcane, and byproducts of the starch industry
3	Solid waste generated by cities and towns and sludge from sewage	i. Organic fraction of MSW (OFMSW)	Organic fraction of MSW (OFMSW): made up of biodegradable parts of products, including food remnants, yard and garden waste, and paper products
		ii. Sewage sludge (bio-solids)	Sewage sludge (bio-solids): sewage sludge is highly unstable and the varied composition of the sludge is difficult to stabilize for anaerobic digestion with high biogas and nutrient-rich digestate production for agriculture use
4	Biomass and energy crops	i. Energy crop	Energy crop—crops harvested for biomass production include corn, sorghum, switch grass and waist
		ii. Crop residue	Crop residue—after harvesting energy crops, the components that weren't needed were left behind, including wheat stubble, cornstalks, and grass clippings
5	Agro-industrial and industrial wastes	i. Pulp and paper industry residue	Pulp and paper industry residue—paper mill wastewater treatment sludge and pulp are the two of the most challenging waste

(continued)

Table 3.2 (continued)

Sl. No.	Types of feedstocks	Sub-types	Examples
		ii. Oil and grease residue	Oil and grease residue—waste from the manufacturing of plant oils, fat traps and food containing fats, and oils to boost the production of biogas
		iii. Agro-industrial byproducts	Agro-industrial byproducts—products that are not used in the manufacturing of food but are instead byproducts of creating and processing food like hells and waste from the processing of fruits and vegetables
6	Streams of neglected organic waste	i. Green waste	Green waste—waste materials from the maintenance of yards such as grass blankets, tree stems, and leaf piles
		ii. Waste from fish processing industry	Waste from fish processing industry—fish offal, heads, and frames from fish processing processes are included
		iii. Algal biomass	Algal biomass—algae grown in sewage or isolated ponds specifically to produce biomass themselves

filtration, and flocculation, which have high energy requirements and might be costly for perhaps large-scale operations to be effective (Ward et al. 2014).

(e) **Biofuel Production Efficiency**: Energy-intensive procedures for removing lipids from microalgae and turning them into biofuels, such as biodiesel, include drying, extracting, and transesterification. Enhancing the conversion process and raising lipid yields were essential for producing biofuel more profitably and efficiently.

Overall, microalgae provide potential uses for wastewater treatment and production of sustainable biofuels which must be balanced against the challenges and environmental considerations highlighted in this review if they are to progress to full commercial utilization. Overcoming some of the limitations mentioned above to maximize the benefits of microalgae as promising source at a large scale is highly needed and only be facilitated via research, innovation, and policy facilitation (Okolie et al. 2022).

There are a number of technical challenges that must be successfully addressed in order for biofuel production in wastewater to become both economically and environmentally sustainable. Biofuels produced from wastewater provide a double benefit of resource recovery and wastewater treatment. The challenges are due to the complexity of wastewater and energy-intensive processes. Minot solved the ten following difficulties of technical problems for biofuel production from wastewater with simplest solutions to achieve.

(a) **Variability in Composition of Wastewater**: Wastewater's composition varies depending on its source (agricultural, municipal, or industrial) and the methods used for treatment (Corato et al. 2022). This variation has a significant impact on the production and growth of microorganisms used to produce biofuels, such as bacteria or microalgae. The presence of inhibitors (such as heavy metals and organic pollutants), variations in pH and salinity, and nutrient levels all have an impact on the rates of microbial growth, lipid accumulation, and biofuel production.

(b) **Selection of Microorganisms and Acclimatization**: Determining the right microorganisms with various wastewater characteristics is crucial. Microalgae, bacteria, and fungi are the most studied microorganisms for biofuels; they all have distinct development environments and metabolic routes (Corato et al. 2022; Luque et al. 2021). The creation of robust species—those that can withstand extremes in effective toxicity—that can effectively utilize wastewater nutrients and still produce high lipid content or route-specific biofuels, like biohydrogen, is a major problem. Several genetic and metabolic engineering techniques are being researched to increase these bacteria' productivity.

(c) **Nutrient Limitations and Imbalances**: Despite wastewater containing a lot of the nutrients (such as nitrogen and phosphorus) is needed for microbial growth, there may be nutritional imbalances or deficits. The nutrients cause microbial overgrowth and oxygen (O_2) depletion in bioreactors, which can lead to algal blooms or other operational problems. However, nutrients may limit the productivity of biomass and the yields of biofuels. Maximizing nutrient recovery from wastewater and regulating nutrient inputs are necessary for sustainable biofuel production (Luque et al. 2021).

(d) **Pretreatment of Wastewater and Compatibility**: In order to guarantee the removal of contaminants, heavy metals, toxins, and organic pollutants that may impede microbial development or the production of biofuel, wastewater must be properly pretreated (Ruan et al. 2019). By using certain pretreatment techniques including filtration, coagulation, activated carbon adsorption, and advanced oxidation processes (AOPs), the wastewater's quality is further improved to the point where the great majority of the requirements for the microorganisms are compromised. There is a cost and complexity associated with these pre-treatment activities.

(e) **Biomass Harvesting and Dewatering**: Microorganism harvesting and solids extraction from liquid wastewater are prohibitively expensive operations that require increased energy efficiency (Luque et al. 2021). Utilizing techniques including centrifugation, filtration, flocculation, and electrocoagulation, biomass is harvested and dewatered. Regrettably, these techniques frequently need specialized equipment, a lot of energy, or chemicals. Current harvesting technology and more efficient, low-energy harvesting methods are common areas of research.

(f) **Conversion Processes are Energy-Intensive**: Due to a number of procedures involved in its pretreatment, drying, lipid extraction, transesterification (biodiesel), hydrolysis, and fermentation (bioethanol/biohydrogen), the conversion of biomass to biofuels is energy-intensive. When it comes to sustainability, efficient conversion is preferred in order to conserve resources and energy while maintaining the lowest feasible levels of pollution. However, several cutting-edge methods, such as supercritical fluid extraction, ultrasound assisted extraction, and various enzymatic processes, show great promise and still require development and scaling up (Ruan et al. 2019; Priya et al. 2022).

Although there are promising prospects in the generation of bioenergy and resource recovery using wastewater, the technological bottlenecks must be addressed for its widespread use. Overcoming some of the current challenges will require integrating advances in microbial engineering, bioprocess optimization, and integrated system design to unlock the promise of wastewater as a renewable feedstock for biofuels. Further research, the cooperation with businesses and politicians, and investment in innovative technology are yet areas of development to allow sustainable biofuel production from wastewater.

3.6 Present Challenges and Future Perspectives

Green biorefineries, particularly based on utilization of microalgae in recent times, have acquired attention for treatment of diverse wastewater generated from diverse industries via technological advancements in order to produce various types of biofuels as well as other important biotechnological, industrial, and environmental applications (Suparmaniam et al. 2019). However, there are still several significant challenges associated with biofuel production that need to be addressed, which include lower biomass productivity, harvesting of microalgal biomass, and energy consumption, as well as higher production cost for cultivation of microalgal biofuel on a large scale. To overcome these key challenges, wastewater treatment is a promising technique that employs microalgae cultivation, which directly or indirectly limits both carbon and nutrient economy. However, microalgae production on commercial basis, especially for the production of biofuel, offers constraints due to its excessive operational cost and efficient harvesting technique. Hence, potential and advanced perspectives toward cultivation of algal biomass production need to be explored for potential application in different sectors. As wastewater is often loaded with different types of pollutants including heavy metals, effective and high output-based extraction technology like supercritical fluid-based techniques pose as ideal process for wastewater valorization involving algal species. Besides, advanced and modern bioengineering tools also could be of additional benefit toward extraction of important bioactive compounds and energy.

Acknowledgements The authors acknowledge their Department and Institute for providing essential facilities and support.

Author Contributions PR, ED, SS, and SK have contributed to the conception of the study, designing, writing, and preparation of figures and tables of the manuscript. RP contributed significantly in the final editing and revision of the manuscript.

Consent for Publication All authors have consented to the manuscript for publication.

References

Abdo SM, El-Enin SAA, El-Khatib KM, El-Galad MI, Wahba SZ, El Diwani G, Ali GH (2013) Preliminary economic assessment of biofuel production from microalgae. Renew Sustain Energy Rev 55; Sci Total Environ 463–464:541–551

Arampantzi C, Minis I (2017) A new model for designing sustainable supply chain networks and its application to a global manufacturer. J Clean Prod 156:276–292

Ayala-Parra P, Liu Y, Field JA, Sierra-Alvarez R (2017) Nutrient recovery and biogas generation from the anaerobic digestion of waste biomass from algal biofuel production. Renew Energy 108:410–416

Banks CJ, Heaven S (2013) 6—Optimisation of biogas yields from anaerobic digestion by feedstock type. In: Wellinger A, Murphy J, Baxter D (eds) The biogas handbook, Woodhead Publishing series in energy. Woodhead Publishing, pp 131–165

Bhatt NC, Panwar A, Bisht TS, Tamta S (2014) Coupling of algal biofuel production with wastewater. Sci World J 210504; Challenges and opportunities. Int J Hydrog Energy 42(12):8450–8461

Cambero C, Sowlati T, Pavel M (2016) Economic and life cycle environmental optimization of forest-based biorefinery supply chains for bioenergy and biofuel production. Chem Eng Res des 107:218–235

Corato UD, Bari ID, Viola E, Pugliese M (2022) Assessing the main 159:323–344

Deepa P, Sowndhararajan K, Kim S (2023) A review of the harvesting techniques of microalgae. Water 15:3074

Elsayed M, Ran Y, Ai P, Azab M, Mansour A, Jin K, Zhang Y, Abomohra AE-F (2020) Innovative integrated approach of biofuel production from agricultural wastes by anaerobic digestion and black soldier fly larvae. J Clean Prod 263:121495

Ghosh P, Shah G, Sahota S, Singh L, Vijay VK (2020) Chapter 7—Biogas production from waste: technical overview, progress, and challenges. In: Singh L, Yousuf A, Mahapatra DM (eds) Bioreactors. Elsevier, pp 89–104

Guiot SR, Frigon JC (2012) Anaerobic digestion as an effective biofuel production technology. In: Hallenbeck P (ed) Microbial technologies in advanced biofuels production. Springer, Boston, MA; HAYATI J Biosci 24(4):163–167. ISSN: 1978-3019 (2017)

Hadhoum L, Aklouche FZ, Loubar K, Tazerout M (2021) Experimental investigation of performance, emission and combustion characteristics of olive mill wastewater biofuel blends fueled CI engine. Fuel 291:120199

Hashemi B, Sarker S, Lamb JJ, Lien KM (2021) Yield improvements in anaerobic digestion of lignocellulosic feedstocks. J Clean Prod 288:125447

Hoang AT, Sirohi R, Pandey A et al. Biofuel production from microalgae: challenges and chances. Phytochem Rev 22:1089–1126

Huang GH, Chen F, Wei D, Zhang XW, Chen G (2010) Biodiesel production by microalgal biotechnology. Appl Energy 87(1):38–46

Jayaseelan M, Usman M, Somanathan A, Palani S, Muniappan G, Jeyakumar RB (2021) Microalgal production of biofuels integrated with wastewater treatment. Sustainability 13:8797

Khan S, Siddique R, Sajjad W, Nabi G, Hayat KM, Duan P, Yao L (2021) Biodiesel production from algae to overcome the energy crisis. Sustain Energy Rev 144:110987

Kothari R, Pathak VV, Kumar V, Singh DP (2012) Experimental study for growth potential of unicellular alga Chlorella pyrenoidosa on dairy waste water: an integrated approach for treatment and biofuel production. Bioresour Technol 116:466–470

Kothari R, Pandey AK, Kumar S, Tyagi VV, Tyagi SK (2014) Different aspects of dry anaerobic digestion for bio-energy: an overview. Renew Sustain Energy Rev 39:174–195

Kumar M, Sun Y, Rathour R, Pandey A, Thakur IS, Tsang DCW (2020) Algae as potential feedstock for the production of biofuels and value-added products: opportunities and challenges. Sci Total Environ 716:137116

Kumari A et al (2022) Wastewater and solid waste as feedstock for energy production. In: Guldhe A, Singh B (eds) Novel feedstocks for biofuels production. Clean energy production technologies. Springer, Singapore

Luque R, Herrero-Davila L, Campelo JM, Clark JH, Hidalgo JM, Luna D, Marinas JM, Romero AA (2021) Departamento de Química Orgánica, Universidad de Córdoba, Campus Universitario de Rabanales, Edificio Marie Curie (C3), E-14014, Córdoba, Spain. Web: www.uco.es/senecagre encat. Green Chemistry Centre of Excellence, Chemistry Department, The University of York, Heslington, York, UK YO10 5DD Seneca Green Catalyst S.L., E-14011, Córdoba, Spain

Maneein S, Milledge JJ, Nielsen BV, Harvey PJ (2018) A review of seaweed pre-treatment methods for enhanced biofuel production by anaerobic digestion or fermentation. Fermentation 4:100

Mehrgan MS, Shekarabi SPH (2022) Electrochemical harvesting of the marine microalgae, Nannochloropsis oculata: effect on approximate composition, fatty acid profile, and metals biosorption. Iran J Fish Sci 21:33–47

Mir MA, Hussain A, Verma C, Dubey S (2016) Design considerations and operational performance of anaerobic digester: a review. Cogent Eng 3(1). Opportunities of integrated biorefining from agro-bioenergy co/by-products and agroindustrial residues into high-value added products associated to some emerging markets: a review

Mo J, Yang Q, Zhang N, Zhang W, Zheng Y, Zhang Z (2018) A review on agro-industrial waste (AIW) derived adsorbents for water and wastewater treatment. J Environ Manag 227:395–405

Muller EE, Sheik AR, Wilmes P (2014) Lipid-based biofuel production from wastewater. Curr Opin Biotechnol 30:9–16

Okolie JA, Epelle EI, Tabat ME, Orivri U, Amenaghawon AN, Okoye PU, Gunes B (2022) Waste biomass valorization for the production of biofuels and value-added products: a comprehensive review of thermochemical, biological and integrated processes. Process Saf Environ Prot

Plácido J, Capareda S (2016) Conversion of residues and by-products from the biodiesel industry into value-added products. Bioresour Bioprocess 3:23

Priya, Deora PS, Verma Y, Muhal RA, Goswami C, Singh T (2022) Biofuels: an alternative to conventional fuel and energy source. Mater Today: Proc 48(Part 5):1178–1184

Quéméneur M, Marty Y (1994) Fatty acids and sterols in domestic wastewaters. Water Res 28(5):1217–1226

Rawat I, Bhola V, Kumar RR, Bux F (2013) Improving the feasibility of producing biofuels from microalgae using wastewater. Environ Technol 34(13–14):1765–1775; Renew Sustain Energy Rev 88:326–346 (2018)

Rodionova MV, Poudyal RS, Tiwari I, Voloshin RA, Zharmukhamedov SK, Nam HG, Zayadan BK, Bruce BD, Hou HJM, Allakhverdiev SI. Biofuel production

Ruan R, Zhang Y, Chen P, Liu S, Fan L, Zhou N, Ding K, Peng P, Addy M, Cheng Y, Anderson E, Wang Y, Liu Y, Lei H, Li B (2019) Chapter 1—Biofuels: introduction. In: Pandey A, Larroche C, Dussap C-G, Gnansounou E, Khanal SK, Ricke S (eds) Biomass, biofuels, biochemicals, Biofuels: alternative feedstocks and conversion

Shah MP (2024a) Microbial remediation of hazardous chemicals from water & industrial wastewater treatment plant. Springer Nature

Shah MP (2024b) Development in waste water treatment research and processes: applied technologies for clean up of environmental contaminants. Elsevier

Shah MP (2024c) Aerobic and anaerobic microbial treatment of industrial wastewater. CRC Press

Shah MP (2025) Extraction of precious metals from industrial waste using microbial technology. Springer Nature

Sivaramakrishnan R, Suresh S, Kanwal S, Ramadoss G, Ramprakash B, Incharoensakdi A (2022) Microalgal biorefinery concepts' developments for biofuel and bioproducts: current perspective and bottlenecks. Int J Mol Sci 23:2623

Skaggs RL, Coleman AM, Seiple TE, Milbrandt AR (2018) Waste-to-energy biofuel production potential for selected feedstocks in the conterminous United States. Renew Sustain Energy Rev 82(Part 3)

Subhash V, Rajvanshi M, Raja Krishna Kumar G, Sagaram US, Prasad V, Govindachary S, Dasgupta S (2022) Challenges in microalgal biofuel production: a perspective on techno economic feasibility under biorefinery stratagem. Bioresour Technol 343:12615

Suparmaniam U, Lam MK, Uemura Y, Lim JW, Lee KT, Shuit SH (2019) Insights into the microalgae cultivation technology and harvesting process for biofuel production: a review. Renew Sustain Energy Rev 115:109361

Torres A, Padrino S, Brito A et al (2023) Biogas production from anaerobic digestion of solid microalgae residues generated on different processes of microalgae-to-biofuel production. Biomass Convers Biorefinery 13:4659–4672

Ward AJ, Lewis DM, Green FB (2014) Anaerobic digestion of algae biomass: a review. Algal Res 5:204–214

Chapter 4
Role and Importance of Biopolymers in Wastewater Treatment and Purification

Abhishek Konar, Pankaj Singh, Pranabesh Ghosh, Jasmine Kaur, Sanjukta Chakraborty, and Puspita Ganguly

Abstract Currently, wastewater treatment is the greatest challenge and is developing daily critical environmental problems. Due to the tremendous increase in the total world's population, rapid increase in industries and metropolitan cities, inadequate water treatment, and lack of efficient and cost-effective treatment processes, we are generating huge quantity of waste water. Approximately, worldwide two billion people do not have access to clean and safe drinking water. This problem is observed in both rural and urban areas of developing countries. Industries release wastewater which contains highly toxic compounds such as heavy metals and radioactive elements that badly impact the health of living organisms. In the present scenario, various technologies have been developed to remove these toxic chemicals and ions, including precipitation and co-precipitation, adsorption, flocculation, ion exchange, coagulation, membrane filtration, and solvent extraction, but it is very expensive, consumes a lot of energy, and some technologies pose environmental hazards. For these reasons, scientists are trying to find out for new advanced and cost-effective membrane technology for wastewater treatment and purification. In today's world, biopolymer-based technologies are the most recent techniques

A. Konar (✉) · P. Ghosh
School of Life Sciences, Seacom Skills University, Santiniketan, Bolpur, Birbhum, West Bengal, India
e-mail: abhishek.konar60@gmail.com

P. Singh
Biotechnology Program, Dr. Rammanohar Lohia Avadh University, Ayodhya, Uttar Pradesh, India

J. Kaur
Department of Applied Biology and Environmental Sciences, Centennial College, Toronto, ON, Canada

S. Chakraborty
Department of Computer Science and Engineering, Seacom Skills University, Santiniketan, Bolpur, Birbhum, West Bengal, India

P. Ganguly
Department of Botany, Bethune College, Kolkata, West Bengal, India

© The Author(s), under exclusive license to Springer Nature Singapore Pte Ltd. 2026
S. Dey and M. P. Shah (eds.), *Microbial Applications in Industrial Wastewater Treatment*, Advances in Wastewater Research,
https://doi.org/10.1007/978-981-96-9128-9_4

based on using microorganisms to purify wastewater. The use of biopolymer-based nanocomposite materials to clean wastewater has made water recycling feasible, eco-friendly, and cost-effective. Cellulose is a highly abundant biopolymer derived from wood, plants, and microbes. Currently, temperature, pH, and agitation are all regulated during the synthesis of cellulose nanofibers (CNFs). Recently, there has been a lot of interest in the bacterial synthesis of CNFs, or bacterial cellulose (BC) (from *Thermus thermophilus*, *Pseudomonas stutzeri*, and a plethora of more extremophiles). The hydrophilicity, biodegradability, renewability, and mechanical characteristics of these newly formed natural biopolymers are extraordinary. The proposed study will be made of biopolymer using bacterial cellulose (BC) and excellent photo-thermal activity for wastewater treatment and purification that can help cost-effective wastewater treatment to reduce water pollution levels.

Keywords Biopolymers · Wastewater treatment · Membrane technology · Water pollution · Photo-thermal activity

4.1 Introduction

In today's world industrial production of non-biodegradable artificial polymer material need to be replaced by sustainable biomaterial for the sake of the environmental hazards. That's why the world search for various biomaterials which can replace artificial biopolymer and used for multiple purposes including wastewater treatment and production of purified drinking (Iguchi et al. 2000). Biopolymers are essential component in waste water treatment. Additionally, biopolymer utilized in various treatment processes, including trickling filters membrane bioreactors and rotating biological contact units (Houghton and Quarmby 1999).

Deposition of several heavy metals including zinc, silver, mercury, arsenic, nickel, lead, cadmium, and copper responsible for water pollution. Living organisms accumulate heavy metal and create chronic diseases in the human being body. There are six types of heavy metals ranging in atomic weight from 63.5 to 200.5 and with a gravity greater than 5.0 (Russo et al. 2021). In developing countries, mining operations, batteries, paper, fertilizer and pesticides industries, etc., are directly or indirectly discharging non-biodegradable and toxic heavy metals such as copper, zinc, cadmium, lead, nickel, mercury, and chromium water. Zinc and copper are in animal metabolism but high amount consumption causes human health problems including stomach cramps, anemia, skin irritations, vomiting and nausea, or even death (Oyaro et al. 2007). According to the U.S. EPA, cadmium is a potential carcinogen of human that can cause kidney and heart damage and even death. Lead can affect basic cellular processes, brain dysfunction, CNS (Central nervous system), kidney, heart, liver, and reproductive system damage (Russo et al. 2021).

In the twenty-first century, the world is facing freshwater crisis and it is one of the most important challenges. Day by day, lack of freshwater is increasing due to rapid growth of industry, industrial polymer uses, increase in metropolitan cities, etc. The most valuable natural asset on the Earth is water and human life depends on its purity. Water shortage is a critical global challenge due to rapid industry growth like food production industry, paper industry, fossil fuel processing industry, hospital operations, chemical synthesis, and other industries, such as jewelry, aerospace, mining, and agriculture (Esfahani et al. 2019). Globally, water contamination from different pollutants is the most challenging issue. So, researchers and scientists are trying to solve this problem. The past few years have seen national and international researchers focusing on developing a variety of innovative biopolymer–nanocomposite membranes for wastewater treatment (Spoială et al. 2021).

Nowadays, various methods have been developed to remove toxic chemicals and ions, including precipitation, co-precipitation, adsorption, coagulation, flocculation, ion exchange, membrane filtration, and solvent extraction. Additionally, these methods utilize poisonous synthetic reagents, once again non-biodegradable. They are expensive, so the treatment has high input costs (Singh et al. 2022a).

The divergence of heavy metal ions has been achieved through different techniques, including membrane filtration, chemical precipitation, ion exchange, and electrochemical treatment. Mainly adsorption is used for wastewater treatment, because it has many advantages, such as high efficiency and easy to operate. But requiring high energy with low metal removal property represents disadvantages of adsorption methods. On the other hand, ion-exchange technique can be used for low-concentration metal solutions and is distinguished by its strong sensitivity to pH in the aqueous phase. However, electrochemical treatments have a high cost due to electricity consumption for water treatment (Russo et al. 2021).

The treatment of wastewater, membrane technology has grown remarkably in the last few decades. To solve this problem, various membrane technologies are being continuously investigated for water purification and wastewater treatment. Currently, distillation techniques, like multistage flash distillation (MSF) and reverse osmosis (RO), require large energy for the distillation process. Developing countries are facing a lack of large power plants in rural areas that cannot provide continuous power supply for distillation of water purification. As a result, rural people suffer various waterborne diseases like cholera, typhoid, hepatitis, gastroenteritis, diarrhea, and worm infections (Spoială et al. 2021; Kim et al. 2012; Sharma et al. 2022).

Nanoparticles consisting of nanofiber polymers (polyurethane, polyethylene oxide, polylactic acid), metal oxides (ZnO, Al_2O_3, MgO, TiO_2, SiO_2, Fe_2O_3, and zeolite), metals (Copper (Cu), Silver (Ag), etc.), and carbon materials (carbon nanofibers, graphene, carbon nanotube, etc.) (Spoială et al. 2021; Kim et al. 2012). But the disadvantage of using Ag and Cu as a nanoparticle in bacterial nanocellulose (BNC) is increase in production costs. Therefore, scientists are trying to find out for new low-cost nanocomposites for biopolymer membrane technology for waste water treatment.

In the developing world, water scarcity and the pollution of water are continuously threatening people's health as civilization develops. So, worldwide research is focusing on in this issue. Bacterial nanocellulose embedded reduced graphene oxide (RGO) can absorb a range of contaminants and can remove suspended particles, including pathogens. Globally, BNC also achieved water purification by using different types of membranes like Forward Osmosis (FO), Microfiltration (MF), Reverse Osmosis (RO), Ultrafiltration (UF), and Nanofiltration (NF) (Sharma et al. 2022; Xu et al. 2022). Bacterial cellulose is ultrafine which do not have hemicellulose, pectin, and lignin, having 20–100 nm diameter, is termed nanocellulose (Wan et al. 2006; Shezad et al. 2010). BNC can be used to liberate toxic metal with life-threatening microbes from wastewater as well as it can be used for purifying water and wastewater treatment (Jiang et al. 2018, 2017). Another paper having BNC developed from banana peel using Gram-negative bacteria have great potential for water purification (Shah 2025; Sijabat et al. 2019).

Chitosan, agar, pectin, agarose, alginate, gelatin, cellulose, xanthan, and other readily available natural biopolymers have been used for various goal in current years because of their biodegradability and less toxicity properties. As versatile biopolymers, many of them can be used as bio-adsorbents. Their dissolvability in water makes them excellent for removing hazardous contaminants from water. Because of their beneficial qualities—such as biocompatibility, water solubility, biodegradability, no toxicity, and ease of availability at a low cost—biopolymers based on carbohydrates have found extensive use water purification. However, their main drawback is their subpar mechanical performance. Biopolymers have been initiated for use in water treatment in conjunction with a variety of filler materials, including natural fibers, various pigments, antioxidants, and fillers of inorganic nano- or micromaterials, due to their renewable, biodegradable, sustainable, and non-toxic properties (Biswal et al. 2024).

A biopolymer can destabilize large colloidal particles during coagulation or flocculation by enhancing their ionic strength and reducing their zeta potential in order to reduce the thickness of the diffuse part of the electrical double layer. In another way, biopolymers have functional carboxyl and hydroxyl groups that link with contaminants adsorbing counterions and neutralizing particle charges. Additionally, biopolymers are non-toxic, biodegradable, and contain no metals, which facilitates the recycling of retained nutrients compared to traditional synthetic polymers (Pandey 2020). Liu et al. reported that biopolymer PLA nanoporous fibrous mats are reported to separate oil–water mixtures sufficiently (Shah 2024a; Liu et al. 2019).

In the developing world, water scarcity and pollution of wastewater are continuously threatening people's health as civilization develops. In this development period, there are different water purification technologies like FO, RO, UF, NF, and MF. These technologies are cost-effective but high in energy consumption and misuse of excess water (Das et al. 2024). In view of this concern about wastewater treatment and water purification, and removal of contaminants from industrial waste water by biopolymers, additionally focus on faster process, cost-effective, eco-friendly, and conserve both energy and another resource. Not only did biopolymer reduce cost and electricity but also have various attractive features like good water solubility, low toxicity, and easy usable, highly colloidal stability, large surface is and easy recovery.

4.2 Biopolymers

Biopolymers are nanofibrils derived from natural sources or synthesized biologically with universal nano-building (Ling et al. 2018). The polymers, that are generated from plants, animals, bacteria, and fungi, are known as natural polymers (Shah 2024b; Singh et al. 2022b). They are composed of monomeric units linked through covalent bonds, forming long chains. Their biocompatibility, biodegradability, and environmental friendliness make them suitable for wastewater treatment applications. A significant amount of organic and inorganic compounds can be found in industrial wastewater, including metals and hazardous compounds. Biopolymers are gaining attention in wastewater treatment due to their sustainable nature, effectiveness in pollutant removal, and ability to replace conventional synthetic materials. They serve as coagulants, flocculants, adsorbents, or membrane materials, contributing to the efficient removal of poisonous substance such as heavy metals, dyes, and organic matter. Below is a detailed exploration of their types, sources, and structures (Ling et al. 2018; Khan et al. 2021).

4.2.1 Types of Biopolymers

The prefix bio means biodegradable. As a result of natural decomposition of biopolymers, organic by-products such as CO_2 and H_2O are left behind, which are not harmful to the environment. Biopolymers have been considered an alternative to petroleum-based plastics because they are easily biodegradable and renewable (Othman 2014). Additionally, super thickening effect with heavy metal absorbent property. Biopolymers are classified based on their chemical structure and the type of monomers forming their backbone. In general, biopolymers include polysaccharides, lipids, polyhydroxyalkanoates, proteins, and other naturally occurring polymers. The major types are as follows (Shah 2024c; Annu et al. 2024; Badran et al. 2023; Bekchanov et al. 2024).

4.2.1.1 Polysaccharides

Polysaccharides are eco-friendly, low-cost biopolymers. These polymers include carrageenan, xanthan gum, curdlan, chitosan, cellulose, hyaluronic acid, and alginate. They are widely used in adsorption and coagulation processes due to their functional groups, which interact with pollutants. Today, they are highly efficient and low-cost, making them ideal for use in any phase of petroleum recovery, including well drilling and wastewater treatment (Xia et al. 2020; Nasrollahzadeh et al. 2021a). Sustainable polysaccharides are obtained from plant animal, algae, and microorganisms (Fig. 4.1) (Nasrollahzadeh et al. 2021a).

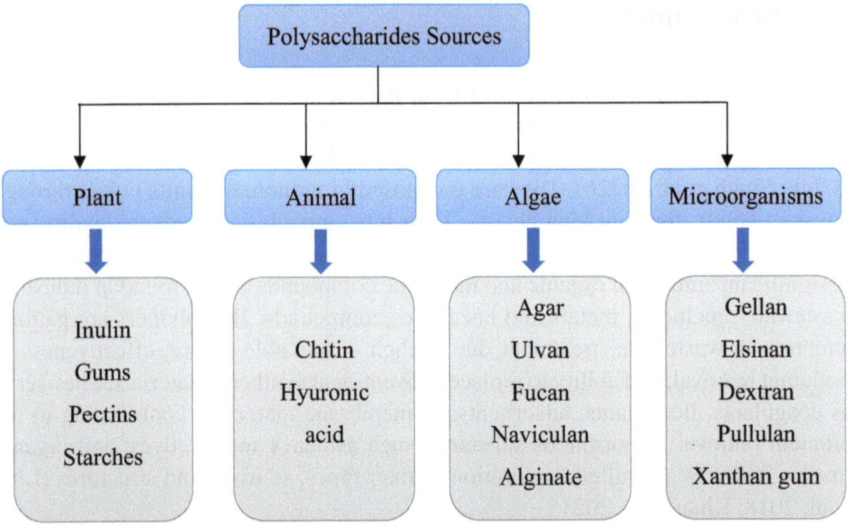

Fig. 4.1 Versatile source of natural polysaccharides polymers

4.2.1.2 Proteins

Proteins like keratin, gelatin, and casein are used for their might to form complexes with heavy metals and organic pollutants. They are particularly effective in bio-adsorption techniques. A protein is a polymer where the 20 amino acids are linked by peptide bonds (Numata 2020; Zubair and Ullah 2021; Renault et al. 2009; Elgarahy et al. 2023). Additionally, researchers investigated the effectiveness of wool keratin proteins in removing Pb(II) from water. A soy protein-based biopolymer has strong absorbent property from industrial wastewater. Other proteins, gelatin and albumen, have also been used to extract dyes and heavy Cd(II), As, and TiO_2 metals from wastewater (Zubair and Ullah 2021).

4.2.1.3 Lignin-Based Polymers

Lignin ($C_{81}H_{92}O_{28}$) is a phenolic natural carbohydrate biopolymer found in plant cell walls. Lignin mainly consists of three constituent monomers, guaiacyl (4-hydroxy-3-methoxyphenyl, G), p-hydroxyphenyl (4-hydroxyphenyl, H), and syringyl (4-hydroxy-3,5-dimethoxyphenyl, S). Cellular glue, lignin protects and connects both cellulose and hemicellulose. Its adsorption properties make it suitable for removing dyes and heavy metals (Zubair and Ullah 2021; Atangana and Oberholster 2020).

4.2.1.4 Synthetic Biopolymers

These are engineered biopolymers, such as polylactic acid (PLA), polyethylene succinate (PES), and poly(hydroxyalkanoates) (PHAs) synthesized using bio-based monomers. The majority of synthetic biopolymers and their composites are used in tissue engineering (Singh et al. 2022b; Mtibe et al. 2021). These synthetic polymers can be synthesized or blended to provide a variety of physicochemical and biological properties. Mechanical strength environment-friendly, hole size, bio-compatibility, and porosity are some of the desirable properties. Chemical building blocks are used to make synthetic polymers in the laboratory. In market, synthetic biopolymers have potential applications in management and life-cycle assessments (Mtibe et al. 2021; Reddy et al. 2021).

4.2.1.5 Polyhydroxyalkanoates (PHAs)

Polyhydroxybutyrate (PHB) is a hydroxy fatty acid polyester synthesized by Gram-positive or Gram-negative bacteria. It is a biodegradable polyester. A total of 150 component parts of PHA have been identified. PHAs are not soluble in water. When PHAs buried in soil, this compound can decompose into water and carbon dioxide in seven months. Various methods are synthesized for PHAs. A new technology is develop by Yield10 Bioscience Company in the agricultural field. In an attempt to circumvent the production cost barrier and trade on a wide scale, they have examined a copolymer. Gram-negative soil bacterium Cupriavidus necator, a naturally occurring carboxylic acid, and a simple sugar were all utilized. Activated sludge and other non-depleting energy sources were used to further reduce. PHAs are employed in membrane filtration systems for water treatment (Samrot et al. 2021).

4.2.2 Sources of Biopolymers

The sources of biopolymers are diverse, ranging from natural organisms to laboratory-synthesized variants. Increasing concerns over water pollution provide a great opportunity for biopolymers to be used in industrial wastewater remediation (Zubair and Ullah 2021). They are categorized as natural and synthetic.

4.2.2.1 Natural Sources

Researchers are increasingly working with biopolymers obtained from various natural sources in large quantities (Barja 2021). Natural sources of biopolymers include plants, animals, and microorganisms (Khan et al. 2021). They are as follows.

Plant Sources:

Cellulose:

Found in plant cell walls, cellulose $(C_6H_{10}O_5)_n$ is the most abundant biopolymer. As an example, wood and cotton fiber contain 40–50% and 90–95% cellulose, respectively. In cellulose, chains of hydroxyl groups and oxygen atoms are linked by hydrogen bonds, give cellulose its crystalline fiber structure (Fig. 4.2a) (Xia et al. 2020; Elgarahy et al. 2023). In addition to plankton and unicellular algae in oceans, cellulose can also be synthesized by different species of bacteria (Barja 2021). Both of these cellulose are nanocellulose and are hydrophilic and hydrophobic with excellent property for wastewater treatment.

Pectin:

Non-toxicity, biodegradability pectin $(C_6H_{10}O_7)$, is a linear heteropolymer, that extracted from land plants such as citrus peels, mango, soy husk, and apple pomace. It consists of alpha-(1 → 4) linked with D-galacturonic acid (GaIA) residues. Backbone of pectin contains various single –COOH and –OH functional groups. However,

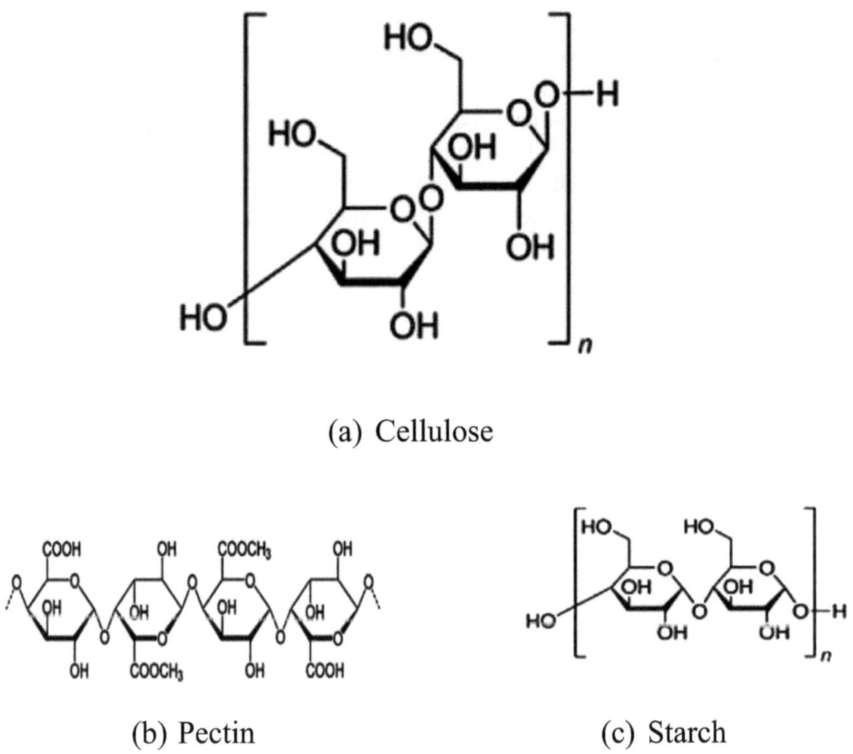

Fig. 4.2 Chemical structures of plant biopolymer (Elgarahy et al. 2023)

pectin, have few inherent restrictions when utilized in specific industries (Fig. 4.2b). Due to the high water solubility and the tendency of the biopolymer to lump and aggregate, pectin solutions are difficult to handle and run (Elgarahy et al. 2023).

Starch:

Biodegradable, low-cost biopolymer starch $(C_6H_{10}O_5)_n$ is derived from agriculture field crops corn, potato, and wheat. These monomers of polysaccharide consist of glucose, starch bears some hydroxyl groups, which can be produced hydrogels. Starch, a naturally occurring plant by-product of biomass, has been investigated for use in bioplastics manufacturing (Fig. 4.2c) (Elgarahy et al. 2023).

Animal Sources:

Chitin and Chitosan:

Chitin and chitosan are extracted from the exoskeletons of crustaceans and insects. Approximately 30% of the dry weight of crustacean shells contains chitin, a cheap and abundant by-product of the food industry (Xia et al. 2020; Zubair and Ullah 2021). Chitin $(C_8H_{13}O_5N)_n$, is a linear chain polysaccharide. Chitin has three forms: alpha-chitin, beta-chitin, and gamma-chitin. The adsorption of various pollutants from water is reported to be achieved using chitin (Elgarahy et al. 2023).

Chitosan showed unique properties like hydrophilic characteristics, biocompatibility, eco-friendly, and high adsorption capacity of toxic substance as a result, they are ideal materials for treating wastewater treatment (Zubair and Ullah 2021). Additionally, Renault et al. reported that chitosan works as a flocculating agent for wastewater treatment and removes heavy metal, lead (Renault et al. 2009). In a study, it was reported that chitosan grafted with polyethylenimine (PEI) and poly(glycidyl methacrylate) could decontaminate heavy metals (Zubair and Ullah 2021).

Keratin:

Plentiful inedible protein keratin found in wool, feathers, and hair. Keratin composed of numerous amino acids make up a polypeptide structure and each one has a distinguishable side chain with a distinct chemical constitution, charge, reactivity, and bonding ability. Keratin has many groups that bind water contaminants and hence it can be used as an adsorbent (Elgarahy et al. 2023).

Sources from Algae:

Alginate, agar, and carrageenan are derived from brown, red algae, and bacteria, respectively. Alginate $(C_6H_8O_6)_n$ is a linear-type copolymer consisting of guluronic acid (G) and mannuronic acid (M) subunits found in brown algae. Agar and alginate are copolymers (Das et al. 2024; Elgarahy et al. 2023). In aqueous conditions, alginate showed negatively charged at $pH > 3.4$, because carboxyl with G and M subunits present. Tasteless agar consists of α (1–4)-3,6-anhydro-L-galactose, and β 9(1–3)-D-galactose units (Elgarahy et al. 2023; Sousa et al. 2021).

Microorganism Sources:

Biodegradable water-soluble Xanthan Gum natural polysaccharide that is produced by *Xanthomonas campestris* and chemically made up of five repeating monosaccharide units (Urlacher and Dalbe 1997). Bacterial cellulose know as bacterial nanocellulose (BNC), which resembles plant cellulose in appearance and is made up of linear glucan molecules joined by hydrogen bonds (Liu et al. 2019; Huang et al. 2014). BNC is synthesized by fermentation process in some Gram-negative bacteria species such as *Pseudomonas, Salmonella, Agrobacterium,* and Gram-positive bacteria *Sarcina ventricu.* BNC is most efficiently synthesized by *Acetobacter xylinum.* The unique nanomorphology of BNC makes it resistant to wet conditions and extremely elastic and conformable (Barja 2021). PHAs are produced by bacteria such as *Ralstonia eutropha* (Elgarahy et al. 2023).

4.2.2.2 Synthetic Sources

Numerous technologies and methods are employed in the synthesis of biopolymers. Usually, synthetic polymers are made up chemically from biomass, including polyethylene succinate (PES), polylactic acid (PLA), and polycaprolactone (PCL) (Elgarahy et al. 2023). Synthetic biopolymers are produced using bio-based monomers via controlled polymerization. These include:

Polylactic Acid (PLA):

This eco-friendly polymer manufactured from lactic acid is derived from fermented carbohydrates. PLA is a member of the aliphatic polyester family, which is often derived from α-hydroxy acids and used in filtration membranes (Avérous 2008). This polymer is more popular than other polymers, also called polylactide. In the 1800s, low-molecular-weight PLA was produced from lactic acid through water distillation process (Sin 2012).

Polycaprolactone (PCL) and Polyethylene Succinate (PES):

Nontoxic, semi-synthetic polymer, aliphatic polyester PCL is bioresorbable and employed in controlled release systems. Hexanoate repeat units make up the aliphatic polyester known as polycaprolactone (PCL) (Woodruff and Hutmacher 2010). PES made from succinic acid and ethylene glycol. PES is known for its biodegradability and mechanical strength (Elgarahy et al. 2023).

Modified Natural or Biopolymers:

Specifically, this review described three kinds of modified low-cost biopolymers: Bio-PE, Bio-PP, and Bio-PET. Bio-PP and Bio-PE were first produced by Braskem. As with other polymers like PVC and PS, ethylene serves as the initial monomer for Bio-PE synthesis and is typically extracted through distillation from petroleum feedstock. First Bio-PE was produced in 2010 and, at that time, it was world's largest polymer. From glucose, bio-isobutanol can be dehydrated with butylene, then polymerized to make bio-PP. Bio-PET is produced by various methods including iso-butanol, muconic acid, and the limonene method. Iso-butanol is a method that begins with sugar, a building block for many chemicals such as isooctane, ethyl tertbutyl ether, methyl methacrylate, and other alkanes (Sousa et al. 2021). Carraher et al. and Matthiesen et al. proposed and reported that using a combination of chemical and biorefining processes, cis-Muconic acid is separated from sugar. Tetrahydro terephthalic acid (THTA) is produced by ethylene addiction reactions following a series of cis–cis to trans–trans transition steps of muconic acid. Bio-PTA is then produced by dehydrogenation (Siracusa and Blanco 2020; Matthiesen et al. 2016; Colonna et al. 2011).

4.2.3 Structure of Biopolymers

The special qualities of biopolymers, including their chirality, hydrophilicity, degradability, and capacity for chemical modification, are astounding. The structure of biopolymers is fundamental to their function and effectiveness in wastewater treatment (Elgarahy et al. 2023; Sousa et al. 2021). High-molecular-weight biopolymers exhibit better flocculation and adsorption due to their extended chains and increased interaction sites (Russo et al. 2021; Gede Widhiantara 2023).

4.2.3.1 Linear and Branched Polymers

It has long chains of monomers with a straight forward structure, offering flexibility and a high surface area for pollutant interaction (Fig. 4.3). It is commonly used in flocculation and adsorption. Examples: Cellulose and alginate (Russo et al. 2021; GedeWidhiantara and I. 2023).

Side chains or branches grow from the main chain of branched polymers as shown in Fig. 4.3. Branched polymers removes dissolved organic matter excellently because its has extra branches which enhance interaction and adsorption of pollutants. From an architectural perspective, branched polymers, or poly(macromonomers), that are made available through the living polymerization of well-defined macromonomers, are the most intriguing class since they offer the highest density branched polymers with precisely controlled structures, carrying one branch on each repeating unit. Examples: starch and modified chitosan (Russo et al. 2021; Gede Widhiantara 2023; Hirao et al. 2005).

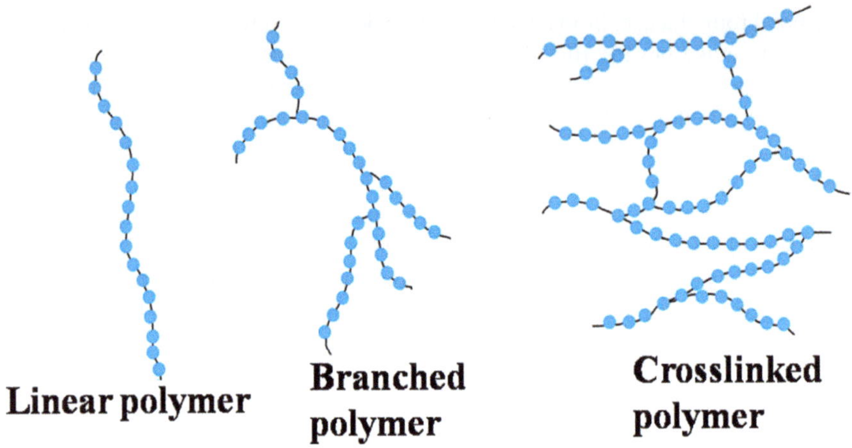

Fig. 4.3 The structure of linear, branched, and cross-linked polymers

4.2.3.2 Cross-linked and Complex Structures (Tertiary and Quaternary)

Cross-linked has interconnected polymer chains which create a network that enhances mechanical stability and porosity. The structure is shown in Fig. 4.3. It is used in filtration of toxic compounds and act as adsorbents for heavy metals. Examples: Cross-linked alginate and chitosan (Russo et al. 2021; Gede Widhiantara 2023).

It has folding chain into 3D conformations patterns which create specific active sites for selective pollutant binding. It is used in binding specific contaminants like dyes and toxic organics. Quaternary structures occur in complex proteins, influencing their binding affinity for pollutants. Examples: Proteins like gelatin (Russo et al. 2021; Gede Widhiantara 2023).

4.2.3.3 Composite Biopolymer Structures

It is the integration of nanomaterials which enhances properties such as adsorption capacity and antimicrobial activity. It is utilized in advanced water treatment technologies. The example of composite biopolymer is chitosan–silver nanocomposites (Russo et al. 2021; Gede Widhiantara 2023). Biopolymers, with their diverse structures and sources, offer a sustainable and efficient solution for wastewater treatment. Their versatility in adsorption, flocculation, and filtration processes is driven by their unique structural properties and environmental compatibility. Research into improving the mechanical and chemical properties of biopolymers will further expand their applicability in industrial-scale water purification systems. By harnessing both natural and synthetic biopolymers, wastewater treatment can be made more eco-friendly, effective, and sustainable (Russo et al. 2021; Elgarahy et al. 2023; Gede Widhiantara 2023).

4.3 Applications of Biopolymers in Sustainable Wastewater Treatment

The effluent contamination water is now being biologically treated using both aerobic and anaerobic processes with the aid of chemical compound. The primary processes used in physicochemical treatment of effluents are membrane-based treatment methods, oxidation to eliminate organic contaminants, adsorption of soluble on solid substrate, and flocculation and coagulation of suspended solids. Recently, biopolymers have been evidenced for contamination water treatment due to their non-toxicity, sustainability, biodegradability, and renewable qualities. To generate special composites with improved properties, biopolymers can also be mixed with a range of buttress components like natural fibers with colors, antioxidants, and micro- and nanoparticles of inorganic fillers (Udayakumar et al. 2021).

4.3.1 Removing Suspended Solids

Coagulation and flocculation are crucial processes in the initial of drinking contamination water treatment that remove organic and inorganic pollutants such as TDS (Total dissolved solids), TS (Total solids), and TSS (Total suspended solids), and these are the main methods to use biopolymers and biopolymer composites in treatment procedures (Udayakumar et al. 2021). Conventional chemical flocculants, such as Poly-aluminum chloride (PAC) were employed in water treatment to eliminate suspended solids (SS), aluminum (Al^{3+}), and chemical oxygen demand (COD) from contaminated water. Chitosan composite flocculants, which were created by combining chitosan, silicate, and poly-aluminum chloride, were found to be more effective in eliminating impurities than standard flocculants. It was discovered that chitosan composite flocculants had better economic and environmental advantages than conventional flocculants used for organic and inorganic separation. The biopolymer and its composites that are utilized as flocculants and coagulants in the water treatment process are included in the following Table 4.1 (Udayakumar et al. 2021).

Table 4.1 Different types of polymers used for removing suspending solids in different types of waste water

Biopolymers	Type of wastewater	Removal parameter	References
Chitosan	Acid dyes	Dye AB292 (52%)	Iguchi et al. (2000)
Polyacrylamide	Waste water that is artificially turbid	Turbidity (98.24%)	Honarkar and Barikani (2009)
Sodium alginate	Synthetic dye waste water	Color (90%)	Wang et al. (2005)
Starch of rice	Palm plant oil mill sewerage	Total suspended solids (84.1%)	Thakur et al. (2014)
Alum and tannin	Water of synthetic	Turbidity (99.3%)	Ouagne et al. (2010)
Anionic sodium carboxymethylycellulose	Drinking water	Turbidity (93%)	Klein et al. (2010)

4.3.2 Reducing Heavy Metals

Biopolymers may potentially be useful in applications involving the detection of heavy metals. There is a lot of study being done on the elimination of toxic metals from water and wastewater because of the significant acute and chronic toxicity concerns. Heavy metal pollutants have been addressed using coagulation flocculation, electrochemical techniques, ion exchange, membrane filtration, flotation, chemical precipitation, and adsorption, with adsorption emerging as the most effective way. The widely accessible and reasonably priced biopolymers have recently drawn a lot of interest as eco-friendly removal heavy metal adsorbents (Palanisamy et al. 2019).

The carrageenan, and agar (natural polysaccharides) extracted from the cell walls of red algae, were utilized to form thin films atop tin-doped indium oxide using the layer-by-layer self-assembly technique. Polyaniline was added in layers that alternated with the polysaccharides of interest. Carrageenan and agar enhanced the conducting polymer's electrochemical stability in an acidic solution. For the detection of Cr(VI), the most promising method was the tin-doped indium oxide–agar–polyaniline system (Farias et al. 2015).

The hydrolyzed zein protein was converted by graft copolymerizing acrylic acid monomers to create zein protein-derived hydrogels for heavy metal cleanup. The capacity of the hydrogel to chelate the heavy metal ions was assessed. The outcome show that hydrogels showed a remarkable capacity for ion of copper chelation at pH 4.5. Several functional groups on polyacrylic acid and hydrolyzed zein protein were responsible for this adsorption efficiency (Aluigi et al. 2007). Interestingly, it is possible to extract and appropriately utilize the biopolymers found in activated sludge for the heavy metals remove. Biopolymers with different constituent proportions were separated and used in Cu(II) ion adsorption tests after conditioning with cetyl trimethyl ammonium bromide and linear alkylbenzene sulfonate. Protein content was raised by the detergent cetyl trimethyl ammonium bromide, but polysaccharide and nucleic acid concentrations were effectively raised by linear alkylbenzene sulfonate. The discovering suggest that any biomass with a maximum protein content may act as a metal ion adsorbent. Karthikeyan et al. reported that TiO_2 nanocomposite has an ability to eradicate both Gram-positive (*Streptococcus pneumoniae*) and Gram-negative (*Pseudomonas aeruginosa*) bacteria as well as generate ROS that irreversibly harm bacteria. When the pH and catalyst concentration are changed, the TiO_2 nanocomposite can eliminate both cationic and anionic dyes (Zhou et al. 2016).

4.3.3 Eliminating Organic Matter

Moving bed biofilm carriers based on polylactide-based products are useful for the removal of atrazine (an S-triazine type of herbicide), oseltamivir (an antiviral medication), and bisphenol A (a material primarily used in the production of epoxy resin and polycarbonate plastic). The cumulative evolution of $14CO_2$ from samples incubated with freely moving carriers was significantly higher than in the control in a study involving wastewater that had been injected with 14C-labeled contaminants. The clearance efficiency was increased by seeding the biofilm carriers with microbial strains. One could consider such biopolymer-based carriers inoculated with degradative microorganisms to be a fresh idea in xenobiotic cleanup (Accinelli et al. 2012). Recently, BNC has been made using carbon nanocomposite to reduce the cost of water purification as well as wastewater treatment. Not only did carbon nanocomposites reduce cost but also have various attractive features like good water solubility, low toxicity, and easy usable, highly colloidal stability, large surface, and easy recovery.

When treating drinking water and wastewater for the first time, coagulation and flocculation are essential procedures that eliminate organic and inorganic contaminants such total solids (TS), total suspended solids (TSS), and total dissolved solids (TDS). When using biopolymers and biopolymer composites in treatment procedures, these are the main methods used (Udayakumar et al. 2021).

4.3.4 Improving Water Quality

Given the growing worries about water pollution, materials made from biopolymers have a lot of potential for wastewater of industrial remediation in the future. In contrast to its equivalents for wastewater purification, materials of biopolymer with higher removal proficiency and less expensive technologies are anticipated to see a rise in demand. In the upcoming years, sustainable and renewable materials for industrial wastewater remediation will be made possible by the industrial-scale advancements in biopolymer-based materials (Thomas et al. 2020).

Chemically modified cellulose filter paper with EDTA has been used to treat water for the removal of Ni(II), Ag(I), Pb(II), Zn(II), Cd(II), Sn(II), and Cu(II), with removal efficiencies ranging from 90 to 95%. Moreover, lignin, alginate, and starch were employed to treat wastewater. To enhance its ability to adsorb Pb(II), Cd(II), and Hg(II) from aqueous solutions, calcium alginate was decreased with PEI and modified with graphene oxide. A naturally occurring biopolymer that has garnered a lot of interest for water treatment is lignin. Poly-aluminum chloride was used to create lignin-derived flocculant, and its efficacy for wastewater containing dyes was evaluated (Aluigi et al. 2007). The findings of this study suggest that SO_4 with or without lignin-derived flocculant significantly reduced flocculation qualities. The presence of Ca and Mg improves flocculation properties and color removals. Proteins have been researched for their potential to remediate industrial effluent and enhance water quality (Saha et al. 2019). Compared to the commercial adsorbing materials available in the market, biopolymer-based products have a bright future because of all these qualities (Thomas et al. 2020). MF, RO, FO, NF, and UF are the methods that is used to purify water globally. Nanoparticles (NPs), nanotubes, nanofibers, or nanosheets are used to create BNC nanocomposite membranes (PNC) using a variety of techniques, such as phase inversion, interfacial polymerization (IP), physical coating, including phase inversion (PI), layer-by-layer assembly, electrospinning, and cross-linking (Kim et al. 2012) (Fig. 4.4).

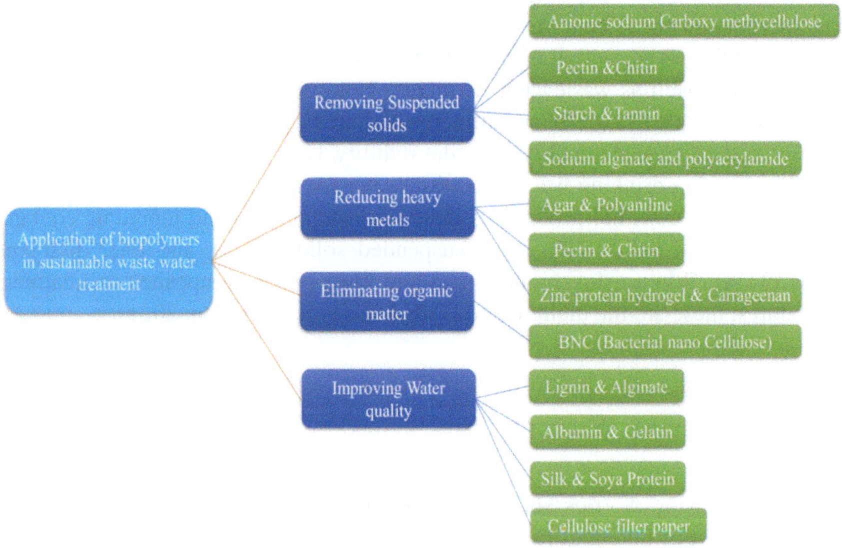

Fig. 4.4 Flow chart of the biopolymers used for wastewater treatment

4.4 Techniques Involved in Wastewater Treatment Using Biopolymers

4.4.1 Adsorption

A surface phenomenon, adsorption, is a metabolically passive process wherein the solute, or adsorbate species, migrates to the adsorbent's surface from the bulk liquid phase. This process is effective for moving pollutants out of wastewater and concentrating them on the surface of the adsorbent (Elgarahy et al. 2023).

4.4.2 Flocculation

To encourage the solid–liquid separation of suspended colloids, flocculation is utilized (Dao et al. 2016). Flocculation is an efficient method of treating turbidity, heavy metal ions, dyes, algae, other industrial and agricultural wastewater, and sludge dewatering, among other forms of water pollution (Jiang et al. 2021). One type of substance used in water treatment is flocculant. It can create bigger flocs by aggregating the colloidal materials and particulate matter suspended in the liquid, encouraging the particles to settle out of the stable suspension (Salehizadeh et al. 2018).

4.4.3 Biopolymer Membranes

Polymeric membranes are said to be the best membranes for water purification applications since membrane purification technology eliminates the need for any extra chemicals or hazardous by-products, allowing for improved industrial design. More importantly, membrane separation may remove contaminants selectively by changing the pore sizes and membrane topologies (Wen et al. 2019). Pressure is the most important factor in membrane separation technology. Pressure-driven membrane techniques are generally classified as FO, RO, NF, MF, and UF depending on the membrane pore size and operating procedures (Mansoori et al. 2020). The UF method is typically employed to eliminate viruses, colloids, and macromolecules; the NF technique is mainly utilized to eradicate hardness, heavy metals, and dissolved organic substances; the MF process is frequently applied to remove suspended molecule and microorganisms; and RO is utilized for desalination, water reclamation, and the generation of ultrapure water (Goh and Ismail 2018).

4.4.4 Sedimentation

The unit process of sedimentation is the separation of suspended particles from a suspension by gravitational settling. Water turbidity must be lowered to the lowest levels feasible in water purification facilities. For water added to filters, a turbidity of 1.5 units is often regarded as ideal. In water treatment operations, chemicals are added to cause floc production since plain sedimentation often takes a long time to be effective. The flocs then settle quickly. Sedimentation in the initial settling tank at a wastewater treatment facility lessens the strain on water treatment units. In secondary settling tanks, sedimentation is crucial because it eliminates a significant portion of the suspended particles from the treated wastewater that would otherwise be transferred into the effluent and subsequently into the streams where the effluent is often released (Shammas et al. 2005). Sedimentation can minimize water treatment costs and the need for added chemicals, but it cannot efficiently settle tiny particles or eliminate germs (Tiehm et al. 1999).

4.4.5 Coagulation

Coagulation is a popular solid–liquid separation method that removes all types of suspended and dissolved materials as well as organic and colloidal pollutants from industrial effluent (Renault et al. 2009). During the coagulation process, charge neutralization weakens colloidal matter and causes finely split or scattered particles to coalesce into huge masses. These massive aggregates use an adsorption process to absorb dissolved organic materials, and straightforward filtering and sedimentation may remove them with ease (Sharma et al. 2006).

4.5 Challenges and Future Prospects for the Use of Biopolymers in Wastewater Treatment

The synthesis of a biopolymer involves biological, thermal, and chemical processes, all of which are influenced by various operating circumstances. The drawback of all thermochemical methods is that they are energy-intensive, chemical processes that result in undesirable compounds; in contrast, a biological method may offer a technologically viable substitute. The nature of the substrate and the microbial strains employed impact how well biological processes for biopolymer synthesis work (Elgarahy et al. 2023).

In terms of four key areas, future research studies are strongly advised to advance the wide-ranging applications of biopolymers in wastewater treatment: (i) expanding the sources of biopolymers, enhancing extraction technologies, and finding new resources; (ii) by identifying the modulating agents and altering the functional parameters to optimize the process, and (iii) realistic applications conditions for cost–benefit evaluation and process standardization (Elgarahy et al. 2023).

4.5.1 Extremophiles

Extreme conditions encompass pressure, physical extremes, including temperature, and radiation, as well as geochemical extremes such as pH, salinity, oxygen tension, and desiccation. Extremophiles are a diverse group of microorganisms that not only withstand extreme conditions but also often necessitate these environmental extremes for their survival and growth (Kazak et al. 2010). Exopolysaccharides (EPSs) serve as emulsifiers and bioflocculants, and they facilitate the sorption of inorganic ions that are essential metabolic elements for bacteria (Sutherland 2002). Additionally, EPSs made by extremophiles have a lot of promise for treating wastewater and addressing the current water shortage. Extremophilic bacterial EPSs have been proposed as a simple, sustainable, and affordable substitute for affordable environmental bioremediation due to the exceptional ecological and commercial value of EPS (Banerjee et al. 2021). Halophiles are an extremophilic group that live in hypersaline settings, such salt lakes, salt pans, salt marshes, or saline soils, where the average salt content is between 0.2 and 5.1 M (Amoozegar et al. 2019).

Reducing groups and charged functional groups are features of EPS biopolymers that aid in binding the polymer to other charged moieties, including metal ions, and enable efficient removal of metal pollutants from wastewater (Durán et al. 2011). In order to lower the harmful heavy metals in the metal-contaminated environmental wastes, bacterial EPS is frequently used in the creation of nanoparticles (Banerjee et al. 2021).

The growth rates and PHA concentrations of extremophilic bacteria may be lower than those of mesophiles. The reduced PHA yield necessitates improving the bioprocess's efficiency through genetic engineering and culture. The majority of bacteria cannot proliferate in environments with high cell densities (Możejko-Ciesielska et al. 2023).

4.6 Advantages of Using Biopolymers in Wastewater Treatment

Cellulose based biopolymer have strong adsorption efficiency of pollutants from wastewater. Because biopolymer have huge porous structure of hydrogel and the availability of large number of hydrophilic groups, including –OH, –COOH, and $-NH_2$. Moreover, biopolymer adsorbents synthesized from cellulose have excellent adsorption efficiency, non-toxicity, renewability, and functionality to remove harmful contaminants from wastewater streams. Cellulose-based biopolymer adsorbents are excellent for removing harmful contaminants from wastewater because of their adsorption efficiency, renewability, non-toxicity, and functionality (Bhaladhare and Das 2022).

Recently, BNC has been made using carbon nanocomposite to reduce the cost of water purification as well as wastewater treatment. Not only did carbon nanocomposites reduce cost but also have various attractive features like good water solubility, low toxicity, and easy usable, highly colloidal stability, large surface is and easy recovery. Some carbon-based nanomaterials are carbon nanotubes, carbon quantum dots, graphene, graphene oxide (GO), or reducing graphene oxide (RGO), and they have loosely π electrons that can be highly absorbed light at different wavelengths, and when the electron is excited on the ground state, heat is generated (Irshad et al. 2021; Nasrollahzadeh et al. 2021b).

Numerous biopolymers, including tannin, chitosan, alginates, starch, cellulose, gums, and mucilage, have been studied for their flocculating and adsorbing capabilities in wastewater treatment. They are effective in removing pollutants from wastewater because of their high selectivity for harmful substances and metals. However, study is ongoing to develop highly effective and environmentally friendly bio-adsorbents and flocculants to replace conventional flocculants (Ling et al. 2018). When compared to alternative scenarios, coagulation–flocculation is generally considered to be an effective and economical approach for the removal of NOM.

4.7 Conclusions

Water is the most important part of the life, but today we are facing major challenges regarding supply of purified water. The ecological balance greatly depends on the fast removal of toxic pollutants from wastewater. Although there are numerous wastewater treatment methods, but they are costly, energy-intensive, and occasionally hazardous to the environment. So, there is an urgent requirement to develop fast and cost-effective method for wastewater treatment. Developed countries are using biodegradable members for removal of microbes, toxic metal to resolve the problem of water crisis. Biopolymer-based wastewater treatments are better technology in compare to other available materials in terms of their commendable capacity, affordability, and biocompatibility. The use of various biopolymers and their compounds in wastewater engineering reclamation technologies, such as membrane technology, coagulation/flocculation, adsorption, and photocatalysis, has been the subject of several fascinating investigations. The findings of this chapter suggest that biopolymers may be used for the treatment wastewater. Therefore, we can accomplish the objective of water purification with the aid of these biopolymers. Along with improving the use of technology, this procedure also becomes far more economical and efficient. Numerous biopolymers can be used for the removal of organic materials, dyes, and heavy metals from wastewater.

Acknowledgements This work was equally contributed by all authors. The authors are also thankful to Seacom Skills University, Dr. Rammanohar Lohia Avadh University and Centennial College for all kinds of support to complete this work.

Conflict of Interest The author declares no conflict of interest.

References

Accinelli C, Saccà ML, Mencarelli M, Vicari A (2012) Application of bioplastic moving bed biofilm carriers for the removal of synthetic pollutants from wastewater. Bioresour Technol 120:180–186
Aluigi A, Zoccola M, Vineis C, Tonin C, Ferrero F, Canetti M (2007) Study on the structure and properties of wool keratin regenerated from formic acid. Int J Biol Macromol 41(3):266–273
Amoozegar MA, Safarpour A, Noghabi KA, Bakhtiary T, Ventosa A (2019) Halophiles and their vast potential in biofuel production. Front Microbiol 10:1895
Annu, Mittal M, Tripathi S, Shin DK (2024) Biopolymeric nanocomposites for wastewater remediation: an overview on recent progress and challenges. Polymers 16(2):294. https://doi.org/10.3390/polym16020294
Atangana E, Oberholster PJ (2020) Modified biopolymer (chitin–chitosan derivatives) for the removal of heavy metals in poultry wastewater. J Polym Environ 28:388–398
Avérous L (2008) Polylactic acid: synthesis, properties and applications. In: Monomers, polymers and composites from renewable resources. Elsevier, pp 433–450
Badran AM, Utra U, Yussof NS, Bashir MJK (2023) Advancements in adsorption techniques for sustainable water purification: a focus on lead removal. Separations 10(11):565. https://doi.org/10.3390/separations10110565
Banerjee A, Sarkar S, Govil T, González-Faune P, Cabrera-Barjas G, Bandopadhyay R et al (2021) Extremophilic exopolysaccharides: biotechnologies and wastewater remediation. Front Microbiol 12:721365
Barja F (2021) Bacterial nanocellulose production and biomedical applications. J Biomed Res 35(4):310
Bekchanov D, Mukhamediev M, Yarmanov S, Lieberzeit P, Mujahid A (2024) Functionalizing natural polymers to develop green adsorbents for wastewater treatment applications. Carbohyd Polym 323:121397. ISSN 0144-8617, https://doi.org/10.1016/j.carbpol.2023.121397. https://www.sciencedirect.com/science/article/pii/S0144861723008627
Bhaladhare S, Das D (2022) Cellulose: a fascinating biopolymer for hydrogel synthesis. J Mater Chem B 10(12):1923–1945
Biswal T, Sethy TR, Sahoo PK (2024) Sustainable removal of contaminants by biopolymers. In: Bio-based polymers: farm to industry. Volume 2: Current trends and applications. American Chemical Society, pp 197–214
Colonna M, Berti C, Fiorini M, Binassi E, Mazzacurati M, Vannini M, Karanam S (2011) Synthesis and radiocarbon evidence of terephthalate polyesters completely prepared from renewable resources. Green Chem 13(9):2543–2548
Dao VH, Cameron NR, Saito K (2016) Synthesis, properties and performance of organic polymers employed in flocculation applications. Polym Chem 7(1):11–25
Das I, Ghosh S, Ghosh P (2024) Future of domestic water purifiers in the Indian market. Int J Res Manag 6(2):305–309
de Farias EA, dos Santos MC, de Dionísio N, Quelemes PV, Leite JR, Eaton P, da Silva DA, Eiras C (2015) Layer-by-layer films based on biopolymers extracted from red seaweeds and poly aniline for applications in electrochemical sensors of chromium VI. Mater Sci Eng B 200:9–21

Durán N, Marcato PD, Durán M, Yadav A, Gade A, Rai M (2011) Mechanistic aspects in the biogenic synthesis of extracellular metal nanoparticles by peptides, bacteria, fungi, and plants. Appl Microbiol Biotechnol 90:1609–1624

Elgarahy AM, Eloffy MG, Guibal E, Alghamdi HM, Elwakeel KZ (2023) Use of biopolymers in wastewater treatment: a brief review of current trends and prospects. Chin J Chem Eng

Esfahani MR, Aktij SA, Dabaghian Z, Firouzjaei MD, Rahimpour A, Eke J, Escobar IC, Abolhassani M, Greenlee LF, Esfahani AR, Sadmani A (2019) Nanocomposite membranes for water separation and purification: fabrication, modification, and applications. Sep Purif Technol 15(213):465–499

Gede Widhiantara I (2023) The role of biopolymers as therapeutic agents: a review. 13(1):042–055

Goh PS, Ismail AF (2018) A review on inorganic membranes for desalination and wastewater treatment. Desalination 434:60–80

Hirao A, Hayashi M, Loykulnant S, Sugiyama K, Ryu SW, Haraguchi N et al (2005) Precise syntheses of chain-multi-functionalized polymers, star-branched polymers, star-linear block polymers, densely branched polymers, and dendritic branched polymers based on iterative approach using functionalized 1, 1-diphenylethylene derivatives. Prog Polym Sci 30(2):111–182

Honarkar H, Barikani M (2009) Applications of biopolymers I: chitosan. Chem Mon 140:1403–1420

Houghton JI, Quarmby J (1999) Biopolymers in wastewater treatment. Curr Opin Biotechnol 10(3):259–262

Huang Y, Zhu C, Yang J, Nie Y, Chen C, Sun D (2014) Recent advances in bacterial cellulose. Cellulose 21:1–30

Iguchi M, Yamanaka S, Budhiono A (2000) Bacterial cellulose—a masterpiece of nature's arts. J Mater Sci 35(2):261–270

Irshad MS, Arshad N, Wang X (2021) Nanoenabled photothermal materials for clean water production. Glob Chall 5(1):2000055

Jiang Q, Kacica C, Soundappan T, Liu KK, Tadepalli S, Biswas P, Singamaneni S (2017) An in situ grown bacterial nanocellulose/graphene oxide composite for flexible supercapacitors. J Mater Chem A 5(27):13976–13982

Jiang Q, Ghim D, Cao S, Tadepalli S, Liu KK, Kwon H, Luan J, Min Y, Jun YS, Singamaneni S (2018) Photothermally active reduced graphene oxide/bacterial nanocellulose composites as biofouling-resistant ultrafiltration membranes. Environ Sci Technol 53(1):412–421

Jiang X, Li Y, Tang X, Jiang J, He Q, Xiong Z, Zheng H (2021) Biopolymer-based flocculants: a review of recent technologies. Environ Sci Pollut Res 28:46934–46963

Kazak H, Oner ET, Dekker RF (2010) Extremophiles as sources of exopolysaccharides. In: Carbohydrate polymers: development, properties and applications, pp 605–619

Khan A, Malik S, Ali N, Bilal M, El-Shazly M, Iqbal HM (2021) Biopolymer-based sorbents for emerging pollutants. In: Sorbents materials for controlling environmental pollution. Elsevier, pp 463–491

Kim ES, Hwang G, El-Din MG, Liu Y (2012) Development of nanosilver and multi-walled carbon nanotubes thin-film nanocomposite membrane for enhanced water treatment. J Membr Sci 394:37–48

Klein AP, Beach ES, Emerson JW, Zimmerman JB (2010) Accelerated solvent extraction of lignin from aleuritesmoluccana (candlenut) nutshells. J Agric Food Chem 58:10045–10048

Ling S, Chen W, Fan Y, Zheng K, Jin K, Yu H et al (2018) Biopolymer nanofibrils: structure, modeling, preparation, and applications. Prog Polym Sci 85:1–56

Liu, Liu Z, Niu J, Tian D, He J (2019) Electrospunpolysulfone/poly (lactic acid) nanoporous fibrous mats for oil removal from water. Adsorpt Sci Technol 37(5–6):438–450

Mansoori S, Davarnejad R, Matsuura T, Ismail AF (2020) Membranes based on non-synthetic (natural) polymers for wastewater treatment. Polym Test 84:106381

Matthiesen JE, Carraher JM, Vasiliu M, Dixon DA, Tessonnier JP (2016) Electrochemical conversion of muconic acid to biobased diacid monomers. ACS Sustain Chem Eng 4(6):3575–3585

Możejko-Ciesielska J, Ray S, Sankhyan S (2023) Recent challenges and trends of polyhydroxyalkanoate production by Extremophilic bacteria using renewable feedstocks. Polymers 15(22):4385
Mtibe A, Motloung MP, Bandyopadhyay J, Ray SS (2021) Synthetic biopolymers and their composites: advantages and limitations—an overview. Macromol Rapid Commun 42(15):2100130
Nasrollahzadeh M, Sajjadi M, Nezafat Z, Shafiei N (2021a) Polysaccharide biopolymer chemistry. In: Biopolymer based metal nanoparticle chemistry for sustainable applications. Elsevier, Amsterdam, The Netherlands, pp 45–105
Nasrollahzadeh M, Sajjadi M, Iravani S, Varma RS (2021b) Carbon-based sustainable nanomaterials for water treatment: state-of-art and future perspectives. Chemosphere 263:128005
Numata K (2020) How to define and study structural proteins as biopolymer materials. Polym J 52(9):1043–1056
Othman SH (2014) Bio-nanocomposite materials for food packaging applications: types of biopolymer and nano-sized filler. Agric Agric Sci Procedia 2:296–303
Ouagne P, Bizet L, Baley C, Bréard J (2010) Analysis of the film-stacking processing parameters for PLLA/flax fiber biocomposites. J Compos Mater 44:1201–1215
Oyaro N, Ogendi J, Murago EN, Gitonga E (2007) The contents of Pb, Cu, Zn and Cd in meat in Nairobi, Kenya
Palanisamy K, Jeyaseelan A, Murugesan K, Palanisamy SB (2019) Biopolymer technologies for environmental applications. In: Nanoscience and biotechnology for environmental applications, pp 55–83
Pandey J (2020) Biopolymers and their application in wastewater treatment. In: Emerging eco-friendly green technologies for wastewater treatment, pp 245–266
Reddy MSB, Ponnamma D, Choudhary R, Sadasivuni KK (2021) A comparative review of natural and synthetic biopolymer composite scaffolds. Polymers 13(7):1105
Renault F, Sancey B, Charles J, Morin-Crini N, Badot PM, Winterton P, Crini G (2009) Chitosan flocculation of cardboard-mill secondary biological wastewater. Chem Eng J 155(3):775–783
Russo T, Fucile P, Giacometti R, Sannino F (2021) Sustainable removal of contaminants by biopolymers: a novel approach for wastewater treatment. Current state and future perspectives. Processes 9(4):719
Saha S, Zubair M, Khosa MA, Song S, Ullah A (2019) Keratin and chitosan biosorbents for wastewater treatment: a review. J Polym Environ 27:1389–1403
Salehizadeh H, Yan N, Farnood R (2018) Recent advances in polysaccharide bio-based flocculants. Biotechnol Adv 36(1):92–119
Samrot AV, Samanvitha SK, Shobana N, Renitta ER, Senthilkumar P, Kumar SS et al (2021) The synthesis, characterization and applications of polyhydroxyalkanoates (PHAs) and PHA-based nanoparticles. Polymers 13(19):3302
Shah MP (2024a) Microbial remediation of hazardous chemicals from water & industrial wastewater treatment plant. Springer Nature
Shah MP (2024b) Development in waste water treatment research and processes: applied technologies for clean up of environmental contaminants. Elsevier
Shah MP (2024c) Aerobic and anaerobic microbial treatment of industrial wastewater. CRC Press
Shah MP (2025) Extraction of precious metals from industrial waste using microbial technology. Springer Nature
Shammas NK, Kumar IJ, Chang SY, Hung YT (2005) Sedimentation. Humana Press, pp 379–429
Sharma BR, Dhuldhoya NC, Merchant UC (2006) Flocculants—an ecofriendly approach. J Polym Environ 14(2):195–202
Sharma PR, Sharma SK, Hsiao BS (2022) Nanocellulose in membrane technology for water purification. In: Separation science and technology, vol 15, pp 69–85
Shezad O, Khan S, Khan T, Park JK (2010) Physicochemical and mechanical characterization of bacterial cellulose produced with an excellent productivity in static conditions using a simple fed-batch cultivation strategy. Carbohyd Polym 82(1):173–180

Sijabat EK, Nuruddin A, Aditiawati P, Purwasasmita BS (2019) Synthesis and characterization of bacterial nanocellulose from banana peel for water filtration membrane application. J Phys Conf Ser 1230(1):012085
Sin LT (2012) Polylactic acid: PLA biopolymer technology and applications. William Andrew
Singh A, Pal DB, Mohammad A, Alhazmi A, Haque S, Yoon T et al (2022a) Biological remediation technologies for dyes and heavy metals in wastewater treatment: New insight. Biores Technol 343:126154
Singh J, Kumar S, Sharma S (2022b) Biopolymer in wastewater treatment. Biopolymers: recent updates, challenges and opportunities. Springer International Publishing, Cham, pp 323–351
Siracusa V, Blanco I (2020) Bio-polyethylene (Bio-PE), bio-polypropylene (Bio-PP) and bio-poly (ethylene terephthalate) (Bio-PET): recent developments in bio-based polymers analogous to petroleum-derived ones for packaging and engineering applications. Polymers 12(8):1641
Sousa AM, Rocha CM, Gonçalves MP (2021) Agar. In: Handbook of hydrocolloids. Woodhead Publishing, pp 731–765
Spoială A, Ilie CI, Trușcă RD, Oprea OC, Surdu VA, Vasile BȘ, Ficai A, Ficai D, Andronescu E, Dițu LM (2021) Zinc oxide nanoparticles for water purification. Materials 14(16):4747
Sutherland I (2002) A sticky business. Microbial polysaccharides: current products and future trends. Microbiol Today 29:70–71
Thakur VK, Thakur MK, Gupta RK (2014) Graft copolymers of natural fibers for green composites. Carbohydr Polym 104:87–93
Thomas S, Gopi S, Amalraj A (eds) (2020) Biopolymers and their industrial applications: from plant, animal, and marine sources, to functional products. Elsevier
Tiehm A, Herwig V, Neis U (1999) Particle size analysis for improved sedimentation and filtration in waste water treatment. Water Sci Technol 39(8):99–106
Udayakumar GP, Muthusamy S, Selvaganesh B, Sivarajasekar N, Rambabu K, Sivamani S et al (2021) Ecofriendly biopolymers and composites: preparation and their applications in water-treatment. Biotechnol Adv 52:107815
Urlacher B, Dalbe B (1997) Xanthan gum. Thickening and gelling agents for food. Springer US, Boston, MA, pp 202–226
Wan YZ, Hong L, Jia SR, Huang Y, Zhu Y, Wang YL (2006) Synthesis and characterization of hydroxyapatite-bacterial cellulose nanocomposites. Compos Sci Technol 26:1125–1131
Wang LK, Hung Y-T, Shammas NK (2005) Physicochemical treatment processes. Humana Press, Totowa, NJ
Wen Y, Yuan J, Ma X, Wang S, Liu Y (2019) Polymeric nanocomposite membranes for water treatment: a review. Environ Chem Lett 17(4):1539–1551
Woodruff MA, Hutmacher DW (2010) The return of a forgotten polymer—polycaprolactone in the 21st century. Prog Polym Sci 35(10):1217–1256
Xia S, Zhang L, Davletshin A, Li Z, You J, Tan S (2020) Application of polysaccharide biopolymer in petroleum recovery. Polymers 12(9):1860
Xu X, Yang Y, Liu T, Chu B (2022) Cost-effective polymer-based membranes for drinking water purification. Giant 10:100099
Zhou Y, Zhang Z, Zhang J, Xia S (2016) Understanding key constituents and feature of the bio polymer in activated sludge responsible for binding heavy metals. Chem Eng J 304:527–532
Zubair M, Ullah A (2021) Biopolymers in environmental applications: industrial wastewater treatment. In: Biopolymers and their industrial applications, pp 331–349

Chapter 5
Antibiotics and Their Metabolites: Distribution in Waste Water and Potential Bioremediation

Anjan Mondal, Chhandak Mondal, and Asif Hossain

Abstract Antibiotics act as crucial tools in healthcare, veterinary medicine, and agriculture against bacterial infections. However, a major amount of the antibiotics used for the treatment of humans and other domestic animals are spent in waste water through their feces and urine in the form of metabolites or even in unmodified states, thereby acting as emerging pollutants. Antibiotics and their metabolites pose significant environmental concerns due to their potential adverse effects on aquatic ecosystems, agriculture, and human health. The ubiquitous presence of antibiotics often leads to the selection of antibiotic-resistant bacteria in the environment. Thus, understanding the distribution of antibiotics and their metabolites in wastewater and implementing effective remediation strategies for further elimination from the environment are imperative for mitigating their environmental impacts. Different physical (coagulation, filtration, ion exchange, sedimentation, adsorption, etc.) and chemical (chlorination, oxidation, ozonation, etc.) techniques are frequently used to remove antibiotics from the environmental samples. However, bioremediation is a cost-effective, environmentally sustainable method that uses living organisms and their derivatives to eliminate or detoxify antibiotics from a specific environment. Overall, this chapter provides insights into the distribution of antibiotics and their metabolites in wastewater, as well as their toxicity in living organisms, and emphasizes the importance of adopting sustainable bioremediation approaches for the elimination of antibiotics and their metabolites for environmental protection and human health preservation.

Keywords Antibiotics · Metabolites · Bacteria · Strategies · Bioremediation

A. Mondal · C. Mondal · A. Hossain (✉)
Department of Zoology, The University of Burdwan, Golapbag, Burdwan, West Bengal, India
e-mail: ahossain@zoo.buruniv.ac.in

© The Author(s), under exclusive license to Springer Nature Singapore Pte Ltd. 2026

S. Dey and M. P. Shah (eds.), *Microbial Applications in Industrial Wastewater Treatment*, Advances in Wastewater Research,
https://doi.org/10.1007/978-981-96-9128-9_5

5.1 Introduction

Antibiotics are definitely the most effective class of medications for treating microbial infections because of their precise action on the target in both humans and animals. These are categorized into 11 groups according to their chemical makeup and mode of action, such as aminoglycosides, lactams, glycopeptides, macrolides, oxazolidinones, polymyxins, quinolones, streptogramins, sulfonamides, tetracyclines, and others (Zhang and Li 2011). Antibiotics cure bacterial infections by killing and inhibiting their growth. Excessive use of antibiotics results in increased antibiotic-resistant bacteria in both the human body and the environment. The body or microorganisms metabolize antibiotics, forming and releasing antibiotic metabolites as by-products. These metabolites can play a variety of functions, including contributing to the antibiotic's effectiveness, producing adverse effects, and influencing the environment.

Various anthropogenic activities, including antibiotic overuse and dumping, have discharged antibiotics into the natural environment, contaminating the land, water, and atmosphere. In addition to the previously mentioned sources, hospitals and municipal wastewater are among the primary sources of antibiotics. Typically, effluents absorb 10–70% of antibiotics, releasing the remaining ones as unaltered forms or metabolites (Ma et al. 2019). The World Health Organization (WHO) strongly advises people not to overconsume medicines, particularly antibiotics. Antibiotics promote the development of antibiotic resistance genes (ARGs) in the bacterial population in the environment, putting selection pressure on them. The spread of ARGs poses serious threats to human health and other living species (Hinchliffe et al. 2018). Scientists caution that uncontrolled antibiotic resistance could escalate into a serious issue over the next 50 years. Without action, mortality rates might reach 300 million fatalities (Shah 2025; Bunce and Hellyer 2018).

Using antibiotic-tainted water to irrigate agricultural land provides a twofold concern. It not only pollutes groundwater but also endangers critical bacterial populations that fix nitrogen, which is necessary for crop development and productivity (Roose-Amsaleg and Laverman 2016). Antibiotics, while meant to treat diseases and increase livestock productivity, also have unexpected consequences, as reported by Mann et al. (2021). Singer et al. (2016) demonstrates how their introduction into aquatic environments modifies microbial community dynamics and functions, contributing to bacterial resistance. Antibiotics have varying effects on fish depending on their position in the food chain. Herbivorous fish have lower antibiotic levels in their systems than carnivorous ones. This discrepancy results from biomagnification, in which carnivorous fish collect larger quantities of antibiotics when they consume food (Shah 2024a; Liu et al. 2021).

Bacterial resistance has increased significantly during the last decade, culminating in the creation of multidrug-resistant pathogens and superbugs. Addressing this issue necessitates effective antibiotic mitigation. Conventional procedures fall short of removing antibiotic residues, prompting the development of new technologies aimed at long-term and cost-effective remediation. These methods improve removal efficiency through methodical, ecologically friendly techniques that use minimal energy and equipment (Frieri et al. 2017).

Antibiotics are degraded via both biological and non-biological processes. Biological degradation involves microorganisms, whereas non-biological degradation includes mechanisms like sorption, hydrolysis, photolysis, and different oxidation and reduction reactions (Shah 2024b; Massé et al. 2014). Researchers are exploring the use of microbial communities to remediate pollutants in terrestrial and aquatic environments. Bioremediation depends on microorganisms' metabolic capacity for degradation (Dangi et al. 2019). Phytoremediation has also evolved into a promising and ecologically benign approach for dealing with a wide range of environmental pollutants (Pilon-Smits 2005). This strategy takes advantage of plants' inherent ability to absorb and break down contaminants. Kotyza et al. (2010) and Hoang et al. (2013) have investigated plants' ability to eliminate pharmaceutical chemicals from their surroundings.

Enzymatic bioremediation is important in breaking down antibiotics, which helps combat antibacterial resistance. There is a correlation between antibiotic-resistant genes, antibiotic-resistant bacteria, and enzyme synthesis that degrades antibiotics (Sodhi and Singh 2022). Essentially, these enzymes assist in neutralizing antibiotics, perhaps making antibiotic-resistant bacteria susceptible again. This chapter discusses the sources of antibiotics, their effects on soil and aquatic environments, different physicochemical methods, various bioremediation techniques, and concludes with future perspectives.

5.2 Sources of Antibiotics

According to the WHO (2012), approximately 2.6 billion individuals don't have a sense of basic sanitation worldwide, which probably leads to the direct discharge of antibiotics into the environment and surrounding water bodies. According to Xu et al. (2007), the livestock industry and human healthcare consume 1–2 lakh metric tons of antibiotics annually. Antibiotics and their metabolites continue to contaminate our surrounding environment (Fig. 5.1). In agriculture, antibiotics improve crop development, honey production, and fish production. However, the overuse of antibiotics can lead to environmental damage through their discharge. Additionally, the hazard of antibiotic contamination is increased by releasing waste from hospitals, veterinarians, pharmaceutical industries (Obayiuwana et al. 2018), dairy farms, livestock farming (Pruden et al. 2013), and domestic wastes.

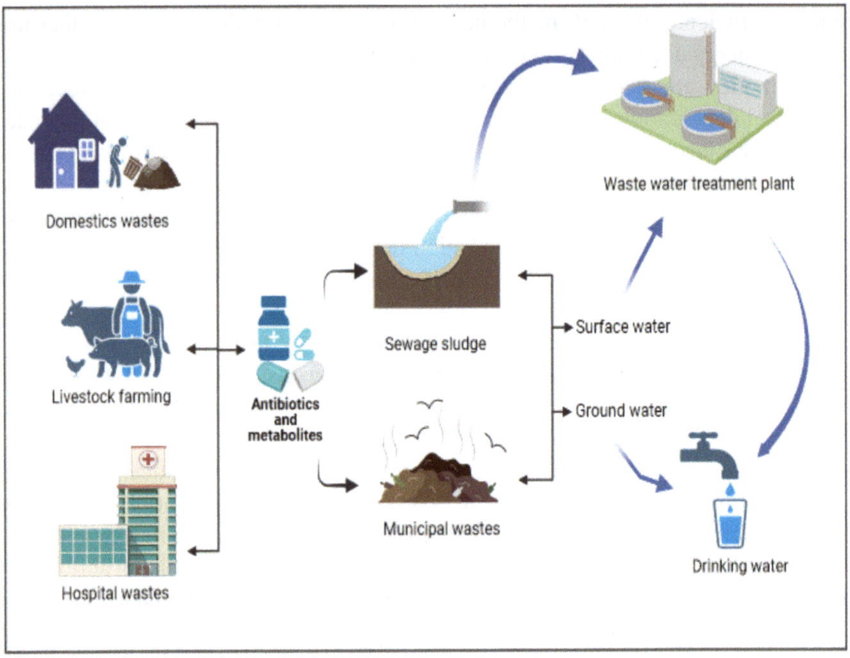

Fig. 5.1 Some sources of the antibiotics and their metabolites (https://biorender.com)

5.3 Hospital Wastes

Healthcare activities produce a wide range of waste, including complex effluents containing both chemical and biological substances. These effluents, originating from various sections of healthcare facilities, are particularly significant sources of emerging contaminants in many developing countries. Among these contaminants are heavy metals, antibiotic resistance genes, resistant bacteria, and antibiotic residues (Sabri et al. 2020; Laffite et al. 2016). Studies have identified elevated levels of various antibiotics in hospital effluent, including metronidazole, ciprofloxacin, oxalinic acid, sulfapyridine, norfloxacin, and trimethoprim, ofloxacin. Typically, conventional methods for treating hospital wastewater are ineffective, leaving a chance for these pollutants to enter the ecosystem, such as rivers, lakes, and seas, either partially or entirely, where they can accumulate over time (Ebrahimi et al. 2020). The discharge of hospital effluents poses a significant health risk. Overuse and misuse of antibiotics have fueled a concerning rise in antibiotic resistance worldwide (Shah 2024c; Fair and Tor 2014). This overuse has led to the selection of multidrug-resistant bacteria. Several studies have documented the environmental impacts of pharmaceuticals, including antibiotics. Antibiotic resistance genes, which can transfer normally, are of particular concern to sensitive microbial species. Infections caused by antibiotic-resistant bacteria are associated with higher mortality and morbidity rates, leading to longer hospital stays (Cosgrove 2006). Several ongoing initiatives aim to evaluate and screen hospitals, investigating their potential as sources of environmental discharge of drugs and other substances that may lead to the emergence of multidrug-resistant bacteria (Escher et al. 2011). In addition to medications, hospitals and surgical facilities utilize a range of substances for medical purposes, such as disinfectants and diagnostic agents.

5.4 Domestic Wastes

Municipal wastewater is the largest source of antibiotics contamination in the environment, although poor treatment of industrial effluents and direct dumping of unwanted drugs are other contributing factors. The exact treatment procedures and processes used determine the final fate of antibiotics arrive at treatment facilities, where they undergo physical or chemical changes. Wastewater treatment plants are often ill-equipped to fully remove these persistent compounds, leading to their release into surface waters and groundwater. This poses a significant threat to aquatic ecosystems and human health. Antibiotics contaminate municipal and surface water systems as a result of inadequate treatment or poor disposal practices by individuals (Sun et al. 2020). Industries like slaughterhouses and poultry farms, along with domestic discards, are the primary sources of these antibiotics. Antibiotics used by humans and other domestic and agricultural animals pass through the digestive system, are ejected, and end up in wastewater treatment facilities (Gao et al. 2012). Municipal wastewater is teeming with a wide array of micro- and macro-organisms, making it a hotbed for the emergence of drug-resistant strains (Kumar and Pal 2018). There are many studies that show the global prevalence of antibiotics in wastewater treatment facilities, indicating that industrialized nations have a wider range of antibiotic concentrations and kinds than developing countries.

5.5 Livestock Farming Wastes

Antibiotics are important in animal production because they treat diseases, maintain health, and promote growth (Boxall et al. 2003; Sarmah et al. 2006). However, their extensive usage raises environmental problems. Animals may excrete a large fraction, ranging from 70 to 90%, of antibiotics unmetabolized after their administration (Kumar et al. 2005). As a result, these residues remain in the environment, influencing water and soil quality (Liu et al. 2015). The discharge of antibiotics through sewage systems, and occasionally through sewage treatment facilities exacerbates the situation (Kümmerer 2009). This contamination has a negative impact on agricultural activities, as antibiotic residues in soil contribute to issues like anoxic denitrification by altering the bacterial populations responsible for this critical process (Pan and Chu 2016). Antibiotic residues in animal-derived foods provide a variety of health risks, including bacterial resistance, toxicity, hypersensitivity responses, cancer, and teratogenicity (Darwish et al. 2013). Recognizing the importance of antimicrobial resistance, the 68th World Health Assembly in May 2015 approved a strategy to reduce the needless use of antimicrobials in both human and animal healthcare. Despite this worldwide attempt, major meat-producing countries, including Argentina, Brazil, China, India, Indonesia, the Philippines, Russia, and South Africa, continue to permit the use of antibiotics as growth promoters (Okocha et al. 2018). Because of a lack of statistics, antibiotic use is highly questionable in low-income nations. Antibiotic usage is projected to rise significantly by 2030, with China, Brazil, India, South Africa, and Russia leading the way (Van Boeckel et al. 2015).

5.6 Impacts of Antibiotics in Soil Environment

Microorganisms in soil carry out a variety of essential functions and contribute to the preservation of the overall health and quality of the soil. They are essential for maintaining soil fertility, releasing nutrients into the soil, and turning over organic materials.

Additionally, a variety of microorganisms act as biological barriers by preventing the spread of infections (Varma and Buscot 2005). Instabilities in soil equilibrium can arise from various abiotic and biotic factors, including bacteriophages, pesticides, heavy metals, toxic hydrocarbons, as well as antibiotics (Cycoń et al. 2019). Antibiotics possessing potent antimicrobial properties in soil are likely to specifically inhibit the growth of soil microorganisms, thus altering the composition of the soil microbial community and potentially impacting the ecological processes within the soil (Molaei et al. 2017). Even exact levels of antibiotics can cause gene flux and genome reorganization in the microbial community in soil, such as the transfer of antibiotic-resistant genes (Grenni et al. 2018). Lots of information is available about the impact of antibiotics on microorganisms and bacterial and fungal mediated soil activities. Research has shown that antibiotics affect microbial populations directly as well as more broadly by changing the makeup of whole microbial ecosystems. To investigate these impacts, a number of methods have been used, including measuring factors including respiration, microbial biomass, and soil organic matter turnover. Furthermore, a variety of techniques have been used to study certain microbial processes such as nitrification, denitrification, sulfate, and iron reduction, methanogenesis, and the activity of enzymes involved in the cycling of carbon, nitrogen, and phosphorus (Liu et al. 2016a; Ma et al. 2016). A significant amount of oxytetracycline leads to a decrease in nitrification rate. Additionally, Ma et al. (2016) observed that applying 30 mg/kg and 100 mg/kg of sulfadiazine, respectively, resulted in certain effects. Furthermore, ciprofloxacin and norfloxacin were discovered to enhance the nitrification rate in soil microcosms, albeit only at the lowest antibiotic concentration tested, which was 1 mg/kg of soil (Yang et al. 2012; Cui et al. 2014). Cui et al. (2014) found that when soil concentration increased, ciprofloxacin degradation was minimized.

5.7 Impacts of Antibiotics in Aquatic Environment

Antibiotics, laxatives, anti-inflammatory medicines, as well as analgesic drugs, are the most often used medications for treatment in hospitals. Antibiotics are the primary source of these compounds that enter municipal drains through a variety of processes. Fermented sludge, activated wastewater, and finally municipal bio-solids. If antibiotics are not eliminated during the purification process of wastewater, they can harm aquatic life once discharged in aquatic life (Meena et al. 2015; Hernández et al. 2019). Some research confirms the presence of bacteria in drinking water, lakes, rivers, groundwater, and sewage treatment facilities. Tetracycline, macrolides, and sulfonamides are the three most well-recognized antibiotic groups now in use. Different concentrations of antibiotics were found in swine farm wastewater: 324.4 g/L is the highest concentration of antibiotics, and 72 g/L is the lowest concentration of antibiotics (Cheng et al. 2018). The bacteria may gradually develop susceptibility to these kinds of antibiotics as a result of the development of these drugs at sub-inhibitory levels, and this tolerance subsequently begins to spread throughout the pathogenic bacteria. In aquaculture, antibiotics are used to control bacterial diseases in farmed fish and shellfish. This can lead to antibiotic residues and resistant bacteria in aquatic environments. Antibiotics exhibit toxicity toward aquatic organisms at high doses and have deleterious effects on developmental processes, digestive system, fertility, and longevity, leading to an ecological crisis. Additionally, it has been shown that sulfonamides and macrolides may harm plant tissues and their photosynthesis process, prevent algal growth and development, and decrease the oxidation rate of atmospheric carbon dioxide (Sosa-Hernández et al. 2021). The research discovered that some ribosomal protein mutations facilitate resistance to aminoglycosides, tetracycline, and macrolides. On the other hand, mutations in RNA polymerase promote resistance to rifampicin (Lupo et al. 2012). In an investigation, more than 35 genetic modifications were found in drug-resistant strains of *M. tuberculosis* and *S. aureus*. According to the sequencing of antibiotic-resistant *M. tuberculosis* strains, the genome of a multidrug-resistant strain exhibited 29 distinct changes, while a very drug-resistant strains genome included 35 mutations (Mwangi et al. 2007). Furthermore, it was found that the gyrA gene in *E. coli* contains seven amino acid residues that promote resistance to antibiotics (Khamenei et al. 2016).

5.8 Remediation

Antibiotics present in the environment are causing detrimental impacts on the environment, making it crucial to establish remediation programs. Absorption of antibiotics and their breakdown products leads to the emergence of drug-resistant bacteria, reducing the potency of antibiotics (Harris et al. 2012). Now a days, the multidrug-resistant bacteria are increasing abruptly because of the excess amount of antibiotics and their metabolites in the environment; as a result, many bacterial diseases are increasing rapidly. To get rid of this situation, remediation or degradation of antibiotics and their metabolites is one of the key solutions for us. Several methods are being explored to remediate antibiotics in a sustainable and cost-effective manner. These techniques enhance removal efficiency by using more sustainable methods and less energy-consuming technology (Frieri et al. 2017). Antibiotic removal is classified into two types: abiotic degradation, which is by oxidation, reduction, photolysis, and hydrolysis, and biotic degradation, which is by fungus, bacteria, algae, and plants.

5.9 Physicochemical Methods

The huge amount of antibiotics discharged into the environment has increased dramatically. Now is the time to remediate these products quickly from the environment (Fig. 5.2). Chu et al. (2021) showed that in the beginning, ionizing radiation was employed to remove toxic microbes in wastewater. Now, the ionizing radiation-coupled approach may effectively reduce antibiotics, ARGs, and antibacterial activity in pharmaceutical waste water. According to Gan et al. (2009), Fenton's procedure is a popular way for removing antibiotics by advanced oxidation. This process is most commonly used for amoxicillin removal. Unlike Fenton's technique, electrochemical oxidation relies on the oxidation caused by an electrochemical battery. Advanced oxidation processes (AOPs) are used to remove antibiotics and their metabolites that are resistant to degradation. This process generates a highly reactive intermediate radical (OH^-) with a significantly larger redox potential than previous methods, which facilitates the oxidation of organic molecules (Deng and Zhao 2015). Akhil et al. (2021) reviewed that, beside the above method, electrocoagulation, photocatalytic degradation, membrane processes, ozonation, and sonocatalytic degradation are used for antibiotics and other pharmaceutical remediation.

5.10 Bioremediation Methods

Bioremediation is a cost-effective, environmentally sustainable method that uses living creatures such as plants, animals, algae, fungi, and bacteria to eliminate or detoxify antibiotics from a specific environment. In the bioremediation process, different types of bacteria, fungi, algae, plants, etc., are used to successfully degrade the antibiotics and also their metabolites.

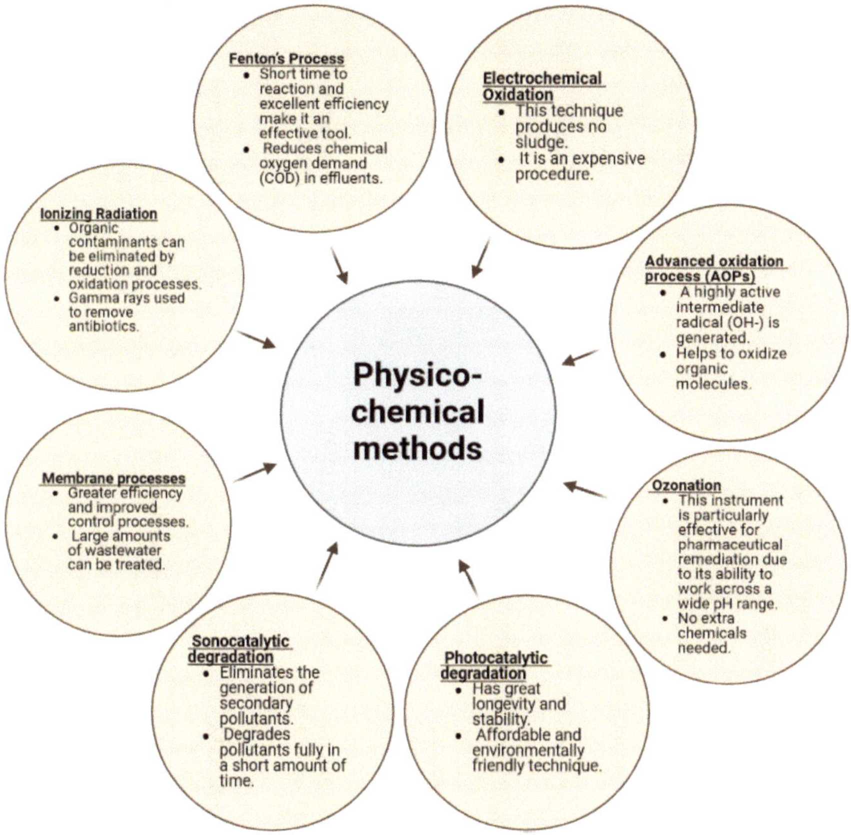

Fig. 5.2 Various physiochemical methods for antibiotics remediation (https://biorender.com)

5.11 Bacterial Remediation

One of the most popular bioremediation processes is bacterial bioremediation, where bacteria come forward to degrade or clean up the antibiotics and their metabolites. Many scientists have recently been involved in this research, identifying different strains of bacteria from various areas and using them to remediate these antibiotics. Al-Gheethi et al. (2015) reviewed that many bacterial techniques, including biosorption as well as enzymatic biodegradation, can remove antibiotics and heavy metals. In Table 5.1, there are many bacterial species that may successfully remove antibiotics in given conditions.

Table 5.1 Some antibiotic degrading bacterial species

Bacterial species	Antibiotic removed	Condition	Removal percentage	Time	References
Stenotrophomonas sp.-WA5	Amoxicillin	150 rpm + M9 minimal media at 30 °C	96%	14 days	Yang et al. (2020) and Bhattacharya and Khare (2022)
Alcaligenes sp.-MMA	Amoxicillin	150 rpm + M9 minimal media at 30 °C	84%	14 days	Sodhi and Singh (2022)
Paraclostridium sp.	Ciprofloxacin	150 rpm + Basal salt media	84%	14 days	Sodhi and Singh (2022)
Shewanella oneidensis-MR-1	Sulfapyridine and sulfamethoxazole	LB + 10 mg/L SPY or SMX at 30 °C	SPY (23.91%); SMX (59.88%)	5 days	Mao et al. (2018)
Shewanella sp.-MR-1	Sulfapyridine and sulfamethoxazole	LB + 10 mg/L SPY or SMX, 30 °C	SPY (23.43%); SMX (63.89%)	5 days	Mao et al. (2018)
Alcaligenes faecalis-CGMCC 1.767	Sulfamethoxazole	MMSM + 1 g/L acetate + 1 g/L glucose + 50 mg/L SMX + 100 ppm vitamin C; 30 °C	80–90%	16 h	Zhang et al. (2016)
Rhodococcus rhodochrous-ATCC 13808	Sulfamethoxazole and sulfamethazine	MMSM + 3 g/L glucose + 31.6 mg/L SMX or 43.4 mg/L SMZ, 26 °C	SMX (20%); SMZ (14%)	SMX 36 days; SMZ (12 days)	Gauthier et al. (2010)
Rhodococcus equi-ATCC 13557	Sulfamethoxazole	MMSM + 6 mg/L SMX + 0.5 g/L glucose, 26 °C	15–29%	120 h	Larcher and Yargeau (2011)
Microbacterium sp.-BR1	Sulfamethoxazole	MSM + 127 mg/L SMX	22–44% mineralization	16 days	Bouju et al. (2012)
		MSM + 127 mg/L SMX	100%	56 h	Birkigt et al. (2015)

(continued)

Table 5.1 (continued)

Bacterial species	Antibiotic removed	Condition	Removal percentage	Time	References
Bacillus subtilis-1556WTNC	Amoxicillin, ampicillin, cephalexin, cefuroxime and ciprofloxacin	Sterilized treated sewage effluents + 0.2 to 5 mg/L amoxicillin or ampicillin or cephalexin orcefuroxime or ciprofloxacin, 35 °C	Amoxicillin (25.03% at 1 mg/ mL); ampicillin (15.59% at 0.8 mg/mL); cephalexin (22.59% at 1 mg /mL); cefuroxime (10.62% at 1 mg/mL); ciprofloxacin (2.45% at 0.6 mg/mL)	12 days	Al-Gheethi and Ismail (2014)
Achromobacter denitrificans-PR1	Sulfamethoxazole	MSM + 152 ppm SMX	100%	14 days	Reis et al. (2014)
	Sulfamethoxazole	MSM + 152 ppm SMX + 590 ppm succinate	100%	1.8 days	
	Sulfadiazine	MSM + 25 ppm SDZ	98%	2.3 days	
	Sulfadimethoxine	MSM + 31 ppm SDM	48%	2.3 days	
	Sulfamethizole	MSM + 28 ppm SMT	98%	2.3 days	
	Sulfapyridine	MSM + 25 ppm SPY	100%	2.3 days	
Methylobacterium sp.-SDZ W2-SJ40	Sulfadiazine	AMS + 0.04% yeast extract + 5 mg/L SDZ	55.2%	12 days	Mulla et al. (2016)
Arthrobacter sp.-D2	Sulfadiazine	MSM + 50 mg/L SDZ + 2 g/L glucose, 37 °C	SDZ (99%) degradation; SDZ (82.5%) mineralization	53 h	Deng et al. (2016)
Acinetobacter sp.-W1	Sulfamethoxazole	MSM + 5–240 mg/L SMX + 5 mg/L DO, 25 °C	100%	24 h	Wang and Wang (2018)
Paracoccus sp.-SDZ-PM2 BSH30	Sulfadiazine	AMS + 0.04% yeast extract + 5 mg/L SDZ; 30 °C	50%	12 days	Mulla et al. (2016)
Kribbella sp. SDZ-3S SCL47	sulfadiazine	AMS + 0.04% yeast extract + 5 mg/L SDZ; 30 °C	60.6%	12 days	Mulla et al. (2016)

SDM: sulfadimethoxine, SMT: sulfamethizole, SPY: sulfapyridine, SDZ: sulfadiazine, SMZ: sulfamethazine, SMX: sulfamethoxazole, LB: Luria-Bertani medium, AMS: ammonium mineral salts, MMSM: minimum mineral salt medium, MSM: mineral salts medium

5.12 Mycoremediation

Mycoremediation is a bioremediation technique that involves using fungi to remove or biotransform the pollutants. Nowadays, many studies are ongoing or already established about the remediation of antibiotics using many fungal species. Čvančarová et al. (2015) studied that norfloxacin (NOR), ciprofloxacin (CIP), and ofloxacin (OF) are biodegraded by the fungal strains, *T. versicolor, Panus tigrinus, Irpex lacteus, Dichomitus squalens, and Pleurotus ostreatus* in liquid medium. Table 5.2 shows some fungal species with their antibiotic degrading capabilities.

5.13 Phytoremediation

Phytoremediation is the technique of using natural or genetically modified plants to remove any material from the environment that is hazardous. Recently, phytoremediation has gained attention as an affordable and sustainable method to remove active pharmaceutical ingredients from contaminated soil and water, surpassing traditional methods.

Brassica juncea is a well-known and promising plant species for phytoremediation of metals. Gahlawat and Gauba (2016) demonstrated that the seeds of *B. juncea* were germinated and cultured for 28 and 24 days under various aspirin and tetracycline concentrations and successfully remediated approximately 90% and 71%, respectively. This study shows that *B. juncea* may tolerate and remediate prescription drugs, including analgesics and antibiotics (Gahlawat and Gauba 2016).

Singh et al. (2018) studied the toxicity of amoxicillin on duckweed (*Spirodela polyrhiza*) during a 7-day exposure. The researchers treated the duckweed with both low (0.0001 to 0.01 mg L^{-1}) and high (0.1 and 1 mg L^{-1}) levels of amoxicillin. Duckweed setups had a much higher reduction in amoxicillin concentration (84.6–100%) compared to the control (62.1–73%), indicating phytodegradation as a key mechanism for removing antibiotics from water, in addition to hydrolysis and photodegradation (Singh et al. 2018).

Alfalfa (*Medicago sativa*) is one type of grass species that is used to remove sulfamethazine (SMN) antibiotics, potentially under hydroponic conditions. Alfalfa grown hydroponically in a solution supplemented with 10 mg/L of SMN antibiotic and detected the amount of SMN, i.e., 8.58 μg/kg in the root zone, 1.89 μg/kg in the top portion, 1.30 μg/kg in the middle one-third, and 0.38 μg/kg in sap and also observed the UV spectra of parent SMN and translocated SMN which are spread throughout the body parts. Using hydroponic alfalfa to absorb SMN has the potential to remove antibiotics from wastewater discharges (Kurwadkar et al. 2017).

Pteris vittata is an ideal plant for remediation because of its high biomass and accumulation capacity. In a seven-day trial, Li et al. (2015) studied that *Pteris vittata*

Table 5.2 Some antibiotic degrading fungi (Jagtap 2020)

Fungal species	Antibiotic	Metabolites	References
Trichoderma viride	Ciprofloxacin and norfloxacin	4-Hydroxy-3-oxo-4 vinylcyclopent-1-enyl ciprofloxacin; 4-hydroxy-3-oxo-4 vinylcyclopent-1-enyl norfloxacin	Parshikov et al. (2002)
Irpex lacteusb, Trametes versicolor	Ciprofloxacin	Desethylene-N-ciprofloxacin; 7-amino-1-cyclopropyl-6-fluoro 4-oxo-1,4-dihydroquinoline-3 carboxylic acid; N-acetylciprofloxacina; Desethylene-N acetylciprofloxacin; N-formylciprofloxacin; desethylene-N formylciprofloxacin; hydroxymethyl-N-ciprofloxacin	Čvančarová et al. (2015)
	Ofloxacin	9-Fluoro-10-[(2-formamidoethyl) amino]-3-methyl-7-oxo-2,3 dihydro-7H-[1,4] oxazino[2,3,4-Ij] quinoline-6 carboxylic acid[b]; 9-fluoro-10-(4-formylpiperazin 1-yl)-3-methyl-7-oxo-2,3 dihydro-7H-[1,4] oxazino[2,3,4-Ij] quinoline-6 carboxylic acid; desethylene Norfloxacin; 10-[(2-acetamidoethyl) amino]-9-fluoro-3-methyl-7-oxo 2,3-dihydro-7H-[1,4] oxazino[2,3,4-Ij] quinoline-6-carboxylic Acid; 10-amino-9-fluoro-3 methyl-7-oxo-2,3 dihydro-7H-[1,4] oxazino[2,3,4-Ij] quinoline-6 carboxylic acid; 10-(4-acetylpiperazin-1-yl)-9 fluoro-3-methyl-7-oxo-2,3 dihydro-7H-[1,4] oxazino[2,3,4-Ij] quinoline-6 carboxylic acid[a]; desmethyl-N-ofloxacina	

(continued)

Table 5.2 (continued)

Fungal species	Antibiotic	Metabolites	References
	Norfloxacin	Desethylene-N-norfloxacina; 7-amino-1-ethyl-6-fluoro-4 oxo-1, 4-dihydroquinoline-3 carboxylic acid; desethylene-N-acetylnorfloxacin; N-formylnorfloxacin	
Pleurotus ostreatus	Ciprofloxacin	–	Singh et al. (2017)
Trametes versicolor	Ofloxacin	–	Gros et al. (2014)
Gloeophyllum striatum	Ciprofloxacin, Enrofloxacin	–	Wetzstein et al. (1997, 1999)
Trametes versicolor	Norfloxacin and ciprofloxacin	–	Prieto et al. (2011)
Aspergillus terreus FZC3	Gentamicin	–	Liu et al. (2016b)
Penicillium restrictum and *Trametes versicolor*	Sulfamethoxazole, erythromycin, and tetracycline Carbamazepine	–	Fakhri et al. (2021) Silva et al. (2019)

effectively removed tetracyclines from a waste water solution in five days. For hyper-accumulation abilities *Pteris vittata* absorbs all of the TCs in the solution. The *Pteris vittata* specimens grew nicely and had no toxic indications (Li et al. 2015).

5.14 Phycoremediation

A bioremediation process where algae play a key role in removing the pollutants from wastewater is called phycoremediation. Algae are photoautotrophic aquatic organisms that give a cost-effective, non-invasive, and safe solution for removing antibiotics from the environment. There are many algae discovered where they successfully removed or degraded antibiotics. *Chlorella pyrenoidosae*, a green alga, effectively eliminates antibiotics like cefradine, cephalexin, cefixime, and ceftazidime (Yu et al. 2017). A biomass of *Phaeodactylum tricornutum* microalga is very useful for removing oxytetracycline successfully (Santaeufemia et al. 2016). Yu et al. (2017) studied that ceftazidime and 7-amino cephalosporanic acid are degraded at 92.70% and 96.07%, respectively, by an algal species, *C. pyrenoidosa.* A group of algal species (*Chlorella vulgaris, Scenedesmus obliquus, Ourococcus multisporus, Chlamydomonas mexicana, and Micractinium resseri*) are used to remove enrofloxacin (Xiong et al. 2017).

5.15 Conclusive Remarks and Future Perspectives

Antibiotics in the environment are of major concern worldwide, as they increase the possibility of bacteria developing resistance to these medications. Because of its widespread use, antibiotic contamination has become a serious problem. Wastewater treatment facilities (WWTPs) are currently unable to completely remove antibiotic residues from water sources. Therefore, considering the current global conditions, there is an urgent need for further research into the capacities and capabilities of WWTPs to address this issue. The toxicity of antibiotics and their metabolites in aquatic systems, agricultural systems, and other systems is a major concern. Due to their short lifespan, the hydrophobic and lipophilic properties of antibiotics contribute to their stability in the environment. Beside the physical and chemical processes, bioremediation may act as a very cost-effective and sustainable way of degrading the antibiotics by bacteria, algae, fungi, and many plant species. Additionally, progress in genetic engineering and the synthetic biology toolbox creates fresh chances for the creation of widely used applications that use organisms that break down antibiotics for bioremediation. Further extensive study is required to enhance technology in order to achieve a cleaner environment and a safer future for future generations.

References

Akhil D, Lakshmi D, Senthil Kumar P, Vo DVN, Kartik A (2021) Occurrence and removal of antibiotics from industrial wastewater. Environ Chem Lett 19:1477–1507

Al-Gheethi AA, Ismail N (2014) Biodegradation of pharmaceutical wastes in treated sewage effluents by Bacillus subtilis 1556WTNC. Environ Process 1:459–481

Al-Gheethi AA, Lalung J, Noman EA, Bala JD, Norli I (2015) Removal of heavy metals and antibiotics from treated sewage effluent by bacteria. Clean Technol Environ Policy 17:2101–2123

Bhattacharya A, Khare SK (2022) Utilizing the ß-lactam hydrolyzing activity of ß-lactamase produced by *Bacillus cereus* EMB20 for remediation of ß-lactam antibiotics. Int Biodeterior Biodegrad 168:105363

Birkigt J, Gilevska T, Ricken B, Richnow HH, Vione D, Corvini PFX et al (2015) Carbon stable isotope fractionation of sulfamethoxazole during biodegradation by *Microbacterium* sp. strain BR1 and upon direct photolysis. Environ Sci Technol 49(10):6029–6036

Bouju H, Ricken B, Beffa T, Corvini PFX, Kolvenbach BA (2012) Isolation of bacterial strains capable of sulfamethoxazole mineralization from an acclimated membrane bioreactor. Appl Environ Microbiol 78(1):277–279

Boxall AB, Kolpin DW, Halling-Sørensen B, Tolls J (2003) Peer reviewed: are veterinary medicines causing environmental risks? Environ Sci Technol 37(15):286A-294A

Bunce JT, Hellyer P (2018) Antibiotic resistance and antibiotic prescribing by dentists in England 2007–2016. Br Dent J 225(1):81–84

Cheng DL, Ngo HH, Guo WS, Liu YW, Zhou JL, Chang SW et al (2018) Bioprocessing for elimination antibiotics and hormones from swine wastewater. Sci Total Environ 621:1664–1682

Chu L, Wang J, He S, Chen C, Wojnárovits L, Takács E (2021) Treatment of pharmaceutical wastewater by ionizing radiation: removal of antibiotics, antimicrobial resistance genes and antimicrobial activity. J Hazard Mater 415:125724

Cosgrove SE (2006) The relationship between antimicrobial resistance and patient outcomes: mortality, length of hospital stay, and health care costs. Clin Infect Dis 42(Supplement_2):S82–S89

Cui H, Wang SP, Fu J, Zhou ZQ, Zhang N, Guo L (2014) Influence of ciprofloxacin on microbial community structure and function in soils. Biol Fertil Soils 50:939–947

Čvančarová M, Moeder M, Filipová A, Cajthaml T (2015) Biotransformation of fluoroquinolone antibiotics by ligninolytic fungi–metabolites, enzymes and residual antibacterial activity. Chemosphere 136:311–320

Cycoń M, Mrozik A, Piotrowska-Seget Z (2019) Antibiotics in the soil environment—degradation and their impact on microbial activity and diversity. Front Microbiol 10:412419

Dangi AK, Sharma B, Hill RT, Shukla P (2019) Bioremediation through microbes: systems biology and metabolic engineering approach. Crit Rev Biotechnol 39(1):79–98

Darwish WS, Eldaly EA, El-Abbasy MT, Ikenaka Y, Nakayama S, Ishizuka M (2013) Antibiotic residues in food: the African scenario. Jpn J Vet Res 61(Supplement):S13–S22

Deng Y, Zhao R (2015) Advanced oxidation processes (AOPs) in wastewater treatment. Curr Pollut Rep 1:167–176

Deng Y, Mao Y, Li B, Yang C, Zhang T (2016) Aerobic degradation of sulfadiazine by *Arthrobacter spp.*: kinetics, pathways, and genomic characterization. Environ Sci Technol 50(17):9566–9575

Ebrahimi SM, Dehghanzadeh Reyhani R, Asghari-JafarAbadi M, Fathifar Z (2020) Diversity of antibiotics in hospital and municipal wastewaters and receiving water bodies and removal efficiency by treatment processes: a systematic review protocol. Environ Evid 9:1–9

Escher BI, Baumgartner R, Koller M, Treyer K, Lienert J, McArdell CS (2011) Environmental toxicology and risk assessment of pharmaceuticals from hospital wastewater. Water Res 45(1):75–92

Fair RJ, Tor Y (2014) Antibiotics and bacterial resistance in the 21st century. Perspect Med Chem 6:PMC-S14459

Fakhri H, Shahi A, Ovez S, Aydin S (2021) Bioaugmentation with immobilized endophytic *Penicillium restrictum* to improve quorum quenching activity for biofouling control in an aerobic hollow-fiber membrane bioreactor treating antibiotic-containing wastewater. Ecotoxicol Environ Saf 210:111831

Frieri M, Kumar K, Boutin A (2017) Antibiotic resistance. J Infect Public Health 10(4):369–378

Gahlawat S, Gauba P (2016) Phytoremediation of aspirin and tetracycline by *Brassica juncea.* Int J Phytorem 18(9):929–935

Gan S, Lau EV, Ng HK (2009) Remediation of soils contaminated with polycyclic aromatic hydrocarbons (PAHs). J Hazard Mater 172(2–3):532–549

Gao P, Munir M, Xagoraraki I (2012) Correlation of tetracycline and sulfonamide antibiotics with corresponding resistance genes and resistant bacteria in a conventional municipal wastewater treatment plant. Sci Tot Environ 421:173–183

Gauthier H, Yargeau V, Cooper DG (2010) Biodegradation of pharmaceuticals by *Rhodococcus rhodochrous* and *Aspergillus niger* by co-metabolism. Sci Total Environ 408(7):1701–1706

Grenni P, Ancona V, Caracciolo AB (2018) Ecological effects of antibiotics on natural ecosystems: a review. Microchem J 136:25–39

Gros M, Cruz-Morato C, Marco-Urrea E, Longrée P, Singer H, Sarrà M et al (2014) Biodegradation of the X-ray contrast agent iopromide and the fluoroquinolone antibiotic ofloxacin by the white rot fungus Trametes versicolor in hospital wastewaters and identification of degradation products. Water Res 60:228–241

Harris SJ, Cormican M, Cummins E (2012) Antimicrobial residues and antimicrobial-resistant bacteria: impact on the microbial environment and risk to human health—a review. Hum Ecol Risk Assess Int J 18(4):767–809

Hernández FELIX, Calısto-Ulloa N, Gómez-Fuentes C, Gómez M, Ferrer J, González-Rocha G et al (2019) Occurrence of antibiotics and bacterial resistance in wastewater and sea water from the Antarctic. J Hazard Mater 363:447–456

Hinchliffe S, Butcher A, Rahman MM (2018) The AMR problem: demanding economies, biological margins, and co-producing alternative strategies. Palgrave Commun 4(1)

Hoang TTT, Tu LTC, Le NP, Dao QP (2013) A preliminary study on the phytoremediation of antibiotic contaminated sediment. Int J Phytorem 15(1):65–76

Jagtap UB (2020) Bioremediation strategies for removing antibiotics from the environment. In: Antibiotics and antimicrobial resistance genes: environmental occurrence and treatment technologies, pp 319–337

Khameneі B, Diab R, Ghazvini K, Bazzaz BSF (2016) Breakthroughs in bacterial resistance mechanisms and the potential ways to combat them. Microb Pathog 95:32–42

Kotyza J, Soudek P, Kafka Z, Vaněk T (2010) Phytoremediation of pharmaceuticals—preliminary study. Int J Phytorem 12(3):306–316

Kumar A, Pal D (2018) Antibiotic resistance and wastewater: correlation, impact and critical human health challenges. J Environ Chem Eng 6(1):52–58

Kumar K, Gupta SC, Baidoo SK, Chander Y, Rosen CJ (2005) Antibiotic uptake by plants from soil fertilized with animal manure. J Environ Qual 34(6):2082–2085

Kümmerer K (2009) Antibiotics in the aquatic environment—a review—part I. Chemosphere 75(4):417–434

Kurwadkar S, Struckhoff G, Pugh K, Singh O (2017) Uptake and translocation of sulfamethazine by alfalfa grown under hydroponic conditions. J Environ Sci 53:217–223

Laffite A, Kilunga PI, Kayembe JM, Devarajan N, Mulaji CK, Giuliani G et al (2016) Hospital effluents are one of several sources of metal, antibiotic resistance genes, and bacterial markers disseminated in Sub-Saharan urban rivers. Front Microbiol 7:1128

Larcher S, Yargeau V (2011) Biodegradation of sulfamethoxazole by individual and mixed bacteria. Appl Microbiol Biotechnol 91:211–218

Li X, Yang X, Wang N, Xie Y (2015) Potential of *Pteris vittata* to remove tetracycline antibiotics from aquatic media. Int J Phytorem 17(9):895–899

Liu B, Li Y, Zhang X, Wang J, Gao M (2015) Effects of chlortetracycline on soil microbial communities: comparisons of enzyme activities to the functional diversity via Biolog EcoPlates™. Eur J Soil Biol 68:69–76

Liu C, Tan L, Zhang L, Tian W, Ma L (2021) A review of the distribution of antibiotics in water in different regions of China and current antibiotic degradation pathways. Front Environ Sci 9:692298

Liu J, Zhao Z, Orfe L, Subbiah M, Call DR (2016a) Soil-borne reservoirs of antibiotic-resistant bacteria are established following therapeutic treatment of dairy calves. Environ Microbiol 18(2):557–564

Liu Y, Chang H, Li Z, Zhang C, Feng Y, Cheng D (2016b) Gentamicin removal in submerged fermentation using the novel fungal strain *Aspergillus terreus* FZC3. Sci Rep 6(1):35856

Lupo A, Coyne S, Berendonk TU (2012) Origin and evolution of antibiotic resistance: the common mechanisms of emergence and spread in water bodies. Front Microbiol 3:18

Ma J, Cui Y, Zhang W, Wang C, Li A (2019) Fate of antibiotics and the related antibiotic resistance genes during sludge stabilization in sludge treatment wetlands. Chemosphere 224:502–508

Ma T, Pan X, Liu W, Christie P, Luo Y, Wu L (2016) Effects of different concentrations and application frequencies of oxytetracycline on soil enzyme activities and microbial community diversity. Eur J Soil Biol 76:53–60

Mann A, Nehra K, Rana JS, Dahiya T (2021) Antibiotic resistance in agriculture: perspectives on upcoming strategies to overcome upsurge in resistance. Curr Res Microb Sci 2:100030

Mao F, Liu X, Wu K, Zhou C, Si Y (2018) Biodegradation of sulfonamides by *Shewanella oneidensis* MR-1 and *Shewanella* sp. strain MR-4. Biodegradation 29:129–140

Massé DI, Cata Saady NM, Gilbert Y (2014) Potential of biological processes to eliminate antibiotics in livestock manure: an overview. Animals 4(2):146–163

Meena VD, Dotaniya ML, Saha JK, Patra AK (2015) Antibiotics and antibiotic resistant bacteria in wastewater: impact on environment, soil microbial activity and human health

Molaei A, Lakzian A, Haghnia G, Astaraei A, Rasouli-Sadaghiani M, Teresa Ceccherini M, Datta R (2017) Assessment of some cultural experimental methods to study the effects of antibiotics on microbial activities in a soil: an incubation study. PLoS One 12(7):e0180663

Mulla SI, Sun Q, Hu A, Wang Y, Ashfaq M, Eqani SAMAS, Yu CP (2016) Evaluation of sulfadiazine degradation in three newly isolated pure bacterial cultures. PLoS One 11(10):e0165013

Mwangi MM, Wu SW, Zhou Y, Sieradzki K, de Lencastre H, Richardson P et al (2007) Tracking the in vivo evolution of multidrug resistance in Staphylococcus aureus by whole-genome sequencing. Proc Natl Acad Sci 104(22):9451–9456

Obayiuwana A, Ogunjobi A, Yang M, Ibekwe M (2018) Characterization of bacterial communities and their antibiotic resistance profiles in wastewaters obtained from pharmaceutical facilities in Lagos and Ogun States, Nigeria. Int J Environ Res Public Health 15(7):1365

Okocha RC, Olatoye IO, Adedeji OB (2018) Food safety impacts of antimicrobial use and their residues in aquaculture. Public Health Rev 39:1–22

Pan M, Chu LM (2016) Adsorption and degradation of five selected antibiotics in agricultural soil. Sci Total Environ 545:48–56

Parshikov IA, Moody JD, Freeman JP, Lay JO Jr, Williams AJ, Heinze TM, Sutherland JB (2002) Formation of conjugates from ciprofloxacin and norfloxacin in cultures of *Trichoderma viride*. Mycologia 94(1):1–5

Pilon-Smits E (2005) Phytoremediation. Annu Rev Plant Biol 56:15–39

Prieto A, Möder M, Rodil R, Adrian L, Marco-Urrea E (2011) Degradation of the antibiotics norfloxacin and ciprofloxacin by a white-rot fungus and identification of degradation products. Biores Technol 102(23):10987–10995

Pruden A, Larsson DJ, Amézquita A, Collignon P, Brandt KK, Graham DW et al (2013) Management options for reducing the release of antibiotics and antibiotic resistance genes to the environment. Environ Health Perspect 121(8):878–885

Reis PJ, Reis AC, Ricken B, Kolvenbach BA, Manaia CM, Corvini PF, Nunes OC (2014) Biodegradation of sulfamethoxazole and other sulfonamides by *Achromobacter denitrificans* PR1. J Hazard Mater 280:741–749

Roose-Amsaleg C, Laverman AM (2016) Do antibiotics have environmental side-effects? Impact of synthetic antibiotics on biogeochemical processes. Environ Sci Pollut Res 23:4000–4012

Sabri N, Schmitt H, Van der Zaan B, Gerritsen HW, Zuidema T, Rijnaarts HHM, Langenhoff AAM (2020) Prevalence of antibiotics and antibiotic resistance genes in a wastewater effluent-receiving river in the Netherlands. J Environ Chem Eng 8(1):102245

Santaeufemia S, Torres E, Mera R, Abalde J (2016) Bioremediation of oxytetracycline in seawater by living and dead biomass of the microalga *Phaeodactylum tricornutum*. J Hazard Mater 320:315–325

Sarmah AK, Meyer MT, Boxall AB (2006) A global perspective on the use, sales, exposure pathways, occurrence, fate and effects of veterinary antibiotics (VAs) in the environment. Chemosphere 65(5):725–759

Shah MP (2024a) Microbial remediation of hazardous chemicals from water & industrial wastewater treatment plant. Springer Nature

Shah MP (2024b) Development in waste water treatment research and processes: applied technologies for clean up of environmental contaminants. Elsevier

Shah MP (2024c) Aerobic and anaerobic microbial treatment of industrial wastewater. CRC Press

Shah MP (2025) Extraction of precious metals from industrial waste using microbial technology. Springer Nature

Silva A, Delerue-Matos C, Figueiredo SA, Freitas OM (2019) The use of algae and fungi for removal of pharmaceuticals by bioremediation and biosorption processes: a review. Water 11(8):1555

Singer AC, Shaw H, Rhodes V, Hart A (2016) Review of antimicrobial resistance in the environment and its relevance to environmental regulators. Front Microbiol 7:219380

Singh SK, Khajuria R, Kaur L (2017) Biodegradation of ciprofloxacin by white rot fungus *Pleurotus ostreatus*. 3 Biotech 7:1–8

Singh V, Pandey B, Suthar S (2018) Phytotoxicity of amoxicillin to the duckweed *Spirodela polyrhiza*: growth, oxidative stress, biochemical traits and antibiotic degradation. Chemosphere 201:492–502

Sodhi KK, Singh CK (2022) Recent development in the sustainable remediation of antibiotics: a review. Tot Environ Res Themes 3:100008

Sosa-Hernández JE, Rodas-Zuluaga LI, López-Pacheco IY, Melchor-Martínez EM, Aghalari Z, Limón DS et al (2021) Sources of antibiotics pollutants in the aquatic environment under SARS-CoV-2 pandemic situation. Case Stud Chem Environ Eng 4:100127

Sun S, Yao H, Fu W, Xue S, Zhang W (2020) Enhanced degradation of antibiotics by photo-fenton reactive membrane filtration. J Hazard Mater 386:121955

Van Boeckel TP, Brower C, Gilbert M, Grenfell BT, Levin SA, Robinson TP et al (2015) Global trends in antimicrobial use in food animals. Proc Natl Acad Sci 112(18):5649–5654

Varma A, Buscot F (2005) Microorganisms in soils: roles in genesis and functions. Springer, Berlin Heidelberg

Wang S, Wang J (2018) Biodegradation and metabolic pathway of sulfamethoxazole by a novel strain *Acinetobacter* sp. Appl Microbiol Biotechnol 102:425–432

Wetzstein HG, Schmeer N, Karl W (1997) Degradation of the fluoroquinolone enrofloxacin by the brown rot fungus *Gloeophyllum striatum*: identification of metabolites. Appl Environ Microbiol 63(11):4272–4281

Wetzstein HG, Stadler M, Tichy HV, Dalhoff A, Karl W (1999) Degradation of ciprofloxacin by basidiomycetes and identification of metabolites generated by the brown rot fungus *Gloeophyllum striatum*. Appl Environ Microbiol 65(4):1556–1563

WHO (World Health Organization) (2012) Health through safe drinking water and basic sanitation. http://www.who.int/water_sanitation_health/mdg1/en/index.html. Accessed 31 Oct 2012

Xiong JQ, Kurade MB, Jeon BH (2017) Ecotoxicological effects of enrofloxacin and its removal by monoculture of microalgal species and their consortium. Environ Pollut 226:486–493

Xu WH, Zhang G, Zou SC, Li XD, Liu YC (2007) Determination of selected antibiotics in the Victoria Harbour and the Pearl River, South China using high-performance liquid chromatography-electrospray ionization tandem mass spectrometry. Environ Pollut 145(3):672–679

Yang CW, Liu C, Chang BV (2020) Biodegradation of amoxicillin, tetracyclines and sulfonamides in wastewater sludge. Water 12(8):2147

Yang JF, Ying GG, Liu S, Zhou LJ, Zhao JL, Tao R, Peng PA (2012) Biological degradation and microbial function effect of norfloxacin in a soil under different conditions. J Environ Sci Health B 47(4):288–295

Yu Y, Zhou Y, Wang Z, Torres OL, Guo R, Chen J (2017) Investigation of the removal mechanism of antibiotic ceftazidime by green algae and subsequent microbic impact assessment. Sci Rep 7(1):4168

Zhang T, Li B (2011) Occurrence, transformation, and fate of antibiotics in municipal wastewater treatment plants. Crit Rev Environ Sci Technol 41(11):951–998

Zhang YB, Zhou J, Xu QM, Cheng JS, Luo YL, Yuan YJ (2016) Exogenous cofactors for the improvement of bioremoval and biotransformation of sulfamethoxazole by *Alcaligenes faecalis*. Sci Total Environ 565:547–556

Singh V, Pandey B, Suthar S (2018) Phytotoxicity of amoxicillin to the duckweed [illegible]

Chapter 6
Bioremediation of Wastewater Emphasizing the Role of Microbes in the Removal of Heavy Metals and Radionuclides

Udisha Datta, Ushasi Roy, and Debleena Roy

Abstract The indiscriminate use of heavy metals and radionuclides for anthropogenic purposes has brought significant increase in their presence above the permissible limits causing disturbances in the ecosystem. Industrial effluents are major contributors to terrestrial and aquatic contamination with toxic heavy metals like arsenic, nickel, zinc, copper, chromium, lead, mercury, cadmium, etc., and radionuclides like uranium, thorium, radon, etc., whose high solubility in the aquatic environments attributes to their hazardous bioaccumulative nature. Various approaches used to degrade such pollutants include physical, chemical, and biological methods. Though none of these methods claim complete degradation of the toxic pollutants, biological techniques are comparatively more effective, economically viable, and environment-friendly, mainly reducing the secondary pollution and by-product accumulation caused by physiochemical methods. Using microbes as the platform for remediation and degradation of heavy metal and radionuclide waste is an alternative yet effective method administered in the recent times. Various processes such as enzymatic transformation including dissolution of pollutants by low molecular weight acidic microbial metabolic secretions or reduction of heavy metals coupled to anaerobic oxidation of organic substrates; biosorption of ions to cell wall layers of microbes; and bioconversion and precipitation are used by microbes to degrade or reduce heavy metals into less toxic innocuous forms. This chapter highlights the biological treatment strategies that may replace or augment existing chemical processes and elaborates on various methods in which microbes are used to remove the most common toxic heavy metals and radionuclides found in wastewater.

Keywords Heavy metals · Radionuclides · Wastewater · Microbes · Biosorption · Bioconversion

U. Datta · U. Roy · D. Roy (✉)
PG Department of Botany, Lady Brabourne College, University of Calcutta, Kolkata, India
e-mail: debleena.roy@ladybrabourne.com

© The Author(s), under exclusive license to Springer Nature Singapore Pte Ltd. 2026

S. Dey and M. P. Shah (eds.), *Microbial Applications in Industrial Wastewater Treatment*, Advances in Wastewater Research,
https://doi.org/10.1007/978-981-96-9128-9_6

6.1 Introduction

Wastewater represents the liquid wastes of community life comprising organic and inorganic solid, liquid, and gaseous pollutants in excess of the prescribed or permissible limit stated by the Central Pollution Control Board (CPCB) in India, World Health Organization (WHO) internationally, and other such authorities (Sonune and Ghate 2004).

Wastewater can be majorly domestic and industrial. Domestic sewage comprises colloidal, suspended, and dissolved impurities with a pH ranging 6–9. Sewage generally has a high percentage of human waste and thus contains a large number of microorganisms some of which might be pathogenic. Depending on the products and byproducts produced by different industries, the industrial wastes have varied composition mostly with total oxygen demand (TSS), biological oxygen demand (BOD), and chemical oxygen demand (COD) in tens of thousands mg L^{-1} and pH often beyond 6–9 range (Gray 2004) (Table 6.1).

The treatment of wastewater is a critical measure that should be followed prior to its disposal to major waterbodies keeping in mind the water scarcity challenges amongst the growing imbalance between freshwater availability and consumption. Wastewater treatment is a process of removal of maximum impurities as much as possible or their reduction by decomposition from highly complex, putrescible forms to relatively stable organic forms.

Different physical and chemical treatment methods with different performance characteristics and impacts have been used for the treatment of wastewater such as ion exchange, chemical precipitation, adsorption, reverse osmosis, coagulation, and flocculation (Shah 2025; Seow et al. 2016). Drawbacks to this conventional methods include chemical and energy requirements, and the formation of toxic byproducts and hazardous biosolids that require specific disposal. An attractive alternative to these approaches is offered by the biological wastewater treatment that aims at degrading or dissolving impurities into biological flocs or biofilms (Henze et al. 2008) (Figs. 6.1, 6.2 and 6.3).

Table 6.1 Various sources of heavy metal contamination in the environment

Direct natural sources	Direct anthropogenic sources	Indirect sources
• Volcanic eruptions • Weathering of rocks	• Coal combustion industry • Automobile • Cosmetics • Mining • Plumbing material • Alloys and coatings • Batteries	• Heavy metal bioaccumulation in crops and fishes • Biomagnification

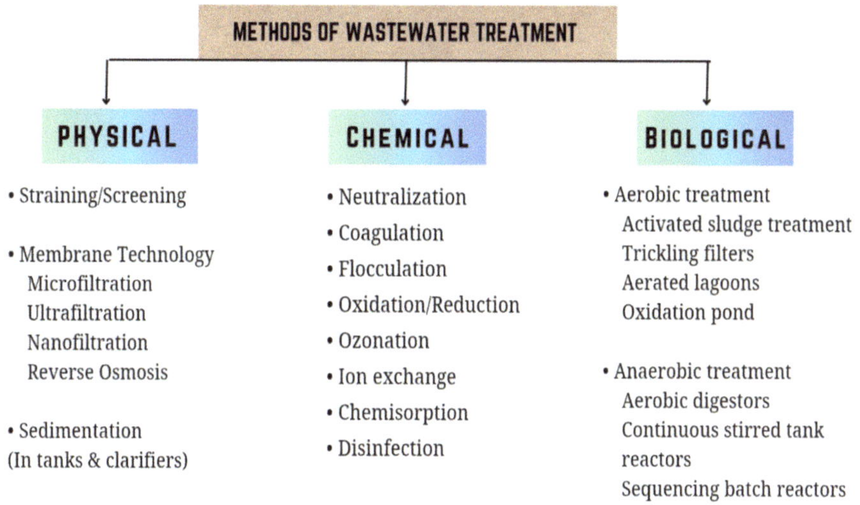

Fig. 6.1 Various methods of wastewater treatment

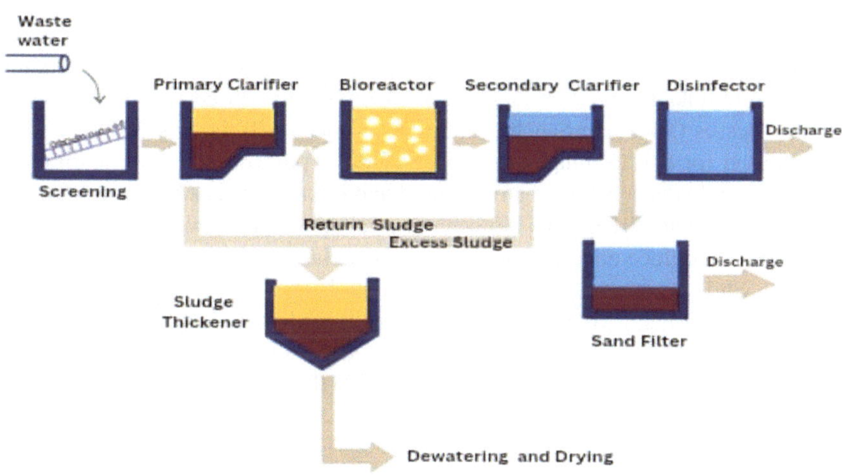

Fig. 6.2 Schematic diagram showing activated sludge treatment

6.2 Microbial Bioremediation of Wastewater

Bioremediation is a biological process that involves minimizing the severity of pollutants through the process of reduction or degradation, where the complex toxic compounds are biosynthesized into simpler less toxic of zero-valent metals, alloys, or inorganic-organic composites, with the involvement of microorganisms (Shah 2024a; Ajona and Vasanthi 2021). Microbial groups such as yeasts, bacteria, algae,

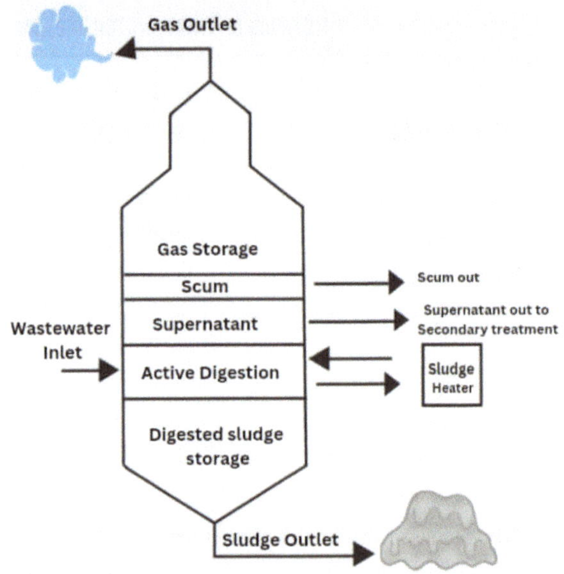

Fig. 6.3 Diagram showing working principle of aerobic digestor

protozoa, and fungi use their enzymes or metabolic pathways to degrade the organic content in pollutants based on their metabolism and growth during the bioremediation processes (Akerman-Sanchez and Rojas-Jimenez 2021), rendering the water to be reused for fish farming, irrigation, and other purposes. The nonsettling and suspended colloidal substances are captured and incorporated as biological floc and biofilm (Rani et al. 2019). This biotechnological approach shows promising potential to removing trace concentrations of heavy metal ions improving the quality of biosolids while producing no toxic waste, and making it a cost-effective process of wastewater treatment. Microbes also possess the ability to degrade a large group of pollutants such as hydrocarbons, dyes, and polychlorinated biphenyls alongside heavy metals (Shah 2024b; Saravanan et al. 2023) (Fig. 6.4).

6.3 Factors Influencing Microbial Bioremediation

Suitable monitoring of certain factors including the presence of nutrients, temperature, oxygen addition or diffusion, pH, heavy metals, presence of toxic compounds, physiology of biodegradative microorganisms, metabolic activity, the effectiveness of contaminants removal mechanisms, and types of microorganisms is necessary to be undertaken to prevent bioremediation failure.

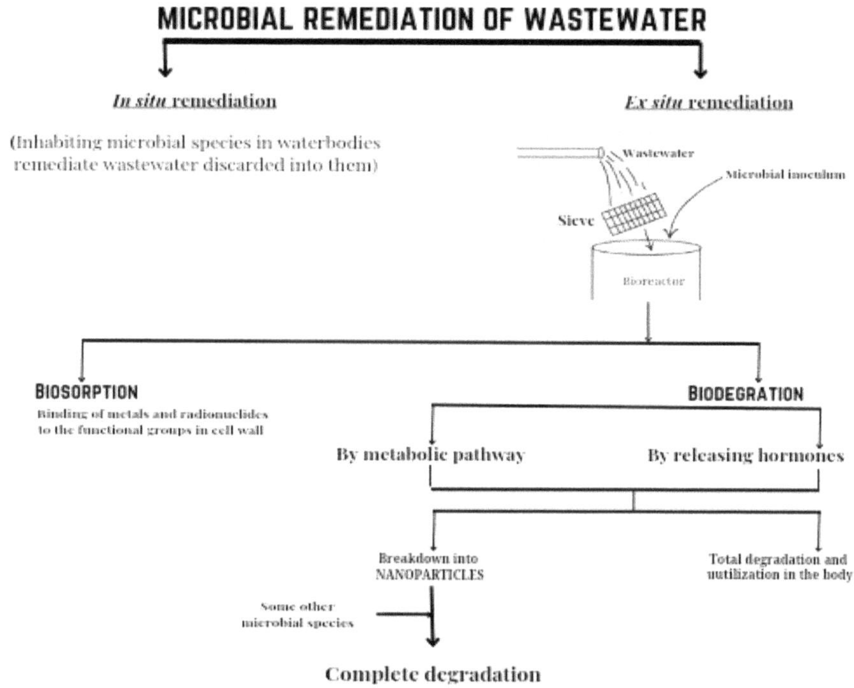

Fig. 6.4 Flowchart depicting in situ and ex situ microbial remediation of wastewater

6.3.1 Concentration of the Contaminant

Toxicity effects are related to concentration of contaminant in direct proportionality. Bacterial enzymes required for degradation are not induced if the contaminant is present in very low concentrations (Adams et al. 2015). The different components of the contaminants interact with each other synergistically showing improvement in the decomposition rates of catabolic enzymes (Mishra et al. 2021).

a. **Nutrient Availability**

Availability of certain essential nutrients is required for the efficient microbial activity as microbes require nutrients such as carbon, phosphorus, nitrogen, calcium, and potassium for their proper growth and survival.

The degradation of the contaminants also depends on the relative concentrations of the available nutrients in the contamination sites; for instance, a few scientific studies on hydrocarbon degradation have shown negative results because of the presence of excess amount of potassium, nitrogen, and phosphorus.

The rate of biodegradation is also determined by bioavailability or accessibility of organic pollutants to microorganisms. Aging reduces this bioavailability because the contaminants are incorporated or absorbed into natural organic matter in course of time (Mishra et al. 2021).

b. **pH and Temperature**

An optimal pH of 6–8 is required for the microbial bioremediation activities. Increased temperature of around 30–50° works best for the bioremediation process as increased temperature enhances the solubility of hydrocarbons and other contaminants in the medium. According to Mueller et al. (1989), a change in temperature range of 15–20° may have countable changes in the biodegradation process (Mueller et al. 1989).

c. **Oxygen Availability**

The bioremediation process can be either anaerobic or aerobic depending on the need for oxygen. The ability of monooxygenase and dioxygenase enzymes to oxidize aromatic rings depends on the first few stages of oxygen aerobic metabolism (Sihag et al. 2014). During anaerobic oxidation of aromatic compounds, electron acceptors are needed in substituted form of ferrous, nitrate, and sulfate. In submerged marine sediments and aquifers, anaerobic processes are primarily responsible for bioremediation (Coates et al. 1996).

6.4 Most Prevalent Heavy Metals and Radionuclides with Their Microbial Bioremediation Processes

a. **Bioremediation of Cadmium**

Cadmium (Cd) is a toxic, non-essential, non-biodegradable heavy metal with a 20 year half-life. It is commonly found in its divalent form. Cd levels are significantly increased by human activities like electroplating, mining, plastic stabilization, and battery production. The United Nations Environment Program reports annual Cd releases into the environment range from 150 to 2600 tons. Natural processes like rock weathering and volcanic eruptions also contribute to environmental Cd pollution. Cd concentrations in rocks range from 0.006 to 8.4 ppm (Pizzino et al. 2017). Cd has adverse effects on major organs like the heart, kidneys, and liver by entering the bloodstream through lungs or intestines and binding to albumin. Cadmium exerts genotoxic and cytotoxic effects through the generation of reactive oxygen species (ROS), which can severely damage DNA and lead to mutations (Shah 2024c; Saluja and Sharma 2014).

Bioremediation using live or dead bacterial biomass has gained significant attention from researchers as a feasible method to eliminate environmental Cd pollutants. Studies have shown the effectiveness of various cadmium-resistant bacteria in removing Cd^{2+} from aqueous media over different time periods. For example, *Escherichia coli* P4 and *Klebsiella pneumoniae* demonstrated removal percentages ranging from 18.8 to 57.4%. *Salmonella enterica* 43C exhibited biosorption efficiencies of 7–22 mM/g Cd^{2+}, while *Microbacterium oxydans* CM3 removed 29–43% of Cd^{2+} from contaminated medium. *Alcaligenes faecalis*, *Pseudomonas stutzeri*, and *Enterobacter* sp. also showed promising bioremediation potentials (Dabir et al.

2019). *Bacillus cereus* S5 and *Bacillus* sp. strains from Brazil were found to have high absorption capacities for cadmium ions. Additionally, *Rhodobacter sphaeroides* bioremediation led to a significant decrease in exchangeable phases of Cd and Zn in soil. Overall, microbial uptake and precipitation of toxic metal ions offer a promising approach for handling industrial wastewater contamination (Teles et al. 2018).

Fungi and yeasts are used to remove toxic metal ions like Cd^{2+} from the environment. Studies have shown the effectiveness of different species in bioremediation. *Pichia hampshirensis* and *Candida tropicalis* were able to remove significant amounts of Cd^{2+} from industrial wastewater. *C. tropicalis* also showed a reduction in Cd^{2+} levels in both controlled and real industrial wastewater conditions (Acosta-Rodríguez et al. 2018). *Trametes versicolor* exhibited tolerance to Cd^{2+} and absorbed cadmium visibly. *Trichosporon asahii* and *P. kudriavzevii* showed promising Cd^{2+} removal abilities in a study. These findings highlight the diverse applications of fungi and yeasts in environmental remediation efforts (Ilyas and Rehman 2018).

Various studies have reported on the adsorption capacities and abilities of different algae species to remove heavy metals from various mediums. Kumar et al. found that metal-resistant *Chlorella vulgaris* could adsorb 97.43 mg/g Cd^{2+} (Kumar et al. 2018). Rehman and Shakoori observed *Chlorella* sp. removing 76–96% of Cd^{2+} after different durations of culturing (Rehman and Shakoori 2004). *Distigma proteus* removed 48–90% of Cd^{2+} over several days. *Euglena proxima* showed resistance to heavy metals and removed over 80% of various metal ions. *Ecklonia maxima*, a marine alga, adsorbed Cd^{2+} up to 83.5 mg/g dry alga. Green algae populations demonstrated varying effectiveness in removing heavy metals, with *Ulva lactuca* removing 99.2% of Cd^{2+} in one study. pH levels were found to impact the metal removal capacities of different algal genera (Ghoneim et al. 2014).

b. **Bioremediation of Lead**

Lead (Pb) contamination is a widespread global issue possessing a significant threat to human health due to its toxicity, oxidative damage, and interaction with nucleic acid and respiratory proteins (Sevak et al. 2021). Bacteria, including lead-resistant bacteria like *Acinetobacter junii* Pb1, *Bacillus subtilis* X3, and others, have developed various mechanisms to survive in lead-contaminated environments. These bacteria have adapted to lead-containing environments by developing resistance through various mechanisms, demonstrating their adaptability and ability to survive in extreme environments (Abdel-Razik et al. 2020). Increased pH from 1 to 6 significantly increases biosorption by *Bacillus pumilus* and *Bacillus cereus*. *Enterobacter* sp. can create an alkaline environment, promoting lead biosorption (Jiang et al. 2020). Live biomass biosorbs lead on *Staphylococcus hominis* strain AMB-2, with living biomass showing higher potential, while Srinath et al. found dead biomass performs better (Rahman et al. 2019). Exopolysaccharide is a crucial component of microbial biofilm, aiding in the survival of bacteria under extreme conditions. It helps in entrapping lead, protecting bacteria from toxic effects, and enhances biosorption and bioremediation.*Bacillus* sp. exhibits tolerance toward lead up to 997.74 mg/L by producing a more compact form of biofilm in the presence of lead. Exopolysaccharide also controls the entry of heavy metals into cells and influences their homeostasis.

Pseudomonas sp. W6 produces exopolysaccharide, which can bind to cationic lead. Exopolysaccharide is a natural reducing agent and a good adsorbent of lead. Its secretion by bacteria contributes to effective lead removal from the environment (Kalita and Joshi 2017). Bioprecipitation and biotransformation are processes that change lead's form and make it less bioavailable for reaction, reducing lead toxicity. Lead is precipitated as lead phosphate by phosphatase enzymes, which can accumulate on cell surfaces or be expelled out. Phosphate-solubilizing bacteria (PSB) play a crucial role in lead bioremediation by making phosphate bioavailable for reaction with lead. Bacillus subtilis X3 bioprecipitates lead in various forms, while Providencia alcalifaciens strain 2EA produces an insoluble brown precipitate. Lead-bioprecipitating bacteria can convert soluble lead nitrate to insoluble lead phosphate as a protective mechanism against lead toxicity (Qiao et al. 2019). Bacteria like *Pannonibacter phragmitetus* BB, *Pseudomonas putida* KNP9, *Pseudomonas psychrophila, Bacillus* sp. MN3-4, and *Bacillus* sp. PZ-1 produce siderophores to alleviate lead toxicity, with variations in siderophore production (Liao et al. 2020).

Two potential fungal strains, *Talaromyces islandicus* and *Aspergillus terreus*, show efficient lead removal under various conditions. These fungi are good candidates for commercial-level Pb removal, with *T. islandicus* being a novel candidate (Sharma et al. 2020).

c. **Bioremediation of Copper**

Copper is one of the essential trace nutrients present approximately in the concentration range of 10–100 mg L^{-1}. But uptake of copper in higher concentration can be toxic affecting the proper functioning of liver, brain and, kidney and even causing death. Copper is present in various industrial effluents. It has been found to impose hazardous impacts on aquatic life causing organ damage in fish and mollusks, interfering with the sense of smell in fish adding up difficulties in premating processes, inhibiting photosynthesis and electron transport chain in marine algae, and affecting lives of other bacteria, diatoms, dinoflagellates, etc. In human, Wilson's disease, Menkes disease, schizophrenia, jaundice, etc. are the most common effects of copper toxicity.

Removal of trace metals from wastewater by physicochemical and electrochemical is a costly and challenging approach. Thus, effective bioremediation of copper metal into nanoparticles can be done by several microscopic and macroscopic living organisms. Use of sulfur-reducing bacteria is an effective process mostly in sulfate and heavy metal-contaminated acid mine drainage. Under anaerobic conditions, these bacteria use sulfate as the terminal electron acceptor for organic substrate metabolism and transform sulfate to sulfide which stably precipitates with heavy metals. This causes increase in the water pH which inhibits growth of the sulfate-reducing bacteria posing as a drawback in this process. An auxiliary reducing agent such as iron (Fe^0) having high reduction potential can improve sulfate reduction and effectively enhance the activity of these bacteria. The anaerobic interaction of Fe^0 and water at low pH oxidizes Fe^0 to ferrous and releases hydrogen which is utilized as the electron donor for sulfate reduction, in turn neutralizing the acidity of the water by hydrogen, resulting in a suitable pH for the growth of sulfate reducing bacteria

(Bai et al. 2013). Apart from sulfate-reducing bacteria, strains of a copper-resistant bacteria, *Bacillus altitudinis*, were observed to respond to biosorption studies indicating the efflux and uptake of Cu in bacterial cells in an ATP-independent manner and positive chemoattraction with removal of Cu respectively (Khan et al. 2022).

Algae-mediated cupric ion (Cu^{2+}) bioremediation has also been investigated with wild-type strain of *Chlamydomonas reinhardtii* that causes efficient reduction of free Cu^{2+} in wastewater to Cu-based nanoparticles in aqueous phase. Several techniques used in the investigation including observations of cell morphology, algal regrowth experiments, fluorescein diacetate hydrolysis, and high-resolution transmission electron microscopy confirmed the presence of well-dispersed, spherical sub-10 nm copper nanoparticle crystallites showed that viability of algae was crucial for efficient Cu^{2+} reduction. UV-vis absorption spectroscopy demonstrated that sulfur-free medium, which enables sustained hydrogen photoproduction in alga, was not the most efficient in NPs formation (Žvab et al. 2021). Other green algae such as *Desmodesmus subspicatus*, *Chloroidium ellipsoideum*, *Scenedesmus almeriensis* can also be considered sustainable biosorbents for copper removal from wastewater although increased copper concentration affects the chlorophyll content, lipid content, and some other growth factors of the algae (Fawzy et al. 2024). Furthermore ciliated protozoan *Paramecium multimicronuleatum* was also found to have bioremediation abilities against copper ions in some scientific studies (Liaqat et al. 2022).

Macroscopic filamentous fungi are advantageous in the wastewater treatment process than those of microscopic alga and bacteria as the mycelial mesh can withstand water flow pressure, or other factors like agitation in bioreactors. Some studies investigated the removal of copper nanoparticles produced by microbial cupric reduction from aqueous solution using dead biomass of the fungus *Tricoderma koningiopsis*, ensuring complete removal of the heavy metals in a cost-effective and environment-friendly way (Salvadori et al. 2014).

d. **Bioremediation of Arsenic**

Arsenic is a toxic metalloid found in various environmental forms including arsenic acids, arsenious acids, dimethylarsinic acid, methylarsenic acid, arsine, and most commonly as arsenates ($AsO_4{}^{3-}$) and arsenites ($AsO_3{}^{3-}$) (referred to as As(V) and As(III) respectively) in aqueous bodies, etc. Its notoriety as a homicidal poison overshadows its historical significance in science, technology, and medicine. Most conventional remediation methods effectively remove high initial arsenic concentrations from water (approx. less than 100 mg/L) but fails in removing the residual arsenic concentrations which exceeds the water quality standard of 0.05 mg/L used in most countries (Mohan and Pittman 2007). Thus bioremediation is the best approach ensuring almost the total clearance of arsenic from contaminated sites. Microorganisms are known to play an important role in the biochemical cycle of arsenic, through its conversion to species with different solubility, mobility, bioavailability, and toxicity. The presence of arsenite oxidizing gene smaller subunit (*aioB*) on chromosomal DNA locus and an efficient arsenic oxidizing potential of a multiple heavy metal-resistant bacterial strain *Brevibacterium* sp. CS2 make it a potential candidate for green chemistry for treatment of arsenic-contaminated wastewater (Silver and Phung 2005). Arsenic and many other heavy metal-resistant bacterial strains like *Micrococcus luteus* AS2 have arsenic-resistant and detoxification genes such as aioB, arsR, arsC1, and ACR3 in its genome (Sher et al. 2023), while a few other genes such as arsB, arsC, and arsH, including arsR, a putative gene responsible for arsenic resistance, are found in bacterial strains of *Thiobacillus ferrooxidans* (Sher et al. 2020). *M. luteus* AS2 has a minimum inhibitory concentration (MIC) of 55 mM against arsenite, which is higher than most of the arsenite-resistant bacteria, proving itself the most efficient for arsenic remediation (Sher et al. 2023). FTIR analysis of these investigated strains under arsenite stress indicates that arsenite interacts with the hydroxyl, carboxyl, and amino groups present on the surface of bacterial cell wall. Some other scientific analysis that showed some other bacteria having an effective MIC for arsenic resistance are *Acinetobacter*, *Agrobacterium*, *Arthrobacter*, *Comamonas*, *Rhodococcus*, *Stenotrophomonas*, and *Pseudomonas* (Sher et al. 2020).

The probability of using certain living microalgae *Chlorella pyrenoidosa*, *Botryococcus braunii*, *Spirulina platensis*, *Anabaena variabilis*, *Cystoseira indica*, *Nostoc muscorum*, etc., and their biosorption potential for phycoremediation of arsenic ions from contaminated water was examined and found to be quite effective (Podder and Majumder 2015).

e. **Bioremediation of Nickel**

Nickel is widely utilized in various industrial processes because of its corrosion-resistant properties, leading to the generation of nickel-containing waste streams (Sharma et al. 2024). Bioremediation of this heavy metal from the industrial wastewater is done by several nickel-resistant bacteria. Bacterial strain *Microbacterium* sp. MRS-1 isolated from electroplating industrial effluent is capable of $NiSO_4$ bioconversion into insoluble unique flower-like NiO nanoparticles, proving it efficient for the reduction of nickel level in it (Sathyavathi et al. 2014). The nickel biosorption ability of autochthonous cyanobacteria *Arthrospira platensis* along with indigenous microbiota was studied in cooking oil industry effluent (Kashyap et al. 2022). Nickel biosorption estimation and plant growth-promoting activities of *Bacillus safensis* (an isolated endophytic strain from *Vigna radiata*) were performed, and high nickel biosorption efficiency of approx. 91.3 % was found at a concentration of 600 mg L^{-1}. While scientific investigations have found few other bacteria such as *Bacillus licheniformis*, *Bacillus subtilis*, *Klebsiella oxytoca*, *Brevibacillus laterosporus*, and *Pseudomonas putida* capable of remediating Ni^{2+} from wastewater, there are also several protozoan species, most commonly of the genera *Aspidisca*, *Trachelophyllum*, *Peranema*, *Euplotes*, etc., that have proved to be efficient bioremediants of nickel (Kamika and Momba 2011).

f. **Bioremediation of Mercury**

Mercury pollution is caused by natural sources like forest fires and volcanic eruptions, and anthropogenic activities like coal combustion and mining. Mercury is a highly toxic heavy metal with varying toxicity patterns. Its bioaccumulative property is responsible for neurotoxic and cytotoxic effects in humans (Kumari et al. 2020). Wastewater is treated for mercury detoxification using bacteria that convert toxic mercury into less toxic forms. Bacteria with *mer* operons have been evaluated for detoxifying mercury compounds. They use mercuric reductase and organomercury lyase enzymes, which catalyze the breakdown of C–Hg bonds in organomercury compounds. These enzymes transport Hg^{2+} into or out of the cytoplasmic membrane (Mahbub et al. 2017). Microorganisms with carrier polymers offer high productivity, constant quality, and cost-effectiveness. Immobilized cells are reusable and protect from environmental conditions. Techniques include gel entrapment, sodium alginate, and calcium alginate. These methods can be performed under ambient conditions, minimize cell damage, and protect cells from external conditions (Oyeagu et al. 2018).

The bacteria *Bacillus cereus* AA-18 can be immobilized using alginate for mercury removal from industrial wastewater, with excellent reusability and stability, making it a potential candidate for biodetoxification of mercury-contaminated aquatic environments (Amin et al. 2022). Mercury-resistant bacteria (MRB) are a group of gram-positive and gram-negative bacteria that counteract and convert mercury (Hg) from toxic to non-toxic forms. Their population is proportional to the scale of Hg pollution at the site. Mechanisms for Hg resistance include formation and binding of the thiol group to different Hg forms, or occurrence of a permeability hurdle. The greatest vital process of Hg resistance is based on mer operons present on plasmids, genomic DNA, Tn21 transposons, or integrons. These operons are highly conserved and code for specific proteins and functions in MRB (Norambuena et al. 2018). Bacteria having *mer* operon include *Shigella flexneri, Pseudomonas aeruginosa, Pseudomonas putida, Pseudomonas stutzeri, Pseudomonas fluorescens, Klebsiella pneumoniae, Morganella morganii, Xanthomonas, Achromobacter, Sphingobium, Acinetobacter calcoaceticus, Serratia marcescens, Mycobacterium marinum, Bacillus* sp. (Mahbub et al. 2016). Researchers have discovered novel mechanisms for Hg bioremediation, including iron-oxidizing acidophilic Hg-sensitive bacteria that reduce Hg^{2+} without mercuric reductase enzyme. These bacteria work in anoxic conditions, reducing Hg^{2+} to volatile Hg° without mercuric reductase. However, they work in anoxic conditions, where concentrations are low. Metallothionein, a low-molecular weight metal-binding protein, can sequester heavy metals in biologically reactive forms. In a study, metallothionein was used to transform an Hg-sensitive E. coli gene, resulting in a 100 μM Hg intracellular accumulation (Le et al. 2016). Three transposons TnMERI1, RC607, and Tc38-2b are isolated from Hg resistance bacteria, *Bacillus megaterium, B. cereus*, and *Exiguo bacterium* sp., which can be used for transposon-mediated *in situ* molecular breeding, requiring further research and soil microbe knowledge (Shahi et al. 2017).

Yeast species, particularly ascomycetous yeasts like *S. cerevisiae*, have been studied for their ability to metabolize toxic heavy metals like Hg. These yeasts can adapt to extreme conditions and serve as better biosorbents than bacteria for removing heavy metal ions from wastewater. The presence of negative groups on the yeast cell surface helps it effectively bind heavy-metal cations. The presence of net-structured yeast floc also facilitates oxygen diffusion and reduces energy consumption. Studies have shown that *S. cerevisiae* can effectively adsorb Hg^{2+} and can be used to clean Hg^{2+} contaminated water. A recent study using a cell surface display system with the merR gene showed that *S. cerevisiae* strains with MerR displayed effectively adsorb Hg^{2+} and could be applied to clean Hg^{2+} contaminated water (Wei et al. 2018). A yeast species named *Rhodotorula mucilaginosa* R1 has been identified with high tolerance to Hg^{2+}, effectively removing it from aqueous solutions. This pigmented yeast can grow in high Hg concentrations and accumulate Hg^{2+} in solution. Yeast also has the unique ability to grow in matrices with high Hg content, protecting and detoxifying the environment (Bahafid et al. 2017).

g. **Bioremediation of Radionuclides**

Radioactive nuclides are used in nuclear power plants, causing massive waste and health risks. Sustainable technologies are needed to manage these wastes. Bioremediation, including microbial, phyto- and myco-remediation, is a promising eco-friendly approach. A sustainable strategy for radioactive nuclide waste management in developing countries is also highlighted.

Uranium, the most radioactive actinide element, poses significant environmental and human health hazards due to its chemical toxicity. Industrial development has led to extensive uranium pollution, particularly from enriched uranium used in energy production, nuclear research, and weapon manufacturing. Reducing U(VI) concentration is crucial for environmental conservation and human health (You et al. 2021). U(VI)-reducing bacteria, primarily occupying three phyla, Proteobacteria, Firmicutes, and Actinomycetes, are the foundation for bioremediation, as demonstrated by 16S rRNA gene PCR amplification, high-throughput sequencing, and PLFA analyses (Zeng et al. 2020). Proteobacteria, a phylum found in uranium-contaminated sites, contains many U(VI)-reducing genera that play a crucial role in reducing uranium toxicity and bioavailability. *Shewanella, Desulfovibrio*, and *Geobacter* are common in radionuclide-contaminated sediments, exhibiting metal resistance and adapting to U(VI) reduction metabolism. *Shewanella* is used as a model organism for U(VI) bioremediation research, while *Desulfovibrio* reduces U(VI) to less soluble mineral uraninite. *Geobacter* is often abundant in uranium-contaminated sites (Kolhe et al. 2018). *Staphylococcus* aureus, a pathogenic bacteria in Firmicutes, can reduce uranyl nitrate by 47% within 1 hour, making it a potential candidate for uranium remediation. *Clostridium acetobutylicum* and Actinobacteria, such as Arthrobacter and Microbacterium, can also reduce uranium in radionuclide-contaminated environments. Acidobacteria, such as *Geothrix*, can tolerate extreme conditions and are associated with Fe(III) and U(VI) reduction (Shukla et al. 2020). U(VI) is primarily reduced through direct enzymatic reduction by bacteria, converting oxidized and soluble forms into reduced U(IV). It could also be reduced through indirect reduction, where bacteria first reduce substrates into secondary metabolites. The U(VI) reduction mechanism is generally a two-electron process, but recent studies have shown that it can be a one-electron enzymatic reduction and re-disproportionated to U(IV). U(V) has been considered a transient and less stable intermediate, but it persists for a considerable lifetime under environmental conditions and exhibits enhanced stability in iron-bearing phases. Further research is needed to resolve the dispute over the single/dual electron reduction mechanism (Vettese et al. 2020). Biosorption is a feasible, eco-friendly, cost-effective, and reusable technique for uranium removal from wastewater. It is influenced by various factors and has the potential to be improved over conventional methods. Genetically engineered biosorbents and modified nanoparticles show maximum biosorption capacity. Different biosorbents, such as *Chlorella salina, Laminaria japonica, Candida utilis, Spirulina platensis, Deinococcus radiodurans* and *Brassica juncea*, exhibit varying uranium uptake potentials due to their pH levels and ionic charges (Kamala et al. 2020). Numerous bacteria can reduce U(VI) outside the cell, either through cytochrome reduction or pili or electron shuttle compounds. U(VI) molecules can also be transported across the membrane. Bioremediation strategies often use in situ methods, with *Geobacter* being a crucial reducing bacteria. *Ex situ* bioremediation, with high catalytic rates and biomass concentrations, is promising for controlling uranium contamination.

Streptomyces sp. CuOff24, a carbohydrate-containing extracellular polymeric substance, was isolated and identified as a strontium ion tolerant marine actinobacterium. The EPS was found to remove 82% of strontium chloride from a solution, with carbohydrate content being the most abundant (Banerjee et al. 2022) (Table 6.2).

Table 6.2 Summary of the microbial strains found resistant to different heavy metals and radionuclides

Heavy metals	Microbial strains found in different research work
Cadmium	*Escherichia coli* P4 *Salmonella enterica* 43C *Microbacterium oxydans* CM3 *Bacillus cereus* S5
Lead	*Acinetobacter junii* Pb1 *Bacillus subtilis* X3 *Pseudomonas* sp. W6 *Providencia alcalifaciens* 2EA *Pannonibacter phragmitetus* BB *Pseudomonas putida* KNP9 *Pseudomonas psychrophila* *Bacillus* sp. MN3-4 *Bacillus* sp. PZ-1
Copper	*Bacillus altitudinis* SMK-08 *Bacillus altitudinis* MT422188
Arsenic	*Brevibacterium* sp. CS2 *Micrococcus luteus* AS2
Nickel	*Microbacterium* sp. MRS-1
Mercury	*Rhodotorula mucilaginosa* R1 *Bacillus cereus* AA-18
Radionuclides	*Streptomyces* sp. CuOff24

6.5 Microbial Bioremediation—An Aid to a Greener Planet

The demand for innovative waste treatment techniques has increased due to strict environmental guidelines and the growing human population. Bioremediation, a cost-effective and environmentally friendly technology, uses microorganisms like algae, fungi, yeasts, and bacteria to remove heavy metals from polluted water and soil. This technology can also reduce industry costs by producing environmentally friendly nanomaterials, promoting cost reduction and sustainable production in the growing nanotechnology sector (Saeed et al. 2022).

Bioremediation is considered an acceptable waste treatment method, as it eliminates future liability and requires no significant disruption to normal activities. Bioremediation can also be carried out on site, reducing potential health and environmental risks. It is also less expensive and especially in situ techniques have no additional costs of equipment installation (Kensa 2011). Geological characteristics, such as pollutant depth and location, should be considered when choosing a bioremediation method (Azubuike et al. 2016).

Bioremediation uses microbial transformation to treat xenobiotics and plastics, and offers a safer and more stable solution to water pollution. This method, which uses fungal, bacterial, or plant species, maintains a steady chemical ecosystem (Shah and Shah 2020).

Heavy metals in wastewater, particularly industrial and domestic sewage, are a major cause of water pollution. These non-biodegradable metals can cause health issues, cancer, and genetic disorders. Conventional methods for removing heavy metals have drawbacks, necessitating the need for alternative methods. Green nanoparticles manufactured from microbes, with their high selectivity and adsorption capacity, are effective sorbents for heavy metal removal (Kaur and Roy 2021).

6.6 Every Boon Comes with Some Banes

Temperature significantly impacts heavy metal remediation efficiency. Extreme temperatures decrease bacterial heavy metal uptake due to affected growth and distorted cell membrane structure. Temperature also affects membrane fluidity, heavy metal biosorption, and overall bacterial metabolism. Therefore, optimal microbial growth at specific temperatures is crucial for maximum HM bioremediation in onsite bioremediation.

pH is crucial for microbial growth and heavy metal removal, as it affects enzymatic activities and influences heavy metal interaction with bacteria. pH also affects the metal ion solubility and immobilization, making it a key factor in heavy metal remediation (Sevak et al. 2021).

Various nutrients are essential for bacterial growth and survival, including glucose, nitrogen sources like tryptone, and yeast. Glucose increases heavy metal uptake, nitrogen increases resistance, and yeast decreases resistance. Different nutrients are required for active uptake of different HM and detoxification.

The presence of other toxic pollutants can affect the bioremediation process. Factors like cation distribution, bacteria type, and environmental conditions can also influence heavy metal tolerance.

Bioremediation faces specificity, scale-up limitations, and demands technological advancement. It requires metabolically capable microbial populations, suitable environmental conditions, and appropriate nutrients and contaminants. Scaling up bioremediation is challenging, and more research is needed to develop modern techniques for sites with composite contaminants in different forms (Sharma and Shukla 2021).

6.7 Conclusion

The development of technologies and industrial manufacturing has led to soil health degradation and pollution. Bioremediation, which involves using microorganisms, has limitations due to their inability to convert toxic heavy metals into non-toxic forms. Genetic engineering is being used to enhance the biodegradation potential of microorganisms. Bioremediation is a cost-effective and organized way to handle contaminated ground water and soil. Environmental biotechnology aims to address these issues by utilizing microbes in bioremediation, promoting their reactions in contaminated biotopes and increasing their degradative potential. Future studies should focus on improving in situ bioremediation technologies using genetically modified microbes or Genetically Engineered Microorganism (GEMs).

References

Abdel-Razik MA, Azmy AF, Khairalla AS, AbdelGhani S (2020) Metal bioremediation potential of the halophilic bacterium, Halomonas sp. strain WQL9 isolated from Lake Qarun, Egypt. Egypt J Aquat Res 46(1):19–25

Acosta-Rodríguez I, Cárdenas-González JF, Rodríguez Pérez AS, Oviedo JT, Martínez-Juárez VM (2018) Bioremoval of different heavy metals by the resistant fungal strain Aspergillus niger. Bioinorg Chem Appl 2018(1):3457196

Adams GO, Fufeyin PT, Okoro SE, Ehinomen I (2015) Bioremediation, biostimulation and bioaugmention: a review. Int J Environ Bioremediation Biodegrad 3(1):28–39

Ajona M, Vasanthi P (2021) Bioremediation of petroleum contaminated soils—a review. Mater Today: Proc 45:7117–7122

Akerman-Sanchez G, Rojas-Jimenez K (2021) Fungi for the bioremediation of pharmaceutical-derived pollutants: a bioengineering approach to water treatment. Environ Adv 4:100071

Amin A, Naveed M, Sarwar A, Rasheed S, Saleem HGM, Latif Z, Bechthold A (2022) In vitro and in silico studies reveal Bacillus cereus AA-18 as a potential candidate for bioremediation of mercury-contaminated wastewater. Front Microbiol 13:847806

Azubuike CC, Chikere CB, Okpokwasili GC (2016) Bioremediation techniques—classification based on site of application: principles, advantages, limitations and prospects. World J Microbiol Biotechnol 32:1–18

Bahafid W, Joutey NT, Asri M, Sayel H, Tirry N, El Ghachtouli N, Sayel NTH (2017) Yeast biomass: an alternative for bioremediation of heavy metals. Yeast-Ind Appl 559:269–286

Bai H, Kang Y, Quan H, Han Y, Sun J, Feng Y (2013) Bioremediation of copper-containing wastewater by sulfate reducing bacteria coupled with iron. J Environ Manag 129:350–356

Banerjee S, Kundu A, Dhak P (2022) Bioremediation of uranium from waste effluents using novel biosorbents: a review. J Radioanal Nucl Chem 331(6):2409–2435

Coates JD, Anderson RT, Lovley DR (1996) Oxidation of polycyclic aromatic hydrocarbons under sulfate-reducing conditions. Appl Environ Microbiol 62(3):1099–1101

Dabir A, Heidari P, Ghorbani H, Ebrahimi A (2019) Cadmium and lead removal by new bacterial isolates from coal and aluminum mines. Int J Environ Sci Technol 16:8297–8304

Fawzy MA, Abdelsalam S, Hafez WA, Fathi AA (2024) Bioremediation potential of microalgae for copper ion from wastewater and its impact on growth and biochemical contents: equilibrium isotherm studies. Mater Res Express 11(5):055404

Ghoneim MM, El-Desoky HS, El-Moselhy KM, Amer A, Abou El-Naga EH, Mohamedein LI, Al-Prol AE (2014) Removal of cadmium from aqueous solution using marine green algae, Ulva lactuca. Egypt J Aquat Res 40(3):235–242

Gray NF (2004) Biology of wastewater treatment, vol 4. World Scientific

Henze M, van Loosdrecht MC, Ekama GA, Brdjanovic D (eds) (2008) Biological wastewater treatment. IWA Publishing

Ilyas S, Rehman A (2018) Metal resistance and uptake by *Trichosporon asahii* and *Pichia kudriavzevii* isolated from industrial effluents. Arch Environ Prot 44:78–84

Jiang Z, Jiang L, Zhang L, Su M, Tian D, Wang T, Sun Y, Nong Y, Hu S, Wang S, Li Z (2020) Contrasting the Pb(II) and Cd(II) tolerance of Enterobacter sp. via its cellular stress responses. Environ Microbiol 22(4):1507–1516

Kalita D, Joshi SR (2017) Study on bioremediation of lead by exopolysaccharide producing metallophilic bacterium isolated from extreme habitat. Biotechnol Rep 16:48–57

Kamala K, Sivaperumal P, Thilagaraj R, Natarajan E (2020) Bioremediation of Sr^{2+} ion radionuclide by using marine Streptomyces sp. CuOff24 extracellular polymeric substances. J Chem Technol Biotechnol 95(4):893–903

Kamika I, Momba MNB (2011) Comparing the tolerance limits of selected bacterial and protozoan species to nickel in wastewater systems. Sci Total Environ 410:172–181

Kashyap S, Chandra R, Kumar B, Verma P (2022) Biosorption efficiency of nickel by various endophytic bacterial strains for removal of nickel from electroplating industry effluents: an operational study. Ecotoxicology 31(4):565–580

Kaur S, Roy A (2021) Bioremediation of heavy metals from wastewater using nanomaterials. Environ Dev Sustain 23(7):9617–9640

Kensa VM (2011) Bioremediation—an overview. J Ind Pollut Control 27(2):161–168

Khan M, Kamran M, Kadi RH, Hassan MM, Elhakem A, Sakit ALHaithloul HA, Soliman MH, Mumtaz MZ, Ashraf M, Shamim S (2022) Harnessing the potential of Bacillus altitudinis MT422188 for copper bioremediation. Front Microbiol 13:878000

Kolhe N, Zinjarde S, Acharya C (2018) Responses exhibited by various microbial groups relevant to uranium exposure. Biotechnol Adv 36:1828–1846

Kumar M, Singh AK, Sikandar M (2018) Study of sorption and desorption of Cd(II) from aqueous solution using isolated green algae Chlorella vulgaris. Appl Water Sci 8(8):225

Kumari S, Jamwal R, Mishra N, Singh DK (2020) Recent developments in environmental mercury bioremediation and its toxicity: a review. Environ Nanotechnol, Monit Manag 13:100283

Le TY, Zimmermann S, Sures B (2016) How does the metallothionein induction in bivalves meet the criteria for biomarkers of metal exposure? Environ Pollut 212:257–268

Liao Q, Tang J, Wang H, Yang W, He L, Wang Y (2020) Dynamic proteome responses to sequential reduction of Cr(VI) and adsorption of Pb(II) by *Pannonibacter phragmitetus* BB. J Hazard Mater 386:121988

Liaqat A, Zahra I, Abbas SZ, Wabaidur SM, Eldesoky GE, Islam MA, Rafatullah M, Shakoori FR, Shakoori AR (2022) Copper bioremediation ability of ciliate paramecium multimicronucleatum isolated from industrial wastewater. Water 14(9):1419

Mahbub KR, Krishnan K, Megharaj M, Naidu R (2016) Mercury inhibits soil enzyme activity in a lower concentration than the guideline value. Bull Environ Contam Toxicol 96:76–82

Mahbub KR, Bahar MM, Labbate M, Krishnan K, Andrews S, Naidu R, Megharaj M (2017) Bioremediation of mercury: not properly exploited in contaminated soils! Appl Microbiol Biotechnol 101:963–976

Mishra M, Singh SK, Kumar A (2021) Environmental factors affecting the bioremediation potential of microbes. In: Microbe mediated remediation of environmental contaminants. Woodhead Publishing, pp 47–58

Mohan D, Pittman CU Jr (2007) Arsenic removal from water/wastewater using adsorbents—a critical review. J Hazard Mater 142(1–2):1–53

Mueller JG, Chapman PJ, Pritchard PH (1989) Creosote-contaminated sites. Their potential for bioremediation. Environ Sci Technol 23(10):1197–1201

Norambuena J, Wang Y, Hanson T, Boyd JM, Barkay T (2018) Low-molecular weight thiols and thioredoxins are important players in Hg(II) resistance in *Thermus thermophilus* HB27. Appl Environ Microbiol 84(2):e01931-e2017

Oyeagu U, Nwuche CO, Ogbonna CN, Ogbonna JC (2018) Addition of fillers to sodium alginate solution improves stability and immobilization capacity of the resulting calcium alginate beads. Iran J Biotechnol 16:e1854

Pizzino G, Irrera N, Cucinotta M, Pallio G, Mannino F, Arcoraci V, Squadrito F, Altavilla D, Bitto A (2017) Oxidative stress: harms and benefits for human health. Oxid Med Cell Longev 2017(1):8416763

Podder MS, Majumder CB (2015) Phycoremediation of arsenic from wastewaters by *Chlorella pyrenoidosa*. Groundw Sustain Dev 1(1–2):78–91

Qiao W, Zhang Y, Xia H, Luo Y, Liu S, Wang S, Wang W (2019) Bioimmobilization of lead by Bacillus subtilis X3 biomass isolated from lead mine soil under promotion of multiple adsorption mechanisms

Rahman Z, Thomas L, Singh VP (2019) Biosorption of heavy metals by a lead (Pb) resistant bacterium, staphylococcus hominis strain AMB-2. J Basic Microbiol 59(5):477–486

Rani N, Sangwan P, Joshi M, Sagar A, Bala K (2019) Microbes: a key player in industrial wastewater treatment. In: Microbial wastewater treatment. Elsevier, pp 83–102

Rehman A, Shakoori AR (2004) Tolerance and uptake of cadmium and nickle by Chlorella sp., isolated from tannery effluents. Pak J Zool 36(4):327–331

Saeed MU, Hussain N, Sumrin A, Shahbaz A, Noor S, Bilal M, Aleya L, Iqbal HM (2022) Microbial bioremediation strategies with wastewater treatment potentialities—a review. Sci Total Environ 818:151754

Saluja B, Sharma V (2014) Cadmium resistance mechanism in acidophilic and alkalophilic bacterial isolates and their application in bioremediation of metal-contaminated soil. Soil Sediment Contam: Int J 23(1):1–17

Salvadori MR, Ando RA, Oller Do Nascimento CA, Correa B (2014) Bioremediation from wastewater and extracellular synthesis of copper nanoparticles by the fungus Trichoderma koningiopsis. J Environ Sci Health Part A 49(11):1286–1295

Saravanan A, Kumar PS, Duc PA, Rangasamy G (2023) Strategies for microbial bioremediation of environmental pollutants from industrial wastewater: a sustainable approach. Chemosphere 313:137323

Sathyavathi S, Manjula A, Rajendhran J, Gunasekaran P (2014) Extracellular synthesis and characterization of nickel oxide nanoparticles from Microbacterium sp. MRS-1 towards bioremediation of nickel electroplating industrial effluent. Biores Technol 165:270–273

Seow TW, Lim CK, Nor MHM, Mubarak MFM, Lam CY, Yahya A, Ibrahim Z (2016) Review on wastewater treatment technologies. Int J Appl Environ Sci 11(1):111–126

Sevak PI, Pushkar BK, Kapadne PN (2021) Lead pollution and bacterial bioremediation: a review. Environ Chem Lett 19(6):4463–4488

Shah MP (2024a) Microbial remediation of hazardous chemicals from water & industrial wastewater treatment plant. Springer Nature

Shah MP (2024b) Development in waste water treatment research and processes: applied technologies for clean up of environmental contaminants. Elsevier

Shah MP (2024c) Aerobic and anaerobic microbial treatment of industrial wastewater. CRC Press

Shah MP (2025) Extraction of precious metals from industrial waste using microbial technology. Springer Nature

Shah A, Shah M (2020) Characterisation and bioremediation of wastewater: a review exploring bioremediation as a sustainable technique for pharmaceutical wastewater. Groundw Sustain Dev 11:100383

Shahi A, Ince B, Aydin S, Ince O (2017) Assessment of the horizontal transfer of functional genes as a suitable approach for evaluation of the bioremediation potential of petroleum-contaminated sites: a mini-review. Appl Microbiol Biotechnol 101(11):4341–4348

Sharma B, Shukla P (2021) Lead bioaccumulation mediated by Bacillus cereus BPS-9 from an industrial waste contaminated site encoding heavy metal resistant genes and their transporters. J Hazard Mater 401:123285. https://doi.org/10.1016/j.jhazmat.2020.123285

Sharma R, Talukdar D, Bhardwaj S, Jaglan S, Kumar R, Kumar R, Akhtar MS, Beniwal V, Umar A (2020) Bioremediation potential of novel fungal species isolated from wastewater for the removal of lead from liquid medium. Environ Technol Innov 18:100757

Sharma S, Sharma M, Kumar R, Akhtar MS, Umar A, Alkhanjaf AAM, Baskoutas S (2024) Recent advances and mechanisms of microbial bioremediation of nickel from wastewater. Environ Sci Pollut Res 31(28):40224–40244

Sher S, Hussain SZ, Rehman A (2020) Phenotypic and genomic analysis of multiple heavy metal—resistant Micrococcus luteus strain AS2 isolated from industrial waste water and its potential use in arsenic bioremediation. Appl Microbiol Biotechnol 104:2243–2254

Sher S, Ishaq MT, Bukhari DA, Rehman A (2023) Brevibacterium sp. strain CS2: a potential candidate for arsenic bioremediation from industrial wastewater. Saudi J Biol Sci 30(10):103781

Shukla SK, Hariharan S, Rao TS (2020) Uranium bioremediation by acid phosphatase activity of *Staphylococcus aureus* biofilms: can a foe turn a friend? J Hazard Mater 384:121316

Sihag S, Pathak H, Jaroli DP (2014) Factors affecting the rate of biodegradation of polyaromatic hydrocarbons. Int J Pure Appl Biosci 2(3):185–202

Silver S, Phung LT (2005) Genes and enzymes involved in bacterial oxidation and reduction of inorganic arsenic. Appl Environ Microbiol 71:599–608

Sonune A, Ghate R (2004) Developments in wastewater treatment methods. Desalination 167:55–63

Teles YV, de Castro LM, Sargentini Junior É, Do Nascimento AP, da Silva HA, Costa RS, do Nascimento Souza RD, da Mota AJ, Pereira JO (2018) Potential of bacterial isolates from a stream in Manaus-amazon to bioremediate chromium-contaminated environments. Water, Air Soil Poll 229:1–10

Vettese GF, Morris K, Natrajan LS, Shaw S, Vitova T, Galanzew J, Jones DL, Lloyd JR (2020) Multiple lines of evidence identify U(V) as a key intermediate during U(VI) reduction by *Shewanella oneidensis* MR1. Environ Sci Technol 54:2268–2276

Wei Q, Yan J, Chen Y, Zhang L, Wu X, Shang S, Zhang H (2018) Cell surface display of MerR on *Saccharomyces cerevisiae* for biosorption of mercury. Mol Biotechnol 60(1):12–20

You W, Peng W, Tian Z, Zheng M (2021) Uranium bioremediation with U(VI)-reducing bacteria. Sci Total Environ 798:149107

Zeng TT, Mo GH, Hu Q, Wang GH, Liao W, Xie SB (2020) Microbial characteristic and bacterial community assessment of sediment sludge upon uranium exposure. Environ Pollut 261

Žvab U, Kukulin DS, Fanetti M, Valant M (2021) Bioremediation of copper polluted wastewater-like nutrient media and simultaneous synthesis of stable copper nanoparticles by a viable green alga. J Water Process Eng 42:102123

Chapter 7
Bioremediation of Emerging Contaminants in Industrial Wastewater: Omics Approach Towards Environmental Safety and Sustainability

Subhra Satahrada, Subrat Kumar, and Ritesh Pattnaik

Abstract The influence of rapid industrialization and human lifestyle excessive disposal of non-degradable pollutants occur particularly in industrial wastewater, consequently posing health risk to both human and ecosystem. Even though many bioremediation techniques can effectively remove dangerous pollutants, they basically require an understanding of the factors influencing the growth, metabolic rate, and native microbial consortia at the areas of contamination. In-depth understanding of the function and functional diversity of microbial cohorts as well as the fundamental processes in environmental bioremediation has been made possible by the recent development of numerous cutting-edge and novel technologies, such as next-generation DNA sequencing, protein and metabolic profiling, and many more sophisticated tools. Metagenomic analysis combined with metabolomics and proteomics provide a collective and comparative analysis of the potential microorganisms, paving way to further active participation in industrial wastewater treatment process. Since most of the potential bioremediation mediating microbes are unculturable under standardized laboratory conditions, or even if they have shown positive results in-situ, their extensive application is limited ex-situ. Such potential drawback can be addressed by analysing collective database of microbial characteristics using multi-omics techniques. The omics frameworks in bioremediation of developing pollutants in industrial wastewater, as well as the efficacy of their holistic strategies for environmental sustainability and the bioeconomy, are therefore summarized and described in the current chapter.

Keywords Environment · Emerging contaminants · Toxicity · Bioremediation · Omics · Sustainability

S. Satahrada · S. Kumar · R. Pattnaik (✉)
School of Biotechnology, Kalinga Institute of Industrial Technology (Deemed-to-be-University), Bhubaneswar, India
e-mail: rpattnaik@kiitbiotech.ac.in

© The Author(s), under exclusive license to Springer Nature Singapore Pte Ltd. 2026
S. Dey and M. P. Shah (eds.), *Microbial Applications in Industrial Wastewater Treatment*, Advances in Wastewater Research,
https://doi.org/10.1007/978-981-96-9128-9_7

7.1 Introduction

The evolution of human lifestyle has led to the massive transformation in multiple sectors like agriculture, value-added industries, automobiles and many more. Expansion of urbanization on a global scale have ultimately triggered the levels of pollution in both land and water habitats. The environment has been impacted from an array of pollutants, including carbon dioxide (CO_2), greenhouse gases like nitrogen oxides (NO), sulfur dioxide (SO_2) and particulate matter, as well as from a range of contaminants, nutrient effluents, spills of petroleum products, pesticides, sludge, and the usage of non-biodegradable materials or throwaway items (Gavrilescu 2010). According to Thomaidis et al. (2012), emerging pollutants (EPs) are a broad category of chemicals that are produced by individuals and consumed all over the world. These chemicals include chemical pesticides, beauty products, personal and household hygiene products, and medications, among others. According to reports, the amount of anthropogenic chemicals produced worldwide rose from a million to 700 million metric tons per year between 1930 and 2020. A wide range of emerging contaminants come from the food and beverage, textile, dye, pharmaceutical, and agro-processing industries, which release large amounts of wastewater that are both biodegradable or non-biodegradable and contain organic and inorganic contaminants (Mishra et al. 2021). In addition to harming plants and people, the emission of such harmful compounds puts the ecosystem at risk.

Numerous physicochemical methods have been employed globally in order to remove pollutants from environment, but they are frequently impractical due to high maintenance and operating costs and the creation of secondary contaminants with high levels of toxicity. Biological remediation is a potential method of pollutant removal that has advantages over other approaches with respect to removal effectiveness, eco-friendliness, and cheap cost of operation and upkeep. Microorganisms like bacteria and fungus are used in bioremediation process to eliminate or alter contaminants.

The emergence of molecular techniques like transcriptomics as well as proteomics, metagenomics and similar omics approaches has created new opportunities and ways for environmental management. Previously dependent on culture dependent technology, these methods improve the study of microbiome organization (Shah 2025; Gutierrez et al. 2018).

7.2 Sources and Types of Emerging Contaminants

The water discarded from processing, cleaning, and manufacturing from industries is regarded as industrial wastewater. This water is laden with chemicals, heavy metals, and contaminants of various types and sources. Without proper treatment, their drainage to the water bodies can cause water pollution increasing the toxicity levels and deteriorating the water quality. These contaminants can be categorized broadly depending on the sources as point sources and non-point sources respectively (Table 7.1). Point sources are recognizable sources that release pollutants at known sites or locations. These sources exhibit a gradient pattern in concentration, with the closest point to the source having the highest concentration of toxins or pollutants while the concentration decreases away from the source (Viman et al. 2010). Point sources are simple to identify and can be checked using conventional methods. wastewater examples of point sources adding pollutants to industrial wastewater are as follows:

(i) Paper and textile factories, oil refineries, paper and pulp manufacturing facilities, chemical factories, and automobile manufacturers are among the industries that frequently release one or more pollutants into their discharged streams (commonly referred as effluents).

Table 7.1 Outline of various types of emerging contaminants, their sources and harmful impacts

Types of emerging contaminants	Sources	Harmful impacts	References
Humic substances	Agriculture based industries	Overloads the organic matter in water	Palanivell et al. (2013)
Phenolic compounds	Cosmetic industries, packaging industries, Textiles and food industries	Increases the toxicity of water	Albuquerque et al. (2021) Anku et al. (2017)
Petroleum products	Petrochemical industries	The presence of cycloalkanes, arenes, and various other agents limits the oxygen supply to the organisms in water, simultaneously degrading the water quality	Cholakov (2010)
Pesticides	Agro-chemical industries	The reactive oxygen species released from these pesticides can cause many health hazards associated with oxidative stress and reactive oxygen pollutants for example, neurodegeneration, coronary ailments, carcinoma, pulmonary diseases etc.	Kaur et al. (2019)
Ionic compounds, surfactants, mordant, soap, chromophores, dyestuff and acetic acids	Dye processing industries	The wastewater generated by wet process from dye industries has a wide range of variables that negatively impact on the overall quality of the water, including chemical oxygen demand (COD), the pH value, TDS, BOD, usage of water, and coloration	Shindhal et al. (2021)

A waterbody receives the effluents that certain factories directly discharge. Some factories treat it themselves before releasing it, while others send their trash to sewage treatment facilities to be cleaned up. The drainage pipelines, discarding ditches and factory smokestacks can be considered as point sources of pollution.

(ii) Using a combined sewer system to blend waste materials with urban runoff is another way that some factories and sewage treatment facilities handle trash. Stormwater that runs off roadways and lawns is referred to as runoff. Pollutants and chemicals are carried by the water as it travels over these surfaces. The sewage system is thus instantly flooded by this dirty, untreated water. These runoffs have high levels of organic and inorganic pollutants loaded in them.

(iii) Heavy rains may be excessive for the coupled sewer system to handle, and a portion of the raw sewage and combined runoff will spill into the closest source of water untreated. Such combined sewer overflow (CSO), is considered a point source pollution, has the potential to cause significant harm to the environment and human health.

Non-point sources of pollution are widely dispersed sources. These sources are not limited to any locations or points but rather keep on getting added to the runoff at several points (Liu et al. 2020). On comparing NPS with PS, NPS is more diluted (Rissman and Carpenter 2015), but they are difficult to be recognized. In contrast to distinct identifiable sources like discharge pipes, nonpoint source pollution is the mixing of contaminants from a wide area. Runoff, or the discharge of contaminated water from sources like gardens, parking lots, or building sites, is typically linked to nonpoint source pollution. When industrial wastewater is discarded into the nearby domestic drainage system without proper treatment or untreated then the point of pollution cannot be detected properly, hence acting as a nonpoint source of contamination. Additionally, elevated rate of microbial growth in household sewage is enriched by industrial waste, which affects the dissolved oxygen (DO) and biological oxygen demand (BOD) variables.

Depending upon the type of industry, manufacturing processes and techniques, and cleaning processes the various types of contaminants are as follows (Table 7.2).

(a) **Organic pollutants**: The organic contaminants or pollutants are high molecular weight compounds mostly released from food processing industries, leather industries, oil refineries, etc. Phenol, alcohols, sugar, oil and fats, proteins, hydrocarbons [polycyclic hydrocarbons PAHs], detergents and organic acids, etc. are included under organic pollutants. Even some suspended solids released from pulp and paper industries are highly organic. These organic contaminants mostly affect the water bodies by changing the BOD levels, thereby making them unfit for use.

(b) **Inorganic pollutants**: Alkalis, mineral acids, inorganic salts, free chloride, ammonia, hydrogen sulphide, salts of chromium, etc. are included under this category. The nitrogenous and phosphorus waste are mainly released from textile

Table 7.2 Outline of various types of refractory emerging contaminants, their sources and harmful impacts

Types of emerging refractory contaminants	Sources	Harmful impacts	References
Nitrogenous wastes	Agro-chemical based industries mostly fertilizer industries, surface runoffs, municipal discharges, Coke oven wastewater discharged from steel industries	Eutrophication, formation of reactive nitrogenous species which hampers the PH and toxicity levels	Chawley et al. (2021)
Heavy metals	Metallurgical industries, for electroplating purposes, paints, pigments, preservatives manufacturing industries, pulp and paper industries	When heavy metals are present in excess of permitted levels, adverse consequences on people, other animals, and the environment as a whole frequently result. carcinogens, harm to the kidneys, lungs, and brain; harm to newborns and young; nausea and vomiting; rashes on the skin; and irritation to the eyes	Kinuthia et al. (2020)
Mineral acids	Chemical factories	At low level their impact is less, but under higher concentration can cause elevation of PH in the water bodies, which ultimately may result in acid rain	Borah et al. (2020)
Radioactive waste	Mining industries, metallurgical industries, industries manufacturing rocket fuels and other elements	The radioactive decay could harmful effects such as genetic mutations in organisms which may lead to cancer, exposure to soil can cause soil infertility, further leads to biomagnification in added into the food	

industries, while the salts are discharged from petroleum industries and agro-based industries. These compounds comprehend the high salinity of industrial wastewater.

(c) **Heavy metals**: The toxic metals released mostly from the paint and dye-making industries released after processing can cause hazardous effects if discharged untreated. Due to multiple applications and relatively poor degradability, these metals bioaccumulate in aquatic lifeforms and may impact food web and trophic structures. Metals like mercury, lead, cadmium, arsenic, etc. are of high concern for their massive usage in various types of industries, accompanied by low degradability.

(d) **Radio-active contaminants**: these contaminants are resistant to biodegradation, while their accumulation in any organism can cause major malfunctioning of the body. Mostly industries involved in mining {e.g. uranium mining} or those processing nuclear fuels dispose of this type of hazardous contaminants. The above-mentioned contaminants are usually found in industrial wastewater running down untreated into the waterbodies. Along with these, many new emerging groups of contaminants are paving their way into our waterbodies further raising alarming conditions. These contaminants or pollutants are mostly refractory.

Refractory pollutants are a complex set of pollutants with poor or very poor degradation capabilities. This group of pollutants are predominantly released from petrochemical industries. Alkanes, olefins, perfluoro-octane sulfonate [PFOS] from metallurgical industries, aromatic hydrocarbons, chloroalkanes, etc. are certain examples of this category. These contaminants are characterized by a low ratio when BOD is compared with the chemical oxygen demand [COD] attributed to its slow microbial degradation. Therefore, this category of contaminants is a significant problem, and efforts must be focused on their bioremediation practices. Gaining knowledge about microbial load or the native microbes at the toxicity site can be a huge advantage to promote as well as to manipulate these microbes for lowering the toxicity level, and to improvise the remediation strategies.

7.3 Presence of Native Microbiome at Contamination Sites

The industrial wastewater running down from the factories is rich in several types of inorganic and organic pollutants. These pollutants are toxic and degrade the water quality. While on the other hand, these contaminants harbor a wide range of microbes resistant to such an environment. These microbes are capable of surviving against the high toxicity level. Hence a vivid analysis of these microbes can add much more to the conventional treatment of wastewater and to the bioremediation strategies as well. The data from the biochemical analysis supplemented with microbial and molecular techniques have suggested the presence of some specific groups of microbes in these water samples discarded from our industries. These microbes vary depending upon the presence of compounds, heavy metals, toxic substances in the effluents released. Depending upon the meta-analysis of the wastewater from industries, certain groups are categorised based on their dominance in the effluent, for example **Proteobacteria** predominate across most of the wastewater samples, taking phylum's abundances into consideration. while **Firmicutes** and **Bacteroidetes** comes next in case of abundancy, respectively. These groups combined made up the major phyla present throughout the wastewater samples. Several bacterial genera like *Comamonas, Acidovorax, Pseudomonas, Acinetobacter, and Arcobacter,* are common in most of the industrial wastewater (Shah 2024a; Palanisamy et al. 2021). This identified core microbiome displayed both positive and negative correlations, demonstrating a symbiotic interaction between the bacteria involved in bioremediation. The assessment of the datasets reveal that genes implicated in xenobiotic metabolism and biodegradation were present.

Most of the wastewater contained unclassified and uncultured bacterial consortia, pointing to a unique microbiome that has not yet been characterized. Industries releasing heavy metals have a certain range of microbial load in them. The most common heavy metals found in industrial effluents include zinc (Zn), copper (Cu), chromium (Cr), mercury (Hg), and cadmium (Cd). The industries of producing goods, electroplating, leather tanning, clothing dyeing, and metal processing commonly use chromium, a hazardous heavy metal. Numerous microorganisms representating *Pseudomonas* sp., *Ochrobactrum*, *Bacillus* sp., *Serratia marcescens*, and *Acinetobacter*, have been found to be present in wastewater that contains heavy metals. These microbes can degrade the inorganic pollutants, furthermore, can be manipulated in the bioremediation processes.

Arsenic being a hazardous contaminant is released from a variety of industries for example, pesticide industries, leather industries, from the smelters of mining industries, even the wood processing industries release arsenic from the used wood preservatives, such as chromated copper arsenate [CCA] (Hayat et al. 2017). This metalloid has been tested for bioremediation by various processes and technologies, but biofilm remains the promising one. Hence the future stands on microbial bioremediation. In this regard certain microbes found in arsenic contaminated water, are as follows: certain methanogens ex. *Methanobacterium bryantii*-MOH strain, *Sulfurospirillum arsenophillum* and *Sulfurospirillum barnesii* etc. while the former acts by volatilization of arsenic (Stolz et al. 2006), the later functions as the dissimilatory arsenate reducing bacteria (Oremland and Stolz 2003).

Plastics, insecticides, and other chemicals produced by industrial activity infiltrate ecosystems in vast quantities and are resistant to biodegradation. These contaminants being refractory in nature poses huge challenge towards the prevailing bioremediation strategies. But at the same time, these pollutants have been attached with a rich diversity of microbiota that lives and degrade these chemicals. Phenolic compounds which are released mostly from dye and printing industries are a matter of concern for their wide applications accompanied with high toxicity levels. Many microbes, fungi have shown effective bioremediation properties on behalf of it. Even certain processes like wet air oxidation methods have given promising results, but the limitation always lies in the by-products so formed during the treatment which may need further processing prior to release. Thereby increasing the overall expenditure. Here microbial treatment can be considered as a better alternative, *Pseudomonas aeruginosa* and *Pseudomonas fluorescence* (Shah 2024b; Agarry et al. 2008) these two microbes comprise the indigenous binary mixed culture which was grown under synthetic phenol water and checked for bioremediation of phenolics. *Corynebacterium* sp., *Proteus* sp., *B. subtilis* and *Staphylococcus* sp. (Ajaz et al. 2004) were isolated from rhizosphere, and were tested for their morpho-cultural and biochemical characteristics. 30 out of the several isolated microbes have shown potential to be implemented in bioremediation of phenolics.

Multiple strategies were attempted to get around the substrate inhibition caused by increasing phenol concentrations, which prevent microbial growth. By gradually raising the phenol content, immobilising cells (biofilm development), or using genetic engineering techniques, microorganisms can be made to adapt to greater phenol concentrations (Banerjee 1997). Nowadays, targeted metabolic engineering has replaced simple adaptation and selection in industrial strain enhancement.

For example, *Rhodococcus erythropolis* strains have been genetically engineered to improve the phenolics' degradation capacity, and the results so obtained suggest that there is about 50% more efficiency in the bioremediation strategies (Zídková et al. 2013). After glucosyl residues, the most prevalent unit of chemical form in nature is the benzene ring. Numerous aromatic compounds are significant environmental contaminants because of their high retention in the environment and their thermodynamic stability (Dagley 1986). Although a wide range of microbes across taxonomic range possess the ability of aerobic degradation of contaminants, most of the study has focused on *Pseudomonas* species along with closely related organisms (Wackett 2003). This is due to their ability to break down a large range of pollutants. Anoxia is prevalent in a variety of contaminated environments, including submerged soils, aquatic sediments, and aquifers. Under such circumstances, biodegradation is carried out by facultative microorganisms or strict anaerobes using alternative electron acceptors, such as nitrates (denitrification processes), carbon dioxide (CO_2) (methanogenesis), Fe(III) (ferric-ion reductions), sulphate (sulphate reductions), as well as different acceptors (chlorate, Mn, Cr, U).

One major obstacle to the bioremediation procedures of PAHs is their low solubility and restricted bioavailability after their incorporation into soil and sediment. Because PAHs mostly go up the food chain and bioaccumulate in adipose tissues, where they may have mutagenic and carcinogenic effects, their intake by benthic creatures poses a serious risk. Bacteria that degrade polycyclic aromatic hydrocarbons (PAHs) may improve the bioavailability of PAHs by excreting biosurfactants, generating exogenous polymeric compounds, or forming biofilms (Johnsen and Karlson 2004). *Pseudomonas aeruginosa* was shown to be increasing the availability of its substrates by producing biosurfactants that enhanced the solubility of PAHs when it was cultured both in phenanthrene and naphthalene (Dziel et al. 1996). The wide variety of microbes work meticulously towards the bioremediation of toxic contaminants, declining their levels in the environment. They are needed to be characterized, conceptualized properly and efficiently to get commendable results in the future. Hence the need of the hour is to know and study the factors hampering or triggering the bioremediation mechanism, so that their manipulation could bring fruitful results.

7.4 Factors Affecting Microbial Bioremediation

The inherent ability of microbes to grow in an environment with high toxicity and pollution level is found to be a boon for the intervention of bioremediation strategies. These microbes utilize the bioaccumulated heavy metals, organic and inorganic contaminants even certain radioactive moieties in their normal metabolic processes. Hence bestows the environment with lower levels of pollution and better chances of survival for the susceptible organisms. These microbial bioremediation strategies are also guided by certain limitations that is questioning their due implementation in mass. The factors affecting microbial bioremediation processes are as follows:

(i) Physiochemical factors
(ii) Environmental factors
(iii) Biological factors
(iv) Bioavailability
(v) Contaminant structure and exposure rate.

(i) **Physiochemical factors:**

Physiochemical factors include environmental chemical and physical variables that influence an organism's metabolic activity. pH level, the redox potential, ion strength, solubility, either the presence or absence of donation of electrons, temperature, age of organometallic ions, and so on are a few examples of these variables.

a. pH:

Most of the microbes involved in bioremediation processes are acclimatized to a certain range of pH, that support their growth and the process of remediation as well. The addition of inorganic salts, acids from the effluents of industries disturbs the prevailing balance of pH, therefore affecting the microbial growth which ultimately disturbs the bioremediation capacity. Even the toxicity resistant microbes ideal for bioremediation are accustomed to a narrow range of pH for their survival, hence keeping a check on the increasing and decreasing pH of the industrial wastewater is crucial for the bioremediation strategies to follow.

b. Redox potential:

The redox potential can be defined as the capacity or tendency of a chemical species to undergo oxidation and reduction processes. It depends on the electron acceptors and donors for regulation. Speaking of microbial growth, biological oxidation is regarded as the energy source for the microbes, thereby creating a reduced environment around. This phenomenon is further implemented to determine the type of microbe growing in a medium, since the redox potential could be a characteristic feature for each type of microbe (Shah 2024c; Abbasian et al. 2015; Reichart et al. 2007). Considering the biodegradation activity oxygen being an electron acceptor is highly required for aerobic processes but in certain cases this oxygen availability is limited due to the oil spills or petrochemicals leakage from industries. These pollutants hinder with the water and air permeability ultimately declining the aerobic degradation. Along with that many enzymes contributing in bioremediation or degradation use oxygen as a substrate hence its declination can cause negative effects (Kumar et al. 2018). Numerous facultative and obligatory anaerobic microbes (bacteria and archaea) have been reported that are able to break down anaerobic hydrocarbon molecules.

Through fermentation, anoxygenic phototrophic processes, or anaerobic metabolism (by means of nitrates, nitrites, sulphate, thiosulfate, carbonate, and metal ions), these microbes provide electrons as well as hydrogen for substrate catabolism activities (Kumar et al. 2018).

c. Solubility:

The solubility of a pollutant is directly related to bioavailability of the substance for the bioremediation to occur. Hence the insoluble contaminants like PAHs, oils, petrochemicals need more time and highly specialized microbes for degradation. The solubility factor is further directly or indirectly affected by temperature as the temperature increases the solubility of organic molecules, specifically PAHs, which increases their bioavailability to microbes. Furthermore, as temperature rises, O_2 solubility decreases, limiting the development and metabolic activities of aerobic microbes (Kumar et al. 2018).

d. Biosorption:

Biosorption is dependent upon the pH in response to the isoelectric point of the solution. Biosorption is the required process for removal of toxic metals as it determines the overall negative and positive charges generating in microbial cells. This defines the action of microbial bioremediation of toxic metals (Srivastava et al. 2014) pH values influence metal ion solubility, which increases as medium pH reduces, directly impacting microbial cell adsorption (Blázquez et al. 2009).

(ii) **Environmental factors:**

The biomass of microbes, population variation, enzyme activity, substrate physicochemical characteristics, molecular composition, and concentration, in addition to a range of environmental parameters, generally impact the process of decomposition (Blázquez et al. 2009). The temperature, moisture content even bioavailability of the nutrient for microbial growth can be considered as environmental factors.

(iii) **Biological factors:**

A particular microbe becomes capable of biodegradation based upon the enzymes it produces, which ultimately is guided by its genetic composition. Hence the biological, molecular code for initiating bioremediation comes within the organism thereby leading us to pay attention towards the biological factors. Under low concentrations of carbon and energy sources, bacteria may opt chemotaxis and ability to degrade xenobiotics as favorable behaviors (Luka et al. 2018). Degradative bacteria need to adhere to surfaces, perhaps by forming biofilms, to access contaminants that have been adsorbed.

Chemotaxis has been discovered to play an important role in biofilm development in several microorganisms, directing a bacterium to swim toward nutrients (hydrophobic pollutants) adsorbed to a surface, followed by flagella attachment (Pandey and Jain 2002). On a second note, many bioremediating microbes need a consortium for proper functioning. The multispecies approach towards biodegradation is found to be more profound and effective. In some cases, certain microbes, were found to use other organic carbon, electron acceptors, and energy sources as co-metabolic substrates and degrade the recalcitrant.

(iv) **Bioavailability:**

To support the metabolic processes, an organism needs energy sources. Similarly, a microbe is dependent upon carbon sources for energy. Hence nutrient availability becomes a crucial point whenever microbial bioremediation is taken into consideration. The accessibility of organic contaminants to microorganisms, or bioavailability, also influences the rate of biodegradation. Although substantial microbial conversion, increased biotransformation is not attained since mass transfer is a constraint (Boopathy et al. 1998).

In this regard, biosurfactants or chemical surfactant can improve the bioavailability of pollutants subjected for microbial bioremediation.

(v) **Contaminant structure and exposure rate:**

Through bioremediation, pollutants are changed into less harmful or non-hazardous compounds. Biotransformation and biodegradation are the terms used to describe these processes, respectively. Biotransformation is the term used to describe any alteration in a compound's atomic or molecular structure brought on by microbes. However, radionuclides and metals cannot be biodegraded. In contrast, microorganisms can interact with these pollutants and alter their state of oxidation by undergoing redox transition processes. Certain bioremediation techniques make the converted metal or radionuclide more soluble, which increases the contaminant's mobility and facilitates its easier removal from the environment. Other approaches could cause immobilization as the changed metal or radionuclide precipitates out of solution. By immobilizing or accelerating the removal of these substances, this creates the possibility for the bioremediation of metallic substances and radioactive substances in the environment (Tabak et al. 2005).

7.5 Transcriptomics, Metagenomics and In-Silico Studies for Analysis of Onsite Microbiome

The central dogma of the living world infers the varsity of information guided and persuaded at the molecular levels, mostly via functional genes. These genes are further designed and designated via suitable codes consisting of nucleotides. The characterised codes and properly aligned sequences forms the foundation for analysing the genetic complexity of an organism. The DNA can be considered as the guide who instructs or lay down the path for formation of the individual. It even stores the required data for the generations to follow. While RNA is the actual pedestrian who walks under the instruction of DNA to give the finesse form to the organism in function. Hence generating data sets and analysing them meticulously is highly essential and inventive for conducting a wholesome analysis on any organism. Even when a group of organisms are to be compared these datasets are regarded as important and feasible parameters to suggest the inter-relation or inter-dependency of one organism over other. The development of transcriptomics and meta-transcriptomics is essential for producing increasingly accurate and definitive findings in any research. The term "transcriptome" refers to all the RNAs that are synthesized in an organism, including sense, antisense, alternatively spliced, alternatively polyadenylated, and RNA-edited transcripts. In other words, transcriptome is the overall account of all the RNAs of an organism, which can be manipulated to obtain inferences regarding the organism's metabolic and molecular processes. While the term transcriptomics can be considered as the study of transcriptome, which gives us the vivid account of the sum of all the RNA transcripts in function (Lowe et al. 2017).

The transcriptomics analysis gives an account of the active and passive metabolic processes depending upon the expression levels of the requisite genes. This kind of analysis is highly beneficial when the genotypic characteristics are questioned. For it gives the in-depth details of an organism's vital genes and their action in the individual's traits. The two well-defined and profusely used technology regarding transcriptomics are:

A. Microarray
B. RNA seq

These techniques help to attain the data sets suggesting the genes in action or to be active with respect to any required metabolic function of the organism. At the same time, the need for high through put data analysis techniques are the absolute requirements since microarray or RNA seq generates huge data to be sequenced and characterised with precision for present and future references.

Microarray

Thousands of genes' expression can be simultaneously detected in a lab setting using a microarray. Microscope slides with hundreds of small spots in precise locations, each containing a known DNA sequence or gene, are called DNA microarrays. These slides are commonly called DNA chips or gene chips. The DNA molecules affixed to each slide identify the messenger RNA (mRNA) transcripts (also called a transcriptome) produced by a group of genes. There are now many different DNA microarrays as well as DNA chip systems and devices for sale. Highly concurrent DNA as well as RNA hybridization screening in microminiaturized forms is made possible by these technologies. The majority of DNA microarray hybridization methods are centered on single nucleotide polymorphism (SNP) screening or gene expression research. Such microarray systems are utilized for forensic and genetic identification, infectious and genetic illness and cancer diagnostics, pharmacogenomic research, and genomic and molecular biology investigations (Heller 2002).

RNA Seq

NGS technologies process the light generated by the sequencing reaction once the correct base is introduced and aligns with the template being sequenced. Accordingly, the image recordings of the light emitted by each parallel sequencing reaction at each sequencing cycle constitute the NGS raw outputs. These image data require a large quantity of storage space because they include terabytes of data. The computational signals from each base's parallel reactions at each synthesis event are then extracted from the photos by filtering them.

The base uses these signals to communicate. NGS data quality and throughput have significantly improved because of improving the precision and consistency of base calling and signal extraction (Marguerat and Bähler 2010). The percentage of reads that correspond to a particular area of the genome is known as the signal in RNA-seq data. Sequencing provides total control regarding which reads are considered in the final analysis and may therefore be linked to the expression signals, in addition to enabling single base pair resolution. RNA-seq data is therefore essentially digital, observable, and quantifiable. According to Marguerat and Bähler (2010), the proper mapping of sequencing reads to corresponding reference genomes or their successful de novo assembly is in fact crucial for the generation of reliable RNA-seq data.

Even though RNA-seq has become the traditional procedure for transcriptome analysis, the technology and tools are still progressing. It needs to be acknowledged that the accordance between results generated by various tools continues to remain inadequate, and the conclusions are significantly affected by parameter settings, notably for low-expression genes. The development of transcriptomes from low amounts of precursor materials and effective transcript recognition from longer reads are two significant standouts in the prevalent application of RNA-seq. The state of the technology in both aspects is constantly evolving (Conesa et al. 2016).

Information about the genes expressed by the population is provided via meta-transcriptomics. Functional annotations of expressed genes can be used to infer the functional profile of a community under situations that are typically reliant on the host's status (Aguiar-Pulido et al. 2016). In these studies, complementary DNA (cDNA) is synthesised from RNA extracted from an environmental sample via reverse transcription and used as a template for highly parallel shotgun sequencing (DeLong 2013). To measure the unique contribution of microbes in biogeochemical and ecological processes, the meta transcriptomics approach for environmental sample research uses rRNA depletion, mRNA amplification, cDNA synthesis, and bioinformatic evaluation (Gifford et al. 2014).

A closer examination of the active members is feasible using RNA sequencing (RNASeq) to monitor the expressed transcripts in a microbiome at a particular time and under specific environmental conditions. Current revelations in mass spectrometry techniques used in proteomics can also shed light on actively expressed proteins, but it works best when combined with reference genomes or metagenomes from which expected peptide masses can be matched. RNASeq can detect, annotate, and map the metabolic pathways having comparatively low expressed genes (Shakya et al. 2019).

On comparing the phylogenetic surveys with metagenomics studies, it is much more evident that, metagenomics provides more investigative and precise records of the organisms since it does not only rely on the 16s rRNA analysis. Metagenomics can provide us with a variety of pertinent information, including evolutionary profiles of community function and structure, potentially novel biocatalysts or enzymes, and genomic correlations between function and phylogeny for uncultured organisms (Thomas et al. 2012). Metagenomics can be used to thoroughly sequence the majority of the available genomes in an environmental sample. This creates a profile of biodiversity in the community that can be connected to studies of the functional makeup of known and potential biological lineages.

The rapid development in the NGS and high throughput sequencing methodology have furthermore accelerated the prevalent metagenomics approach. This type of study furnishes highly reliable datasets, which have not only changed the course of onsite microbial research but also have given insights into the vast environmentally important microbial community. The various tools and techniques with respect to metagenomics analysis are as follows.

A. **Sampling**

Sampling is the first and foremost step for this type of studies. The sampling methodology is highly dependent upon the type of information needed to generate by the following study. The DNA yield is a crucial methodological consideration, particularly when large quantities are required for high throughput sequencing or cloning. Previous research has shown that, despite a lower purity, direct extraction produces more DNA per mass of soil than indirect extraction does (Delmont et al. 2011). DNA can be amplified using the multiple displacement amplification process (MDA). The entire DNA template is targeted with a set of random primers. Due of its exceptional processivity and potent strand displacement activity, the preferred enzyme is the Φ29 DNA polymerase, which is produced from *Bacillus subtilis* bacteriophage 29. Every time the polymerase binds the primer template; it inserts an average of 70,000 nucleotides. By repeatedly copying over the same template the polymerase may simultaneously extend new primers and replace previously extended products. An intermediate branching DNA is produced because of this exponential amplification (Thomas et al. 2012; Lasken 2009).

B. **High throughput sequencing technology**

The development of microbial community studies was greatly influenced by Sanger sequencing technologies in the beginning. Sequence length and sequencing output have varied dramatically since Sanger sequencing. Given that Sanger sequencing can currently extract up to 96 sequences each run with an average length of 650 bp, phylogenetic marker analysis might be possible. Next Generation Sequencing technologies (NGS) are low-cost platforms that allow for the simultaneous sequencing of millions of DNA molecules in various yields and sequence lengths (Escobar-Zepeda et al. 2015). The 454-pyrosequencing technology, which revolutionized the disciplines of genomics and metagenomics, was one of the first advancements in the field of

NGS. Because of the one-by-one nucleotide addition cycle in these technologies, the pyrophosphate (PPi) released during the DNA polymerization stage is transformed into a bright signal. The device encodes the light emission into nucleotide sequences with a base quality score after monitoring the light emission from a plate that contains millions of microwells, each of which contains a distinct DNA fragment (Escobar-Zepeda et al. 2015). After eliminating the adapters, barcodes, and primers, as well as using the Ribosomal database project (RDP) in the Pyrosequencing Pipeline to screen the low-quality reads during the primary phases, it is possible to significantly enhance the number of useable reads (Zhang et al. 2012).

With the advancements of technologies, new age NGS methods are being employed. Using solid-surface PCR amplification, the Illumina/Solexa method immobilises random DNA fragments on a surface to produce clusters of identical DNA fragments. Following that, in a sequencing-by-synthesis procedure, they are sequenced utilizing reversible terminators (Bentley et al. 2008). Further advent high-end techniques like ion torrent sequencing and the third generation nanopore sequencing have made huge contributions in this respect. The Ion torrent personal genome machine, from Life technologies have given some conclusive and evident results in metagenomics (Whiteley et al. 2012).

C. **Procedure for onsite metagenomics analysis**

The ideal procedure onsite studies vary depending upon the inferences to be drawn. But the basic methods can be summed up as follows (Fig. 7.1).

Together with further biochemical studies, the identification of novel genes retrieved from the environment by functional metagenomics may provide exposure to the molecular analysis of a variety of bacterial functions that take place in harsh environments. Moreover, the potential for genetic engineering and targeted evolution

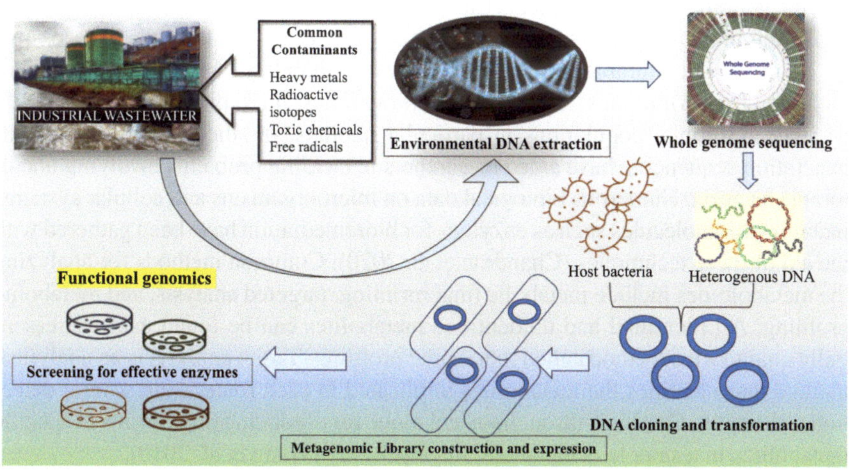

Fig. 7.1 A schematic description of onsite metagenomic analysis for targeted microbiome

to strengthen extremozymes will expand their industrial applications (Mirete et al. 2016).

7.6 Fluxomic and Metabolomics, an Overview of Metabolic Manipulation

The integration of computational and experimental methods to the explanation and prediction of intricate cellular behaviour in biological systems is known as systems biology. Systems biology can be divided into at least two different areas: simulation-based analysis and knowledge discovery, or data mining, depending on how the data that is accessible and is processed computationally. As part of the decoding of genome-wide data, the former builds and analyzes cell network models using specific high-throughput data. The latter, on the other hand, seeks to create dynamic simulations for the evaluation of hypotheses through in silico experiments, providing hypotheses to be verified using in vitro and in vivo research that may be pertinent to the development of drug discovery and preventive medicines.

Every metabolite present in a biological sample at any given time is examined using metabolomics. However, metabolism is an ever-evolving process. The metabolites that comprise the metabolome are ever-changing since they are part of the variety of metabolic processes that take place within cells. These processes combine to produce a complicated system of metabolic pathways that make up the metabolic network, which depends entirely on the fluxome and metabolome. The entire collection of fluxes in a cell's metabolic processes network, also referred to as the fluxome, contains comprehensive data on a variety of biological processes that give cells their unique phenotypic traits. A dynamic representation of the underpinning phenotype is revealed by flux analysis, which documents the relationship between the metabolome and the genome and environment (Cascante and Marin 2008).

Recent breakthroughs in molecular technologies (Fig. 7.2) have made it possible to examine unculturable organisms in natural habitats. Molecular techniques including transcriptomics, metabolomics, proteomics, fluxomic, and genomics help us study microbial populations in particular habitats. In-silico analysis and next-generation sequencing have aided researchers in tackling problems involving unculturable bacteria. Numerous biological data on microorganisms and cellular systems, including biomolecules such as enzymes for bioremediation have been gathered with the aid of these techniques (Chandran et al. 2020). Common methods for analyzing the metabolomes include metabolic fingerprinting, targeted analysis, and metabolite profiling. All identified and unidentified metabolites can be found using an essentially quantitative method called metabolite profiling. Target analysis is an analytical method that identifies the metabolites implicated in each route using several detection techniques. One high-throughput technique for producing fingerprints of specific metabolites in a sample is metabolic fingerprinting (Wang et al. 2010).

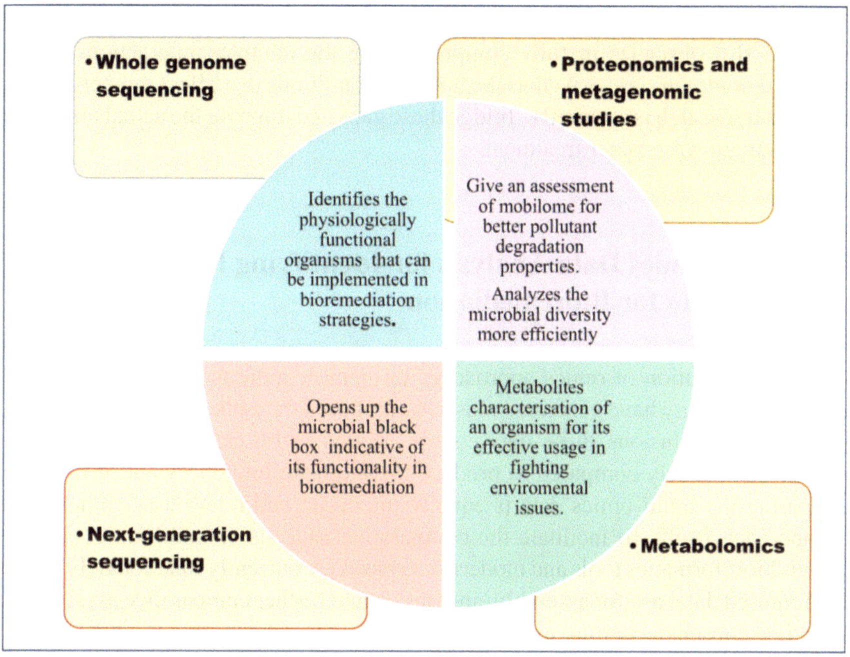

Fig. 7.2 Application of different multi-omics approaches for remediation of emerging contaminants in industrial wastewater

Fluxomic techniques examine metabolic interfaces that contribute to the progression and design of cellular processes. Metabolic reaction rates assess in-vivo enzymatic activity beyond transcriptome, proteomic, and metabolomic data. To characterize stoichiometry and regulation of metabolic reactions, a thorough fluxomic analysis depends on exact quantitative metabolomics data, radioisotope labeling observations, and metabolic network reconfiguration. Using more thorough metabolic models, the integrated approach incorporates tracer-based metabolomic data into software applications to forecast flux distribution. The intracellular fluxes of a broad range of bacteria have been mainly estimated using the 13C-based flux monitoring methodology. Numerous fluxes can be calculated using software programs like 13C-FLUX. According to biochemical, GC–MS, statistical, and genetic research, the marine bacterium Shewanella has co-metabolic pathways for biodegrading radio nuclides, halogenated organic compounds, and hazardous metals. Exposure to various carbon sources, the bacteria demonstrate sensitive metabolic flux (Tang et al. 2007).

By combining fluxomic with additional omics methods, researchers can discover critical metabolic pathways and enhance microbial activity, resulting in increased pollution removal. Fluxomic can assist in screening and improving specific microbial strains which are competent at decomposing refractory chemicals, hence increasing the total efficacy of wastewater treatment systems (Wang et al. 2021).

While fluxomic promises significant advances in biodegradation optimization, there are still obstacles in fully comprehending the complex interactions among microbial communities, as well as the ambient conditions that affect their efficiency. Additional research is required to bridge these gaps and improve the actual utilization of fluxomic in wastewater treatment.

7.7 Multi-omics Data Analysis for Identifying Potential Microbes for Bioremediation

The implementation of omics approaches in identification and characterisation of microbial diversity have generated massive number of datasets. To infer the required results and conclusions these datasets are needed to be carefully examined and even more precisely compared to produce the requisite inferences out of it. Hence comparing the multi-omics data is equally important and relevant as compared to the approach itself. To facilitate the comparative analysis of multi-omics datasets various bioinformatics tools and modern data bases are currently functional (Fig. 7.3). The required database for assembly, annotation and further comparative assessments should be chosen depending upon the expected outcomes we are pertaining.

The scientific community can now do computationally demanding analyses accredited to MG-RAST. With the introduction of this server, a small academic or industrial team may now quickly and cheaply produce high-quality findings for dozens of data sets. The MG-RAST server is dependable, flexible, and easily accessible to all researchers. It has built-in support for numerous data sources and a backend that holds abstract data types. In addition to excellent annotations and powerful computing, MG-RAST offers automated studies of phylogenetic context, gene and protein family identification, and subsystem and metabolic reconstructions. The user can access the analyses in a short period of time, which greatly boosts productivity

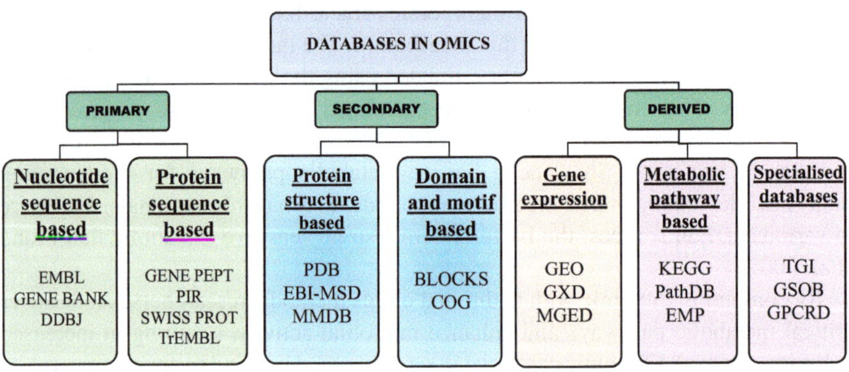

Fig. 7.3 Different databases used in multi-omics data analysis for remediation of emerging contaminants

(Glass et al. 2010). The pollutant-degrading enzymes (PDEs) can now be found on a huge scale owing to high-throughput metagenomic strategies.

The PDE protein sequences, which included both validated and disputed sequences, were compiled using the Uniprot database. 30,925 PDE enzyme sequences were acquired from Uniprot, comprising 14,442 sequences for the main enzyme family in charge of hydrocarbon breakdown and 13,072 sequences supporting plastic degradation. Sequences that were presumptive, hypothetical, or aberrant were eliminated using Python programs that were internally developed. RemeDB was created to extract PDEs from metagenomic data sets. RemeDB accepts the metagenome data set in a variety of formats, including faa, fasta, and fastq. Every processed sequence and input file format is independently checked by a Python script. The user is thereafter guided through the pipeline, which includes RAP Search, HMMER, and Transeq (Sankara Subramanian et al. 2020).

Recently, binning which involves assembling DNA sequences that represent an individual genome and genomes from related organisms emerge as an efficient approach. Based on multivariate statistics of tetranucleotide frequencies and interpolated Markov models, a new binning method was devised. Four metagenomes that are available to the public and one artificial metagenome that has already been published were used to compare its performance to that of existing algorithms. The new method's accuracy was on par with or better than that of the old ones. It was also up to 100 times faster. The entire open-source graphical binning programme was created in Java Swing and is available for download and further development (Strous et al. 2012).

However, to achieve a complete picture of microbial metabolism, researchers have attempted to leverage multiple levels of information to improve understanding of biological systems. Scientists have investigated on how *E. coli* cells react to genetic and environmental changes. This investigation included a number of 'omics' measurements, such as DNA microarray for 4213 *E. coli* genes, qRT-PCR for 85 genes that regulate central carbon metabolism, 2D DIGE for nearly 2000 proteins, and quantitative techniques using liquid chromatography-mass spectrometry/mass spectrometry (LCMS/MS) for 57 proteins participating in central carbon metabolism, CE-TOFMS for 579 metabolites, and 13C-labeled glucose for fluxome. Most of the identified metabolic enzyme gene disruptors that were examined displayed very slight changes in both proteins and mRNA. Metabolite concentrations were stable overall (Ishii et al. 2007).

The statistical integration methods implemented in this respect, are categorized as multivariate, concatenation-based, or transformation-based depending upon how the data are modified prior to the algorithm's usage (Tini et al. 2019). In summary, the analysis demonstrated that three omics integration is superior to paired omics integration in terms of sample classification. This illustrates that to improve our understanding of the mechanisms driving the characteristics of sample subgroups, it is crucial to consider different omics data at once (Tini et al. 2019).

For producing credible multi-omics datasets, metagenomic sequencing techniques can be divided into two categories: read-based and assembly-based. One is by mapping the resultant reads back to reference genomes, read-based metagenomic techniques may evaluate the taxonomic makeup and functional layouts of a complex microbial sample. But, read-based methods significantly depend on reference databases, which may not adequately reflect understudied environments like constructed water systems. Furthermore, this method is unable to assign precise functions to certain microbial consortia or populations. The second approach depends upon the binning of assembled contigs, or scaffolds is the process through which assembly-based methods, such as genome-resolved metagenomics, recover specific genomes de novo from a mixed community. As a result, genomes are produced that might be referred to as population genomes, metagenome-assembled genomes (MAGs), or draft bins since they frequently reflect a mixed consensus sequence of closely linked microbial cells. Using information from meta transcriptomics, metaproteomic, and/or metabolomics to try and explicitly link metabolic guilds and functional activities to microbes, these genomes can then be utilised as a platform for integrated multi-omics analysis (McDaniel et al. 2021).

Stable isotope probing (SIP) cataloging can be used to particularly enhance the phospholipids, DNA, or RNA of variable microbial populations. Metagenomic reference libraries are created by cloning DNA fragments taken from environmental samples in a suitable vector (phage, plasmid, or bacterial artificial chromosome, or BAC) (Mani 2020). The resulting transformants can be evaluated for phylogenetic evidence, conserved genes, the expression of characteristics, etc. Sequence-driven or function-driven analysis can be used to retrieve biological information from metagenomic libraries. Clone sequencing that contains conserved DNA sequences or phylogenetic anchors that are likely to be the genetic origin is subjected to sequence-driven analysis. Finding clones that exhibit their functional activity is the foundation of function-driven analysis (Chandran et al. 2020).

A few unique bacteria that weren't likely to be accessible with traditional culturing techniques have been discovered because of the application of omics-tools to determine the systematic and functional traits of microbial populations from polluted areas (Ojha et al. 2021). Using high throughput omics techniques, the systems biology of microbial consortia in many environments has lately been studied. However, a comprehensive understanding of the factors governing the growth, metabolic rate, structure, dynamics, and activities of the native microbial population at these sites is necessary for the efficient use of these intricate bioremediation strategies.

7.8 Economic Feasibility of Omics Dataset for Addressing Emerging Contaminants

The computational approach towards analysing the biological datasets have added a new perspective to the bioremediation protocols. For the potential microbes are not only screened based on their phenotypic expressions but are also carefully assessed based upon the genotype that lies behind. This approach has widened up the scenario of microbial biodiversity along with their rightful implementation in solving the prevalent environmental issues. Multi-omics screens a microbe based upon the functional genes, proteins produced, even analyse the ones targeted towards pollutant degradation (PDEs). The metabolism of a microbe playing the critical role in this can also be assessed as well as processed to give us the requisite metabolites. Further the metabolomics, fluxomic have significantly improvised the viewpoint of potential microbes to be employed into bioremediation.

The industrial effluents are overloaded with harmful chemicals, heavy metals aromatic compounds and many more. The microbes having the capacity to lower down the toxicity levels can now be assessed on site and can be wisely treated with the omics technologies to get relevant and conclusive results. These datasets need strategies to be properly diagnosed, so that valid inferences could be drawn out of it. Testing for marginal connections between several data modalities is a fundamental method for the merging of data sets. Molecular quantitative trait locus mapping is a well-established example, where numerous association studies are conducted between specific genetic variants and levels of gene expression or epigenetic marks. Considering the variant annotation approach, the basic strategy can be inculcated within, but the inferences so obtained are limited to local which do not illustrate the global overview of the molecular variation between the samples. Hence there is the need for a more effective strategy. The employment of kernel- or graph-based techniques to merge several data types into a uniform homology network between samples is one of the effective approaches (Argelaguet et al. 2018). Another eminent benefit of the omics application to the bioremediation or for environmental microbiome analysis is the supply of fundamental data to set the base for microbiome engineering. The datasets so generated from the multi-omics aspect is well-versed to suggest or can be implemented to design the models for microbiome engineering. Due to the size restriction on building these communities, synthesising microbial communities generally remains a significant difficulty. When communities have more than two or three members, interactions frequently become unmanageably complicated. A truly helpful input for creating synthetic consortia would be model-based identification of important member species that can co-grow to create a certain result (Kessell et al. 2020). Hence in short it can be concluded the application of multi-omics literally fill in the gap between the microbiome analysis in vitro and its application on site, ultimately paving way towards the better application and engineering of bioremediation strategies.

Since omics techniques like genome sequencing, transcriptomics, metabolomics, including fluxomic provide an extensive amount of information about microbial communities and how they interact with pollutants, they have significantly enhanced bioremediation methods. The various attributes of omics applications have been the key to its massive applications with the growing age of scientific research. Noting some key parameters like cost effectiveness, efficiency have brought in the overall benefits of omics approaches. Omics approaches facilitate in situ bioremediation procedures which is typically less expensive than ex situ methods. This lowers the need to excavate and transport polluted materials (Gupta et al. 2020). Because of its eco-friendliness, bioremediation employing omics techniques is frequently more widely accepted, potentially lowering the expenses relating to compliance with regulations and public relations attributing towards its high public acceptance.

Omics technologies enable the pinpointing of those microbial strains and metabolic pathways that are especially effective at degrading pollutants, resulting in more efficient and rapid remediation operations. The vast amount of data provided by omics techniques can be utilized for enhancing bioremediation strategies, lowering trial-and-error costs and increasing success rates. Hence the choice of organism for bioremediation strategies are highly data driven rather than assumptions based. With addressing the basic processes of microbial degradation, omics techniques help to build long-term bioremediation solutions that may reduce future remediation costs, adding onto the factor of sustainable solutions towards venturing science from labs to field applications. Insights gathered from omics studies can aid in the design of preventive measures to avert pollution, resulting in cost savings associated with potential remediation initiatives (Kumar et al. 2022). The initial installation expenses associated with omics technologies can be significant, including the procurement of advanced equipment and employee training, requires high initial investment, this could one of the major challenge of omics applications. Managing and analysing the massive amounts of data produced by omics methods necessitates large investments in bioinformatics infrastructure and skills (Alidoosti et al. 2024).

The overall economic impact of omics approaches in research can be dissipated as follows: firstly, the adoption of omics technologies in bioremediation can create jobs in various sectors, including research, environmental consulting, and bioinformatics. Secondly the growing demand for sustainable and efficient bioremediation solutions, with omics technologies playing a crucial role aid in its expansion and implementation with respect to cutting edge research. In summary, while the initial costs of applying omics techniques in bioremediation may be significant, the long-term economic benefits, such as cost savings, improved effectiveness, and ecofriendly approaches for environmental sustainability.

7.9 Highlights and Challenges of Multi-omics Studies

By enabling high-resolution monitoring of microbial populations and their functional dynamics, omics technologies—which include metagenomics, meta transcriptomics, metaproteomic, & metabolomics—are revolutionizing the treatment of industrial wastewater. Particularly for complex effluents including pharmaceuticals, toxic metals, as well as persistent organic pollutants, these methods enable continuous surveillance of pollutant breakdown, management of microbial consortia, and identification of important degradative pathways. Using microbial nanomaterials, adaptable sludge communities, and algae–bacteria symbioses, unified omics frameworks have made it easier to develop bioengineered systems that improve energy recovery and nutrient removal. Therefore, future work must concentrate on standardizing data integration, enhancing predictive modeling, and encouraging cross-sector partnerships that connect industrial engineering and microbial ecology to fully realize

Table 7.3 A comparative analysis of highlights and challenges in omics approach

Highlights	Challenges
1. Renders careful investigations into the molecular and biochemical processes behind the bioremediation potential of a microbe 2. Requisite proteins for mediating the bioremediation can be studied and even isolated on site given the metagenomics, meta transcriptomics and proteomics strategies 3. The microbial diversity to be exploited for the bioremediation strategies becomes broader with the intervention of omics 4. Several enzymes, proteins involved in the bioremediation mechanism can be diagnosed intensively using omics	1. Requires highly trained professionals for conducting the omics protocols 2. Requirement of highly advanced databases and bioinformatic tools to handle the massive data generated from the omics applications 3. The databases handling the highly important and sensitive omics data is prone to online malwares 4. The techniques and tools implemented with respect to omics are continuously evolving, needing constant upgradation of individual's skills

the potential of omics-driven remediation. The below mentioned table outlines the key highlights and challenges in omics approach in the bioremediation of industrial wastewater (Table 7.3).

7.10 Conclusion

The use of microorganisms in wastewater treatment have emerged as a viable and environmentally sustainable method for the efficient management and breakdown of new contaminants. Since several new enzymes derived from fungus, bacteria, and consortiums have proven successful in treating industrial wastewater, bioremediation techniques as a whole provide an economical and environmentally friendly method. Multi-omics approaches have been used recently and have the potential to completely transform the biological remediation of polluted environments by enabling comprehensive characterization and operation of microorganisms that have not yet been cultured. In order to address the treatment of wastewater contaminated by emerging contaminants, a number of contemporary and sophisticated techniques, including transcriptomics, proteomics, metabolomics, fluxomics, and genomics, have been recognized as cutting-edge technologies.

Acknowledgements The authors acknowledge their department and Institute for providing essential facilities and support.

Author Contributions SS and RP have contributed equally to the conception of the study, designing, writing, preparation of figures and tables of the manuscript. Both RP and SK contributed significantly in the final editing and revision of the manuscript.

Consent for Publication All authors have consented to the manuscript for publication.

References

Abbasian F et al (2015) A comprehensive review of alipHatic hydrocarbon biodegradation by bacteria. Appl Biochem Biotechnol 176(3):670–699

Agarry SE, Solomon BO, Layokun SK (2008) Kinetics of batch microbial degradation of phenols by indigenous binary mixed culture of *Pseudomonas aeruginosa* and *Pseudomonas fluorescence*. Afr J Biotechnol 7(14):2417–2423

Aguiar-Pulido V, Huang W, Suarez-Ulloa V, Cickovski T, Mathee K, Narasimhan G (2016) Metagenomics, metatranscriptomics, and metabolomics approaches for microbiome analysis: supplementary issue: bioinformatics methods and applications for big metagenomics data. Evol Bioinform 12:EBO-S36436

Ajaz M, Noor N, Rasool SA, Khan SA (2004) Phenol resistant bacteria from soil: identification, characterization and genetical studies. Pak J Bot 36(2):415–424

Albuquerque BR, Heleno SA, Oliveira MBP, Barros L, Ferreira IC (2021) Phenolic compounds: current industrial applications, limitations and future challenges. Food Funct 12(1):14–29

Alidoosti F, Giyahchi M, Moien S, Moghimi H (2024) Unlocking the potential of soil microbial communities for bioremediation of emerging organic contaminants: omics-based approaches. Microb Cell Fact 23(1):210

Anku WW, Mamo MA, Govender PP (2017) Phenolic compounds in water: sources, reactivity, toxicity and treatment methods. In: Phenolic compounds-natural sources, importance and applications, pp 419–443

Argelaguet R, Velten B, Arnol D, Dietrich S, Zenz T, Marioni JC et al (2018) Multi-omics factor analysis—a framework for unsupervised integration of multi-omics data sets. Mol Syst Biol 14(6):e8124

Banerjee G (1997) Treatment of phenolic wastewater in RBC reactor. Water Res 31(4):705–714

Bentley DR, Balasubramanian S, Swerdlow HP, Smith GP, Milton J, Brown CG et al (2008) Accurate whole human genome sequencing using reversible terminator chemistry. Nature 456(7218):53–59

Blázquez G, Hernáinz F, Calero M, Martín-Lara MA, Tenorio G (2009) The effect of pH on the biosorption of Cr (III) and Cr (VI) with olive stone. Chem Eng J 148(2–3):473–479

Boopathy R, Manning J, Kulpa CF (1998) Biotransformation of explosives by anaerobic consortia in liquid culture and in soil slurry. Int Biodeterior Biodegrad 41(1):67–74

Borah P, Kumar M, Devi P (2020) Types of inorganic pollutants: metals/metalloids, acids, and organic forms. In: Inorganic pollutants in water. Elsevier, pp 17–31

Cascante M, Marin S (2008) Metabolomics and fluxomics approaches. Essays Biochem 45:67–82

Chandran H, Meena M, Sharma K (2020) Microbial biodiversity and bioremediation assessment through omics approaches. Front Environ Chem 1:570326

Chawley P, Yadav K, Jagadevan S (2021) Nitrogenous wastes and its efficient treatment in wastewater. In: Water pollution and management practices, pp 147–175

Cholakov GS (2010) Control of pollution in the petroleum industry. Pollut Control Technol 3:1–10

Conesa A, Madrigal P, Tarazona S, Gomez-Cabrero D, Cervera A, McPherson A et al (2016) A survey of best practices for RNA-seq data analysis. Genome Biol 17(1):1–19

Dagley S (1986) Biochemistry of aromatic hydrocarbon degradation in Pseudomonads. In: Sokatch J, Ornston LN (eds) The bacteria, vol 10. Academic Press, Orlando, pp 527–555

Delmont TO, Robe P, Clark I, Simonet P, Vogel TM (2011) Metagenomic comparison of direct and indirect soil DNA extraction approaches. J Microbiol Methods 86(3):397–400

DeLong E (2013) Microbial metagenomics, metatranscriptomics, and metaproteomics. Academic Press

Dziel E, Paquette G, Villemur R, Lepine F, Bisaillon JG (1996) Biosurfactant production by a soil Pseudomonas strain growing on polycyclic aromatic hydrocarbons. Appl Environ Microbiol 62:1908–1912

Escobar-Zepeda A, Vera-Ponce de León A, Sanchez-Flores A (2015) The road to metagenomics: from microbiology to DNA sequencing technologies and bioinformatics. Front Genet 6:348

Gavrilescu M (2010) Environmental biotechnology: achievements, opportunities and challenges. Dyn Biochem Process Biotechnol Mol Biol 4:1–36

Gifford S, Satinsky B, Moran MA (2014) Quantitative microbial metatranscriptomics. In: Environmental microbiology: methods and protocols, pp 213–229

Glass EM, Wilkening J, Wilke A, Antonopoulos D, Meyer F (2010) Using the metagenomics RAST server (MG-RAST) for analyzing shotgun metagenomes. Cold Spring Harb Protoc 2010(1):pdb-prot5368

Gupta K, Biswas R, Sarkar A (2020) Advancement of omics: prospects for bioremediation of contaminated soils. In: Microbial bioremediation & biodegradation, pp 113–142

Gutierrez B, Gant-Branum RL, Romer CE, Farrow MA, Allen JL, Dahal N, Nei Y-W, Codreanu SG, Jordan AT, Palmer LD, Sherrod SD, McLean JA, Skaar EP, Norris JL, Caprioli RM (2018) An integrated, high-throughput strategy for multiomic systems level analysis. J Proteome Res 17(10):3396–3408

Hayat K, Menhas S, Bundschuh J, Chaudhary HJ (2017) Microbial biotechnology as an emerging industrial wastewater treatment process for arsenic mitigation: a critical review. J Clean Prod 151

Heller MJ (2002) DNA microarray technology: devices, systems, and applications. Annu Rev Biomed Eng 4(1):129–153

Ishii N, Nakahigashi K, Baba T, Robert M, Soga T, Kanai A et al (2007) Multiple high-throughput analyses monitor the response of E. coli to perturbations. Science 316(5824):593–597

Johnsen AR, Karlson U (2004) Evaluation of bacterial strategies to promote the bioavailability of polycyclic aromatic hydrocarbons. Appl Microbiol Biotechnol 63:452–459

Kaur R, Mavi GK, Raghav S, Khan I (2019) Pesticides classification and its impact on environment. Int J Curr Microbiol Appl Sci 8(3):1889–1897

Kessell AK, McCullough HC, Auchtung JM, Bernstein HC, Song HS (2020) Predictive interactome modeling for precision microbiome engineering. Curr Opin Chem Eng 30:77–85

Kinuthia GK, Ngure V, Beti D, Lugalia R, Wangila A, Kamau L (2020) Levels of heavy metals in wastewater and soil samples from open drainage channels in Nairobi, Kenya: community health implication. Sci Rep 10(1):8434

Kumar V, Thakur IS (eds) (2022) Omics insights in environmental bioremediation. Springer

Kumar V, Shahi SK, Singh S (2018) Bioremediation: an eco-sustainable approach for restoration of contaminated sites. In: Microbial bioprospecting for sustainable development. Springer, Singapore, pp 115–136

Lasken RS (2009) Genomic DNA amplification by the multiple displacement amplification (MDA) method. Biochem Soc Trans 37(2):450–453

Liu Y, Li H, Cui G, Cao Y (2020) Water quality attribution and simulation of non-point source pollution load flux in the Hulan River basin. Sci Rep 10(1):3012

Lowe R, Shirley N, Bleackley M, Dolan S, Shafee T (2017) Transcriptomics technologies. PLoS Comput Biol 13(5):e1005457

Luka Y, Highina BK, Zubairu A (2018) Bioremediation: a solution to environmental pollution—a review. Am J Eng Res 7(2):101–109

Mani I (2020) Metagenomics approach for bioremediation: challenges and perspectives. In: Bioremediation of pollutants, pp 275–285

Marguerat S, Bähler J (2010) RNA-seq: from technology to biology. Cell Mol Life Sci 67:569–579

McDaniel EA, Wahl SA, Ishii SI, Pinto A, Ziels R, Nielsen PH et al (2021) Prospects for multi-omics in the microbial ecology of water engineering. Water Res 205:117608

Mirete S, Morgante V, González-Pastor JE (2016) Functional metagenomics of extreme environments. Curr Opin Biotechnol 38:143–149

Mishra M, Singh SK, Kumar A (2021) Chapter 32—Role of omics approaches in microbial bioremediation. In: Kumar A, Singh VK, Singh P, Mishra VK (eds) Microbe mediated remediation of environmental contaminants. Woodhead Publishing, pp 435–445

Ojha N, Karn R, Abbas S, Bhugra S (2021) Bioremediation of industrial wastewater: a review. In: IOP conference series: earth and environmental science, vol 796, no 1. IOP Publishing, p 012012

Oremland RS, Stolz JF (2003) The ecology of arsenic. Science 300:939–944

Palanisamy V, Gajendiran V, Mani K (2021) Meta-analysis to identify the core microbiome in diverse wastewater. Int J Environ Sci Technol 1–18

Palanivell P, Susilawati K, Ahmed OH, Majid NM (2013) Compost and crude humic substances produced from selected wastes and their effects on Zea mays L. nutrient uptake and growth. Sci World J 2013

Pandey G, Jain RK (2002) Bacterial chemotaxis toward environmental pollutants: role in bioremediation. Appl Environ Microbiol 68(12):5789–5795

Reichart O, Szakmár K, Jozwiak Á, Felföldi J, Baranyai L (2007) Redox potential measurement as a rapid method for microbiological testing and its validation for coliform determination. Int J Food Microbiol 114(2):143–148, ISSN 0168-1605

Rissman AR, Carpenter SR (2015) Progress on nonpoint pollution: barriers & opportunities. Daedalus 144(3):35–47

Sankara Subramanian SH, Balachandran KRS, Rangamaran VR, Gopal D (2020) RemeDB: tool for rapid prediction of enzymes involved in bioremediation from high-throughput metagenome data sets. J Comput Biol 27(7):1020–1029

Shah MP (2024a) Microbial remediation of hazardous chemicals from water & industrial wastewater treatment plant. Springer Nature

Shah MP (2024b) Development in waste water treatment research and processes: applied technologies for clean up of environmental contaminants. Elsevier

Shah MP (2024c) Aerobic and anaerobic microbial treatment of industrial wastewater. CRC Press

Shah MP (2025) Extraction of precious metals from industrial waste using microbial technology. Springer Nature

Shakya M, Lo CC, Chain PS (2019) Advances and challenges in metatranscriptomic analysis. Front Genet 10:904

Shindhal T, Rakholiya P, Varjani S, Pandey A, Ngo HH, Guo W et al (2021) A critical review on advances in the practices and perspectives for the treatment of dye industry wastewater. Bioengineered 12(1):70–87

Srivastava J, Naraian R, Kalra SJS, Chandra H (2014) Advances in microbial bioremediation and the factors influencing the process. Int J Environ Sci Technol 11:1787–1800

Stolz JF, Basu P, Santini JM. Oremland RS (2006) Arsenic and selenium in microbial metabolism. Annu Rev Microbiol 60:107e130

Strous M, Kraft B, Bisdorf R, Tegetmeyer HE (2012) The binning of metagenomic contigs for microbial physiology of mixed cultures. Front Microbiol 3:410

Tabak HH, Lens P, Van Hullebusch ED, Dejonghe W (2005) Developments in bioremediation of soils and sediments polluted with metals and radionuclides–1. Microbial processes and mechanisms affecting bioremediation of metal contamination and influencing metal toxicity and transport. Rev Environ Sci Biotechnol 4:115–156

Tang YJ, Hwang JS, Wemmer DE, Keasling JD (2007) Shewanella oneidensis MR-1 fluxome under various oxygen conditions. Appl Environ Microbiol 73(3):718–729

Thomaidis NS, Asimakopoulos AG, Bletsou AA (2012) Emerging contaminants: a tutorial minireview. Global NEST J 14:72–79

Thomas T, Gilbert J, Meyer F (2012) Metagenomics—a guide from sampling to data analysis. Microb Inf Exp 2:1–12

Tini G, Marchetti L, Priami C, Scott-Boyer MP (2019) Multi-omics integration—a comparison of unsupervised clustering methodologies. Brief Bioinform 20(4):1269–1279

Viman O, Oroian I, Fleseriu A (2010) Types of water pollution: point source and nonpoint source. Aquacult Aquar Conserv Legisl 3

Wackett LP (2003) Pseudomonas putida—a versatile biocatalyst. Nat Biotechnol 21:136–138

Wang L, Lin Y, Ye L, Qian Y, Shi Y, Xu K et al (2021) Microbial roles in dissolved organic matter transformation in full-scale wastewater treatment processes revealed by reactomics and comparative genomics. Environ Sci Technol 55(16):11294–11307

Wang JH, Byun J, Pennathur S (2010) Analytical approaches to metabolomics and applications to systems biology. In: Seminars in nephrology, vol 30, no 5. WB Saunders, pp 500–511

Whiteley AS, Jenkins S, Waite I, Kresoje N, Payne H, Mullan B et al (2012) Microbial 16S rRNA ion tag and community metagenome sequencing using the ion torrent (PGM) platform. J Microbiol Methods 91(1):80–88

Zhang T, Shao MF, Ye L (2012) 454 Pyrosequencing reveals bacterial diversity of activated sludge from 14 sewage treatment plants. ISME J 6(6):1137–1147

Zídková L, Szőköl J, Rucká L, Pátek M, Nešvera J (2013) Biodegradation of phenol using recombinant plasmid-carrying Rhodococcus erythropolis strains. Int Biodeterior Biodegrad 84:179–184

Chapter 8
Enzymatic Degradation of Pollutants

Gaurisha Tripathi, Aliza Irfan, Gauri Raj, and Debasish Kar

Abstract Bioremediation is a green remediation approach that uses biological agents for the degradation or detoxification of pollutants in the environment. Among other pollutant-degrading enzymes capable of working on a wide range of pollutants such as hydrocarbons, pesticides, phenols, dyes, plastics, and pharmaceutical residues including reductases, laccases-reductases, monooxygenases and dioxygenases, Superoxide Dismutase (SOD), hydrolases, esterases, phosphatases, lipases, amylases, lyases, ureases, nitroreductases, PETases, and oxygenases. To facilitate the breakdown and detoxify environmental toxins, this research focuses on the enzymatic breakdown of environmental pollutants using extremozymes. There are enzymes called "extremozymes," which come from microorganisms that survive in extreme conditions. The industrial water treatment process employs microbes that break down these pesticides and remove the heavy metals in water to meet green standards. The review encompasses the classification and diversity of extremophiles, focusing on some enzymes such as leases, laccases, and peroxidases for breaking down numerous compounds. It also gives the molecular adaptations of the extremophiles and discusses the role of the extremozymes in bioremediation for increasing access to valuable metals for industrial use. Different types of extremozymes include thermoenzymes, halophilic enzymes, acidoenzymes, alkyloenzymes, and psychrophilic enzymes. They maintain stability in high temperatures, salinity, and pH, under which most of the enzymes are denatured. Due to the excellent catalytic efficiency of the extremozymes and their ability to act on a broad spectrum of substrates that includes recalcitrant pollutants, they are particularly fit for bioremediation processes at extreme conditions. They are also micro-factories with the ability to synthesize biomolecules through genetic and biotechnological means. It will summarize the discussion of innovative approaches toward the improvement of bioremediation using enzymatic engineering, namely by improving enzyme activity and genetically altering the microbial strains toward increasing productivity.

G. Tripathi · A. Irfan · G. Raj · D. Kar (✉)
Department of Biotechnology, Ramaiah University of Applied Sciences, Bangalore, India
e-mail: debasish.bios@gmail.com

© The Author(s), under exclusive license to Springer Nature Singapore Pte Ltd. 2026

S. Dey and M. P. Shah (eds.), *Microbial Applications in Industrial Wastewater Treatment*, Advances in Wastewater Research,
https://doi.org/10.1007/978-981-96-9128-9_8

Keywords Bioremediation · Extremozymes · Pollutant-degrading enzymes · Extremophiles · Enzymatic engineering

8.1 Introduction

The industrial advancement and technological improvement of this age have made it alarmingly clear that there is an environmental cost to human activity. Pollution by pollutants has become a severe challenge to sustainability and public health as natural ecosystems are degraded. Harmful substances in soil, water, and air comprise pollution, causing ecological imbalance, endangering wildlife, and jeopardizing human well-being. While traditional methods of pollution mitigation through chemicals and physics have been somewhat effective, their limitations, such as the high energy demands and secondary pollution, reveal a need for innovative approaches.

Bioremediation has proven to be one of the most promising ways to meet this global challenge. It utilizes the natural ability of microorganisms and their enzymes to detoxify, degrade, or sequester harmful substances from the environment. The most specific, efficient, and adaptable tool of bioremediation is enzymatic degradation. Enzymes accelerate chemical reactions that reduce complicated contaminants to simpler, less harmful compounds, reducing environmental contaminants.

This paper discusses enzymatic degradation of pollutants with special emphasis on extremozymes, enzymes from extremophilic microorganisms that have been known to thrive at extreme temperatures, salinity, and pH levels (Bala et al. 2022). These extremozymes have captured much attention because they can thrive in extreme conditions, where most other microorganisms perish, making them useful for bioremediation in adverse environments. From the oxidoreductases such as laccases and peroxidases to hydrolases such as cellulases and esterases, it is evident that these enzymes have been versatile in degrading hydrocarbons, pesticides, phenols, dyes, plastics, and pharmaceutical residues.

There are three major categories of pollutants: soil, water, and air. All these have their own characteristics and consequences. Soil pollutants include organic (such as pesticides and petroleum products) and inorganic (like heavy metals and acids) forms that damage the fertility of the soil, harm organisms, and contaminate food chains. Water pollutants encompass industrial chemicals, biological agents, and plastic debris, which threaten both aquatic ecosystems and human health. Similarly, volatile organic compounds and heavy metals are air pollutants that cause respiratory diseases and environmental degradation (ExclusivePapers.com 2014). The diversity and persistence of these contaminants need advanced solutions, where enzymes offer a potent remedy because of their biochemical adaptability.

Extremophiles, which are the microorganisms producing extremozymes, have been developed to thrive and survive in conditions that would usually denature conventional enzymes. Be it in volcanic springs at extremely high temperatures, the polar oceans at extremely low temperatures, or hypersaline lakes with salinity much higher than sea salt, these microorganisms are robust. Their enzymes, having high catalytic efficiency and stability, have paved new horizons in pollution management. Extremozymes such as oxygenases and hydrolases are used for the degradation of insecticides, polycyclic aromatic hydrocarbons, and plastics into non-toxic byproducts.

The role of extremozymes in bioremediation is not limited to degradation. They are also integral to biofilm formation, metal sequestration, and detoxification of heavy metals. For example, laccases from phenolic substances and azo dyes have been broken down by white rot fungi. Hydrocarbons in saline environments can be biodegraded by halophilic bacteria. Such applications underscore the potential of extremozymes to address a broad range of environmental challenges.

Although promising, practical applications of extremozymes in bioremediation suffer from a lack of scalability, cost-effectiveness, and environmental compatibility (Shah 2024a; Olaya-Abril et al. 2024). The only key for the unlocking of this great potential is further research in optimizing enzyme production, stability, and functionality under industrial conditions. Using recombinant DNA technology and other advanced techniques, scientists can unlock biocatalytic capabilities from extremophiles, opening avenues to even more efficient and environmentally friendly approaches to pollution management.

This paper explores classification of pollutants, biochemical properties of extremozymes, and their implications on bioremediation. Furthermore, it also analyses molecular adaption by extremophiles and novel strategies for designing enzymes, which are spearheading advances in this research area. With these efforts, the paper throws more emphasis on how enzymatic decomposition, in all its glory, can become the tool toward achieving a better and healthier future in environmental terms.

8.2 Classification of Pollutants

Pollutants are majorly classified into the following.

8.2.1 Soil Pollutants

The detrimental effects of hazardous substances on the natural composition, structure, and function of soil are referred to as soil pollution. These harmful substances that contaminate the soil are known as soil contaminants. There are numerous ways to categorize soil pollutants, such as by looking at their chemical nature, source, toxicity, persistence, physical form, and biological form (Sharma 2023a, b). Based on their chemical nature, soil pollutants can be classified as either organic or inorganic. Mining, transportation, urbanization, agriculture, and industry are a few of the causes of contaminated soil. Sorting soil pollutants according to their toxicity levels is an additional method of classification. There are several different physical forms of contaminants, such as liquid, vapour, and solid. Both biodegradable and non-biodegradable pollutants can be found in soil.

Based on the chemical nature of soil pollutants:

- Organic soil pollutant: Organic soil pollutants come from natural or human-caused soil contamination, endangering human health, agriculture, and ecosystems. Since these contaminants are organic, they contain carbon. Examples include pesticides (e.g., DDT), herbicides, petroleum products, and polychlorinated biphenyls (PCBs), polyethylene, aliphatic/aromatic, and terephthalate.
- Inorganic soil pollutants: Inorganic soil pollutants are not formed from living organisms and contaminate the soil, hence adversely affecting ecosystem health and its overall quality (Sharma 2023b). Usually originating from natural processes or human activity, these pollutants are substances or elements that are difficult for the environment to break down. They include acids, salts, and heavy metals. These are non-carbon-based substances. Examples include lead, mercury, cadmium, arsenic, and nitrates.

Both organic and inorganic soil contaminants have the potential to seriously harm ecosystems, human health, and the environment. These contaminants have the following main effects:

I. **Soil Contamination**: The soil may get contaminated by organic contaminants over time, which lowers soil fertility and inhibits plant development. Elevated levels of heavy metals have been linked to soil deterioration, which lowers crop yield and damages plant nutrients.
II. **Organism Toxicity**: Animals, insects, and soil microorganisms are all poisoned by a variety of organic contaminants, including pesticides. Heavy metals are toxic to plants and limit crop yields by restricting growth. In animal tissues, they also build up and can cause poisoning and other health problems.
III. **Bioaccumulation**: Certain organic pollutants, including persistent organic pollutants (POPs), have the ability to lodge themselves in plant and animal cells. This can put higher creatures, including humans, at risk of bioaccumulation in the food chain.
IV. **Risks to Human Health**: Prolonged exposure to organic pollutants and heavy metals increases the risk of developing cancer, endocrine disruption, brain damage, kidney failure, developmental problems, and reproductive issues (Shah 2024b; Shetty et al. 2023).
V. **Soil Salinization**: Excessive use of chemical fertilisers and irrigation in agriculture can cause salinization of the soil, which weakens the structure of the soil and lessens its capacity to sustain plant life.

8.2.2 Water Pollutants

Water pollutants are substances that contaminate water bodies, making them unsafe for human consumption and harmful to aquatic ecosystems. These pollutants can be chemical (such as industrial chemicals, insecticides, and heavy metals), biological (including bacteria, viruses, and parasites), or physical (like sediments and plastics). They may come from sewage, industrial discharges, trash, and agricultural runoff, among other causes. A variety of environmental and health problems, including waterborne illnesses, ecosystem deterioration, and disturbance of aquatic habitats, can result from the presence of these contaminants, underscoring the significance of efficient water management and pollution control strategies. The origin, chemical makeup, persistence, and impacts of water contaminants can all be used to classify them into different groups.

Based on chemicals, water pollutants can be classified into the following.

1. **Organic Pollutants**: These include compounds containing carbon, such as pesticides, herbicides, pharmaceuticals, petroleum products, and industrial solvents. They can be toxic, bioaccumulative, and persistent in the environment, posing risks to both human health and aquatic life. Since many organic contaminants can build up in the food chain and are resistant to natural degradation, they are especially dangerous to aquatic and terrestrial life.

   Examples of Organic Pollutants:

   - **Pesticides and Herbicides**: In agriculture, these substances are frequently used to manage weeds and pests. Examples include dichlorodiphenyltrichloroethane (DDT), atrazine, and glyphosate.
   - **Petroleum Hydrocarbons**: Crude oil and its derivatives, such as benzene and toluene, are common organic pollutants in water due to oil spills, leaks from pipelines, and runoff from roads.
   - **Industrial Solvents**: Organic solvents like trichloroethylene (TCE) and tetrachloroethylene (PCE) are used in various industries for cleaning and degreasing.
   - **Polychlorinated Biphenyls (PCBs)**: PCBs are man-made organic compounds that are employed as coolants and in electrical devices.
   - **Pharmaceuticals and Personal Care Products (PPCPs)**: These include drugs, hormones, and cosmetic ingredients that enter water bodies through sewage systems or improper disposal. Even at low concentrations, PPCPs can disrupt the endocrine systems of aquatic organisms and may contribute to antibiotic resistance.
   - **Volatile Organic Compounds (VOCs)**: VOCs like chloroform and formaldehyde can enter water through industrial discharges or atmospheric deposition. These compounds can be carcinogenic and pose risks to human health when present in drinking water supplies.

2. **Inorganic Pollutants**: These are chemical substances that do not contain carbon-hydrogen bonds, such as heavy metals (e.g., lead, mercury, arsenic), acids, salts, and minerals (Admin-Infection 2023). They can cause toxicity and disrupt water chemistry, making water unsafe for consumption and harming ecosystems. When present in water bodies, inorganic pollutants can alter water chemistry, disrupt ecosystems, and pose serious health risks to humans and aquatic organisms.

- **Heavy Metals**:
  (1) Lead (Pb): Lead contamination in water often originates from old pipes, industrial discharge, and mining activities.
  (2) Mercury (Hg): Mercury pollution can result from coal combustion, industrial processes, and mining.
  (3) Cadmium (Cd): Found in batteries and electroplating industries, cadmium can enter water through mining and industrial waste. It accumulates in the liver and kidneys.
  (4) Nitrates and Nitrites: These pollutants are commonly found in fertilisers and can enter water bodies through agricultural runoff. High nitrate concentrations in drinking water can cause methemoglobinemia or "blue baby syndrome" in infants, where oxygen transport in the blood is impaired.

Water pollutants can be classified based on the effects they cause on human health, aquatic life, and the environment. The main categories include

i. **Toxic Pollutants**: These pollutants are poisonous and can cause immediate or long-term health effects, depending on their concentration and exposure duration. They may include heavy metals, pesticides, and industrial chemicals (Manisalidis et al. 2020). Toxic pollutants can lead to cancer, neurological disorders, and reproductive issues. Examples: Mercury (causes neurological damage), Lead (affects the nervous system), and DDT (a pesticide causing bioaccumulation in the food chain).
ii. **Persistent Organic Pollutants (POPs)**: Organic chemical compounds which are not degradable to the environment and stay in the environment for long without biodegrading. POPs are associated with chronic health problems among humans and wildlife including endocrine disruption, damage of immune systems, and reproduction problems. Examples: PCBs; Dioxins; Some pesticides, including DDT.

8.2.3 Air Pollutants

It is commonly known that air pollution poses health concerns. The air we breathe, commonly known as ambient air, can contain a variety of contaminants that are emitted into the indoor and outdoor atmospheres. Solid particles, liquid droplets, and gases can all be found in air pollution, which can come from a number of sources such as industry, automobiles, heating appliances, and tobacco smoke (Shah 2024c; Manisalidis et al. 2020). The health effects of short-term exposure can differ from those of long-term exposure, depending on the specific pollutant. Pre-existing conditions including heart disease, asthma, or chronic bronchitis (also known as chronic obstructive lung disease, or COPD) are made worse by brief exposure. Long-term exposure can lead to potentially catastrophic heart and breathing issues. Let's talk about organic and inorganic pollutants:

Organic Pollutants: Organic air pollutants are substances that include carbon and are released into the atmosphere from a number of sources. These pollutants may have detrimental effects on the environment and human health. The main categories of organic air pollutants include the following.

Volatile Organic Compounds (VOCs): At room temperature, chemicals referred to as volatile organic compounds (VOCs) easily evaporate into the atmosphere. Paints, solvents, cleaning supplies, and even some industrial processes can release pollutants into the air. Common examples include benzene, formaldehyde, and toluene. VOCs can cause headaches, light-headedness, and respiratory issues in addition to being a contributing factor to the creation of ground-level ozone, or smog.

Polycyclic Aromatic Hydrocarbons (PAHs): A class of compounds known as polycyclic aromatic hydrocarbons (PAHs) are created when organic materials like coal, oil, gas, wood, and even tobacco are burned partially. They can be released from vehicle exhaust, industrial emissions, and smoke from wildfires.

Many PAHs are known to be carcinogenic and can cause lung, skin, and other types of cancer.

Inorganic Air Pollutants: Inorganic air pollutants are substances that are not based on carbon and are present in the atmosphere. Natural sources, industrial processes, and combustion all commonly discharge these pollutants into the atmosphere. These pollutants can have negative effects on ecosystems, the environment, and human health. Common inorganic air pollutants include the following.

Ozone (O_3): Three oxygen atoms are combined to form ozone, or O_3. O_2, or basic oxygen, is made up of two oxygen atoms linked together. Because of the extra third atom, ozone is a very reactive and unstable gas. Within the Earth's atmosphere, ozone can be found at both ground level and in the upper atmosphere (Ground-level Ozone Basics | US EPA 2024). Ozone, which is present in the upper atmosphere, protects us from the sun's damaging UV radiation. NO_2 (nitrogen dioxide): The highly reactive gas nitrogen dioxide is produced by the emissions of gas stovetops, unflued gas heaters, and automobiles. Particularly close to busy roadways and indoors where unflued gas heaters are in operation, high amounts are seen.

Carbon monoxide (CO): The deadly chemical carbon monoxide is invisible to the senses of taste, smell, or sight (SilentShadow 2019). Even in the absence of smoke, it is created when fuels like charcoal, gas, and wood are burned.

Sulphur dioxide (SO_2): Sulphur dioxide is a highly reactive gas with a strong, disagreeable smell. It is produced by burning fossil fuels in power plants and other industrial facilities.

8.3 Mechanism Degradation of Pollutants by Enzymes

Enzymatic degradation is the process by which enzymes break down complex molecules into simpler ones. Enzymes function as biological catalysts to speed up the chemical reaction and break down a substance (such as proteins, carbohydrates, and lipids) into smaller, easier-to-manage components. Many biological processes, including digestion, depend on this process. Enzymes like lipases, amylases, and proteases break down food into nutrients the body can absorb. In addition, it is used in industrial settings for waste treatment and recycling biodegradable materials, as well as in environmental processes including the breakdown of contaminants. Many inorganic and organic pollutants, such as pesticides, heavy metals, and persistent organic pollutants (POPs), can be broken down by these enzymes. They are useful in bioremediation processes that cause contaminants to change chemically. Because of their broad specificity, enzymes can react with a variety of compounds that share similar structures, conditions, or substrates (Fig. 8.1).

8.3.1 Steps Involved in Mechanism

1. Pollutant Recognition: Enzymes have a high degree of selectivity toward their substrates, which are frequently contaminants in this setting. Pollutants like hydrocarbons, pesticides, and industrial chemicals are recognized by enzymes

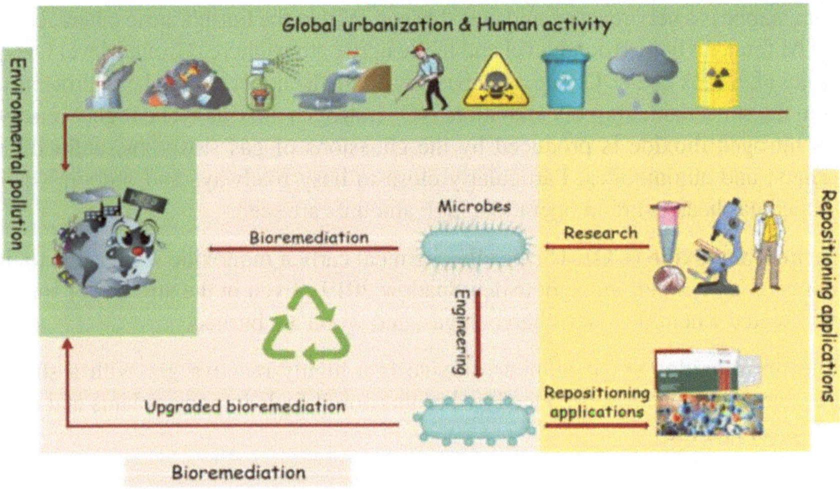

Fig. 8.1 Bioremediation approach. Newly discovered pollutants are frequently discharged into the environment, leading to detrimental effects and health issues (Xu et al. 2021). Diagram illustrating the “expanded” uses of microbial bioremediation for the removal of environmental contaminants

produced by microbes, fungi, or plants, which have evolved to utilize or break down such compounds. Enzymes can bind to a pollutant molecule because of their specific structure, ensuring that only particular types of pollutants are targeted for degradation.

2. Enzyme–Substrate Interaction: Active site binding: The pollutant (substrate) attaches itself to the active site of the enzyme. This interaction frequently follows the induced-fit or lock-and-key models, in which the enzyme modifies its structure slightly to accommodate the pollutant. It leads to the formation of enzyme–substrate complex. Because of this binding, the activation energy required to start a chemical reaction is lowered, accelerating the degradation process.
3. Catalysis of chemical reactions: Redox (oxidation–reduction) reactions are facilitated by enzymes such as dehydrogenases, peroxidases, and oxygenases.

   - **Oxygenases**: By adding oxygen to the pollutant molecules, or oxidation, these enzymes change them into less hazardous or more soluble forms. As an illustration, consider the enzymes monooxygenases and dioxygenases, which add oxygen to aromatic hydrocarbons to transform them into alcohols, ketones, or carboxylic acids (Arora et al. 2009). By introducing water molecules and cleaving bonds in esters, amides, and other compounds, enzymes like hydrolases and esterases hydrolyse contaminants. The contaminant becomes more degradable and less complicated as a result (Fig. 8.2).
   - **Hydrolysis**: By hydrolysing their ester linkages, esterases break down insecticides such as organophosphates (Fig. 8.3).

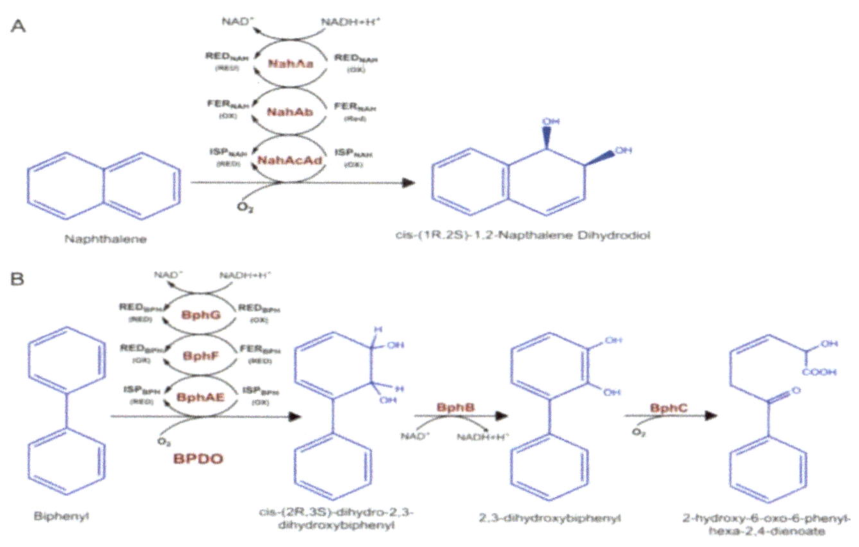

Fig. 8.2 Naphthalene and biphenyl catabolic pathway: **A** naphthalene catalysed by three-component NDO to *cis*-(1*R*, 2*S*)-1,2-napthalene dihydrodiol (Xu et al. 2021). **B** Red, reduction, and oxidation are the four enzymatic processes of the bacterial biphenyl metabolic pathway that are catalysed by BPDO

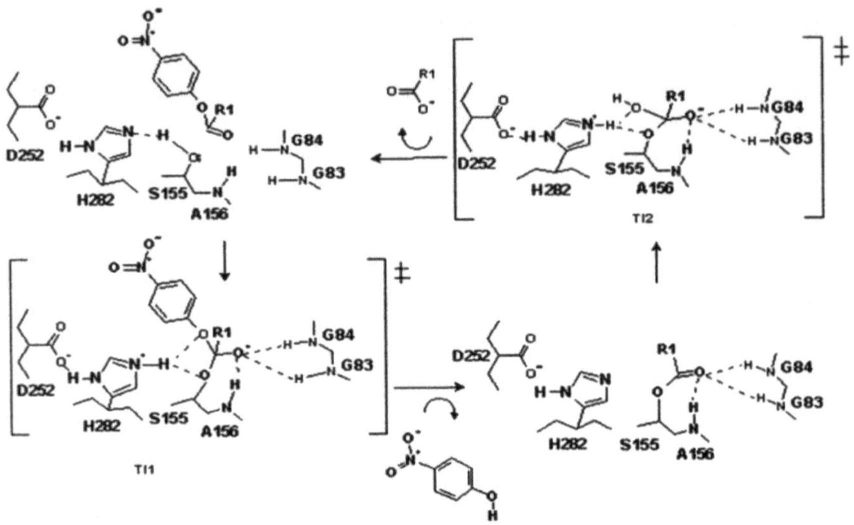

Fig. 8.3 Hydrolysis mechanism of EST2. TI1 and TI2, two intermediate transition phases, are experienced by the reaction. The M211S/R215L complexed with HDSC and the EST2 wild type structure complexed with Hapes, which are mimics of the second intermediate transition state, have been used to assign the hydrogen bonds

- **Dehalogenation**: The toxicity of halogenated contaminants (such as PCBs or halogenated solvents) is decreased by the removal of halogen atoms (such as chlorine and bromine) by enzymes such as reductive dehalogenase.
- **Deamination/Decarboxylation**: When contaminants have their amino or carboxyl groups removed by enzymes, they become more readily degraded.

4. Breaking Down Complex Pollutants: Primary and Secondary Metabolism: Certain pollutants undergo sequential degradation (metabolic pathways). They are first transformed into primary metabolism's intermediate metabolites, which are frequently still dangerous. They are then converted into harmless byproducts like biomass, carbon dioxide, and water by further enzyme reactions (secondary metabolism). As an illustration, during the breakdown of polyaromatic hydrocarbons (PAHs), enzymes such as dioxygenases gradually break down the aromatic rings, resulting in a series of increasingly simpler and non-toxic compounds.
5. Detoxification and Mineralization:

    - Detoxification: Enzymes change harmful pollutants into less harmful or non-toxic forms. For instance, enzymes can chemically alter heavy metals or persistent organic contaminants.
    - Mineralization: Complete mineralization is the consequence of some enzymatic activities that completely convert contaminants into water, carbon dioxide, and inorganic molecules.

8.3.2 Types of Extremozymes

1. **Pectinases**: Fruit processing wastewater is heavily charged with the plant polysaccharide pectin this pectin can make treatment of the effluent problem ridden since it generates methane treatment of this effluent in activated sludge systems is made easy through the breakdown of pectin by pectinolytic enzymes such pectin-rich wastewater can be well pre-treated before its entry into the tank by alkaline-loving bacteria like the species of bacillus.
2. **Celluloses**: Cellulases are enzymes that facilitate breaking down cellulose, one of the major components of plant fibres. There are three main types of cellulases: Endoglucanase: Cuts randomly throughout cellulose fibres, releasing free chain ends.

   Exoglucanase (or cellobiohydrolase): Catalyses the reaction on exposed chain ends to liberate small units called cellobiose.

   β-Glucosidase breaks cellobiose down into glucose, the simplest sugar.

   For instance, a specific type of cellulase, called Puradax HA, is made from an alkalophilic bacterium that belongs to the genus Bacillus, and is an example used in textile industries as a stain-removing and colour-preserving chemical applied on clothing, catalyses to reduce and break indigo dye that is textile colours, then washed and flushed into bodies of water as wastes hence contaminating it.
3. **Laccases**: Enzymes known as laccases use oxygen to accelerate the breakdown of a variety of complex compounds. They are multi-copper enzymes that act with oxygen to convert it to water as they oxidize the substrates (Shah 2025; Janusz et al. 2020). Various types of laccases exist, including high redox potential types, which exist in white rot fungi, and low redox potential types, which exist in moulds.

   The breaking down of harmful compounds involves Laccases; these compounds include polycyclic aromatic hydrocarbons, PAHs. These are chemicals that are carcinogenic in pollutants emanating from vehicle exhausts and industrial combustion. They break the PAH into harmless byproducts such as carbon dioxide and water.

   These enzymes also can be used in degrading textile dyes—azo dyes. There is widespread use of the application of these dyes in the textile business which are poisonous to aquatic and humans' health as well. More effective degradation ability is possessed by Laccases by extremophilic bacteria against these dyes and further reduce their bad impacts in the environment.

   Laccases are used in bioremediation to remove various contaminants such as phenols, polyamines, lignin, and many hazardous dyes. For example, there exist various isolated fungi and bacteria that break down problematic pollutants, making laccases important for a clean environment.

4. **Microbial dioxygenases**: Dioxygenases are enzymes that involve the incorporation of oxygen into another molecule to break up harmful, complex compounds (Iacopino and Licausi 2020). They are very useful for bioremediation, particularly in breaking down aromatic compounds (ring-shaped molecules usually associated with pollutants) into a less complex, safe compound. There are two categories of dioxygenases: intradiol and extradiol enzymes. Both of these use different iron species (Fe III and Fe II) to execute their work.
5. **Monooxygenases**: Monooxygenases are those enzymes that help to degrade the pollutants since they specifically bind to certain kinds of molecules (stereoselectivity) (Tõlgo et al. 2022). Therefore, it is helpful for bioremediation that is both aromatic (ring) and aliphatic (chain) compounds, where one example of methane monooxygenase could break the hydrocarbons from the oils and also in some other pollutant.
6. **Peroxidases**: Peroxidases are enzymes that help clean up pollutants, especially in contaminated solids and wastewater. They are useful in breaking down hydrocarbons, oils, and many phenolic compounds. The most significant types include lignin peroxidases (LiPs) and manganese peroxidases (MnPs), which work with hydrogen peroxide to carry out reactions like cutting chemical bonds and oxidizing certain groups in pollutants. LiPs and MnPs can degrade harmful compounds like polychlorinated biphenyls (PCBs) and various toxic dyes, making them valuable for bioremediation.
7. **Lipases**: Lipases are the enzymes that break oils into smaller components, hence useful in cleaning oily wastewater. They are applied in the treatment of saline wastewater, oilfield waste, and effluents from dairy and tanning industries, rich in oil and proteins, respectively. Lipases are applied in the degradation of polyester wastes and removal of biofilm build-up in cooling water systems. For instance, the Moraxella bacteria obtained from Antarctic seawater, as well as some yeast, produce lipases, which degrade oils effectively.
8. **Esterases**: Hydrolase enzymes known as esterases catalyse the degradation of substances that include ester, carbamates, pyrethroids, and organophosphates. Some specific genes have been shown to play a role in the biodegradation of specific pesticides including cypermethrin, sulfosulfuron, and fipronil. Isolated esterases from thermophilic microorganisms are of great utility in removing harmful substances as well as other toxic compounds from the environment (Merry 2021).
9. **Nitrilase**: Nitrilases are enzymes useful in modifying surfaces, waste treatment, and carboxylic acid production, among others. They work by breaking up nitrile compounds consisting of a carbon–nitrogen triple bond into the respective carboxylic acid and ammonia. Nitrilases can act upon plant-derived nitriles, including cyanolipids and cyanoglycosides.

   There are two main pathways for the degradation of nitriles: the use of nitrilases, which form carboxylic acids and ammonia, and nitrile hydratases (Gong et al. 2012). Nitrilases exist in a wide variety of organisms, including bacteria, fungi, yeast, and plants. They are particularly useful for the environmentally friendly synthesis of amides and carboxylic acids.

Especially useful are thermophilic, or heat-loving nitrilases, since these function better at high temperatures with quicker reactions and the low chance of contamination. *Bacillus pallidus*, *Geobacillus pallidus*, and *Acidovorax facilis* are some of the most thermophilic nitrilase producers. Particularly useful in bioremediation is a nitrilase from the hyperthermophilic archaeon *Pyrococcus* sp. as it has shown efficiency in cyanide degradation.

10. **Tyrosinases**: These are such industries as conversion of coal, plastic, and resin manufacturers, and refining of petroleum. In their production process, there arises the production of harmful phenolic compounds, which interfere with biological treatments of wastewater. Over the past decades from the 1980s, there have been numerous scientists researching possibilities of using enzymes in wastewater treatment, some of whom used horseradish peroxidase (Brandão et al. 2017). Very recently, the β-tyrosinase, a thermophilic enzyme derived from a bacterium referred to as *Symbiobacterium* sp. SMH-1, was useful in transforming the compound produced in phenolic resin manufacture into one known L-tyrosine. The purification of phenolic wastewater using enzyme-based techniques is effective, affordable, and produces valuable byproducts.
11. **Catalase**: In catalase, hydrogen peroxide is a strong oxidizing agent. The usage of hydrogen peroxide in textile and semiconductor industries involves high concentration in wastewater streams, which is detrimental to the environment. Thus, hydrogen peroxide is normally removed by chemicals before its biological treatment. Hydrogen peroxide can be effectively removed from effluent from the textile industry using the catalase enzyme or microorganisms that produce it.

8.4 Metabolic Pathway of Extremozymes

According to reports, there is a dearth of information in the literature regarding the metabolic processes and the enzymatic system involved in pollutant degradation, and the data that is currently available on the rate of pollutant degradation by extremophiles—which thrive in extremely cold, hot, and pressured environments—is phenomenological. The process is described extensively by which halophilic bacteria found in extremely salinized conditions break down xenobiotics. Chia et al. (2024) talked about the biodegradation pathway of aliphatic and aromatic chemicals by halophilic microorganisms and information on the use of microbial enzymes and catabolic genes involved in pollutant degradation. The Marinobacter strain NCE312, which was isolated from a naphthalene-degrading culture inoculated from creosote-contaminated marine sediment, has been described by earlier studies. Naphthalene 1,2-dioxygenase is like pseudomonas. Because naphthalene is a carbon and energy source, this strain may thrive on it. The activity of a catechol 2,3-dioxygenase causes the meta-ring cleavage of catechol to yield yellow compounds. Catechol

2,3-dioxygenase and naphthalene 1,2-dioxygenase are the enzymes that break down naphthalene.

Pseudomonas, Burkholderia, and Neptunomonas naphthalene 1,2-dioxygenase were linked to this dioxygenase in strain NCE312, according to phylogenetic analyses. The horizontal transmission of genes may be the cause of this. According to people, extremophiles may have a role in the biodegradation of pollutants. Several extremozymes, such as phenol oxidase for the breakdown of phenolic chemicals and phosphotriesterase-like lactonase for the breakdown of organophosphate, were described. Using recombinant DNA technology, which introduces catabolic genes from bacteria with high degradation potential, increasing the mechanisms of degradation by extremophiles is possible.

Previous researchers have genetically modified *P. haloplanktis* TAC125 to increase its potential for producing the aromatic oxidative activity that is encoded by the toluene-o-xylene monooxygenase that was identified from the Pseudomonas sp. bacteria. They concluded that toluene-o-xylene can oxidize a wide range of substances, including dimethylphenols, cresols, and phenol. There is a potential for biodegradation by Geobacter sp. This has been accomplished using acetate enrichment, which has so far shown itself to be a successful method for both uranium removal and growth stimulation. In situ, bioremediation and uranium reduction were enhanced by this acetate enrichment procedure.

In recent decades, the use of extremophilic microbes in industrial processes has increased dramatically. Over 3,000 enzymes from extremophiles were identified and isolated. The majority have been applied to industrial and biotechnological purposes (Ingemansson 2024). The organisms that produce extremozymes are less likely to become contaminated, and they are highly stable. Among the enzymes that fall within the category of polymer-degrading enzymes are xylanases, cellulases, pullulanases, amylases, and proteases. There is more literature suggesting that extremozymes may serve as innovative catalysts for activities of economic importance. Chemical, food, pharmaceutical, paper, pulp, and waste-treatment sectors all depend on these enzymes. Additional significant extremozyme products include polyunsaturated fatty acids, suitable solutes, and cyclodextrins. Extremozymes of marine origin have also been reported to be effective tools for bioremediation applications. There are other sources of pollution, including unintentional oil spills, mining operations for oil extraction, and industrial wastes and toxins. These activities are the source of many hydrocarbons, including pesticides, heavy metals, chlorinated hydrocarbons, and polycyclic aromatic hydrocarbons. Extremozymes are capable of detoxifying and eliminating some of the various sorts of pollutants. Some of the extremozymes and how they are used in the process of biodegradation are discussed below.

8.5 Role of Extremozymes in Bioremediation

Bioremediation is aided by the enzymes that bacteria and fungi produce. The enzymes may degrade a wide range of resistant materials in vivo settings, such as when soil enzymes are absorbed and chelated (Thirumalaivasan et al. 2024). Focusing on the function of extremozymes formed from extremophilic microorganisms and their role in the bioremediation and biodegradation of different hazardous chemicals and pesticides is the goal of this review study.

8.5.1 Extremophilic Microorganisms and Their Classification

Microorganisms known as extremophiles are those that flourish in harsh environmental settings. Their ability to adapt to a variety of environmental circumstances has led to the classification of these extremophilic bacteria (Rekadwad et al. 2023). These microbes include acidophiles, alkaliphiles, endoliths, thermophiles, hyperthermophiles, hypoliths, metallotolerant, oligotrophs, piezophiles, psychrophiles, radioresistant, toxitolerant, and xerophiles. The bacteria that have a pH lower than 3.0 are known as acidophiles.

Microbes known as **alkaliphiles** thrive in alkaline environments (pH 9–11) 26.

The organisms known as **endoliths** reside inside rocks, coral, and animal shells.

To grow, an organism must withstand temperatures between 80 and 122 °C.

In frigid deserts, **hypoliths** are organisms that dwell beneath the rocks.

Organisms that can withstand elevated quantities of harmful metals are known as metallotolerant organisms.

Organisms that thrive in environments with little nutrients are known as **oligotrophs**.

Organisms known as **piezophiles** can withstand high hydrostatic pressure.

Psychrophiles are creatures that thrive in temperatures between −10 and 20 °C. They are primarily found in the Arctic and Antarctic oceans and are microorganisms that like cold environments. The enzymes from psychrophiles are characterized by their limited thermal stability at moderate temperatures and high catalytic activity.

Organisms classified as toxicity-tolerant can withstand elevated quantities of harmful substances.

Organisms that thrive in arid environments are known as **xerophiles**.

Treatment of pollutants by the following

- **Thermophiles**: Thermophiles or heat-loving microorganisms can be applied in the cleaning up of contaminated environments especially heavy metals and industrial wastes through their ability to withstand very high temperatures and attachment to heavy metals to remove them from the affected region. Metals like cobalt and chromium can be eliminated through bacteria like *Anoxybacillus amylolyticus* and *Geobacillus thermantarcticus*. Some thermophiles, such as *Bacillus* species, degrade long-chain hydrocarbons as well as other harmful substances which are present in petroleum-contaminated soils (Wiegel 2014).

Some of the thermophiles can degrade pollutants from oil spills, such as alkanes and other hydrocarbons, just like some strains of Geobacillus. Microorganisms like *Pyrobaculum* species and *Thermus scotoductus* can convert radioactive elements during radioactive waste decontamination, thus making it safe.

Another good invention of thermophiles in the treatment of wastewater involves extremozymes. In industrial sectors like the food and pharmaceutical industries, pollutants like heavy metals and dyes are effectively broken down. Certain thermophiles accelerate and require deeper reactions than traditional methods, which allows them to degrade the plastics like polyethene more effectively.

- **Acidophiles**: Acidophiles are strong cleaning agents on habitats polluting heavy metals toxic to plants and animals like Cu and Cd. Microorganisms utilize bioleaching and bio-oxidation to derive these metals from which *A. ferrooxidans* is the good one with clean-up on large-scale suitability in acidic environment. Even in highly acidic environments, for example, it can be presumed that *Acidocella aromatica* and *Acidiphilium symbioticum* absorb harmful vanadium and cadmium elements.

Furthermore, at low pH, certain lactic acid bacteria may also create huge, stable molecules called S-layer proteins, which are very good at binding and removing heavy metals like cadmium (Cd) and lead (Pb). It has been established that the strains of *Lactobacillus plantarum* and *Lactobacillus kefiri* have a significant capacity for the adsorption of specific metals. Acidophiles are those that can clean industrial waste best, containing metals like lead and zinc. Even more efficient than single strains, mixed cultures like A. ferroxidase and A. trioxide can eliminate the presence of deadly metals harmful for humans. They also convert hydrogen sulphide to innocent forms as solid sulphur for effluent systems in industrial wastewater treatments.

Thermoacidophiles are ideal for industrial wastewater treatment since they tolerate acidic and high temperatures. Thus, they are not affected by the extreme acido-heat required for the degradation of contaminants. For example, *Sulfolobus solfataricus* could be used in degrading one of the common pollutants, phenol, at 80 °C and at pH 3.2.

- **Psychrophiles**: Psychrophiles are the microorganisms that love cold. In fact, they are very good at decomposing those contaminants in cold conditions. They can convert hydrocarbons like jet fuel and crude oil into non-toxic forms. While *Arthrobacter* species help in wastewater treatment by decomposition of synthetic compounds, the species *Pseudoalteromonas* and *Oleispira antarctica* are specifically efficient in cleaning up oil spills in cold climates.

Psychrophiles such as *Oceanospirillales*, *Colwellia* and *Cycloclasticus* use pollutants in the case of oil spills including BTEX chemicals and alkanes. Additionally, in marine settings, complex organic materials are consumed by cold-water microbes including *Rhodobacteraceae* and *Flavobacteria*. Some psychrophilic microbes break down plastics, including *Shewanella*, *Moritella* and *Psychrobacter*. They can form biofilms on plastic waste, another important adaptation that has enabled psychrophiles to thrive at cold temperatures. Biofilms allow for slow material degradation and encourage microbial activity. Such microorganisms would degrade the plastic faster because some microbe strains, such as *Pseudomonas* and *Lysinibacillus*, are very efficient at adhesion to plastic surfaces.

- **Alkaliphiles**: Some microorganisms, known as alkaliphiles (organisms that flourish in high pH environments), can break down harmful industrial pollutants. For example, *Pseudomonas pseudoalcaligenes* utilizes cyanide extracted from the waste of gold mining in its nitrogen source to clean up cyanide-impregnated wastewater (Lee et al. 2022). A specific mechanism of an enzyme is used by *Pseudomonas putida* and *Rhodococcus* to degrade acetonitrile, whereas *Natronocella acetinitrilica* can survive with the presence of high salt levels but eliminates nitriles through the utilization of the substrate for carbon and energy.
- **Halophiles**: The salt-loving microorganisms, known as halophiles, are helpful in the purification of saline surroundings. Microorganisms are more effective in degrading hazardous chemicals because they produce sticky substances, referred to as EPS, which help them bind to surfaces and thus create biofilms. The *Halomonas* strain was observed degrading hydrocarbons at the spill site of Deepwater Horizon (Castillo-Carvajal et al. 2014). Other examples of halophiles include *Vibrio parahaemolyticus* and *Bacillus licheniformis*. They synthesize substances that emulsify oils and help degrade contaminants.

8.6 Utilizing Potential Extremophiles to Bioremediate Xenobiotic Substances

It is well known that microorganisms are superior biocatalysts at removing xenobiotics from polluted environments. However, there are several factors that determine biodegradation, including the hydrophobic character of hydrocarbons, their bioavailability, and the environment. Microorganisms have the intrinsic ability to modify the outer membrane and enhance the hydrophobicity of the cell surface to facilitate the uptake of hydrocarbons (Borgne et al. 2008). As reported, PAH biodegradation is low in high salinity environments, and the rates of biodegradation are slow. Biodegradation of petroleum hydrocarbons is enhanced by the huge amounts of biosurfactant produced by extremophiles, e.g., halophilic bacteria. For the biodegradation of contaminants from different settings, the details of extremophilic bacteria are summarized in Table 8.1 (Fig. 8.4).

Table 8.1 Utilization of extremophiles for bioremediation of pollutants

Sl. No.	Pollutants	Microbial species/strain
1	Crude oil	*Streptomyces albiaxialis* grow at 30% of crude oil, NaCl: 3–10%
2	Benzoic acid, cinnamic acid, and 3-pheylpropionic acid	*Haloferax* sp. strain D1227
3	Uranium	*Shewanella putrefaciens* MR-1
4	Paraffin and other petroleum derivatives	*Dietzia maris*, NaCl: 10%
5	Petroleum hydrocarbons	*Bacterial consortia, Marinobacter* sp., *Erwinia ananas* and *Bacillus* sp., NaCl: 0–22% NaCl
6	4-Hydroxybenzoic acid	*Haloarcula* sp. D1
7	Diesel	*Cellulomonas* spp., *Bacillus marisflavi, Dietzia maris* and *Halomonas eurihalina*, NaCl: 0–17.5%
8	Phenol and catechol	*Cellulomonas* spp., *Bacillus marisflavi, Dietzia maris* and *Halomonas eurihalina*, NaCl: 0–17.5%
9	Benzoic, p-hydroxybenzoic, cinnamic, salicylic, phenylacetic, phenylpropionic, p-coumaric, ferulic and p-aminosalicylic acids	*Halomonas organivorans*
10	Hydrocarbon and organic pollutants	*Marinobacter, Chromohalobacter, Salinococcus*, and *Halobacillus*
11	Ferulic acid	*Halomonas elongate*
12	Acetonitrile and propionitrile	*Natronocella acetinitrilica*
13	Crude oil	*Ochrobactrum* sp.
14	Fuel oil	*Acinetobacter* strain, culture collection (CC): JSC Biocentras, Lithuania, NaCl: 35%
15	Crude oil	*Fusarium* sp., NaCl: 35%
16	Crude oil	*Phanerochaete chrysosporium*, Iranian Research Organization for Science and Technology (IROST)

A recent study investigated the population of microorganisms using a 16S rRNA gene from geyser-discharge water effluent, finding that the most prolific species of bacteria that contributed to the degradation of the environment through the reduction of arsenite were *Deinococcus-Thermus, Chloroflexi*, and *Chlorobi*. Other related species of bacteria, such as *Geobacillus*, may also be involved in degrading organic contaminants. At 60 °C, other members of the same group, such as *Geobacillus thermoleovorans* T80, degraded about 70% of the hexadecane. The hyperthermophilic archaea *Pyrococcus* sp. M24D13 was also isolated and studied by Dennett and

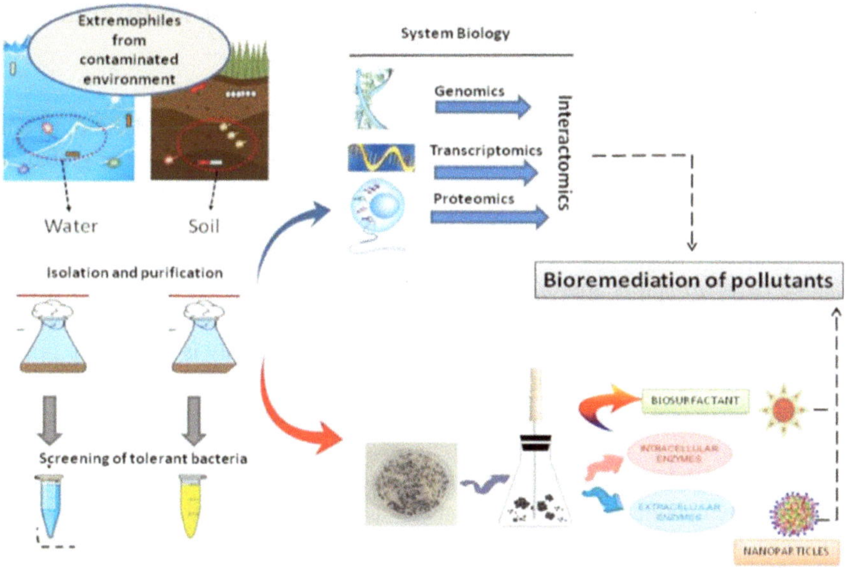

Fig. 8.4 Extremophile isolation and its utilization in bioremediation

Blamey as a novel thermostable cyanide-degrading nitrilase. The treatment of waste streams polluted with cyanide may be a potential application for this thermostable enzyme (Table 8.2).

Table 8.2 Exploitation of extremozymes for biodegradation

Sl. No.	Microbes	Enzymes	Target pollutants
1	*Bacillus safensis* (CFA-060	Oxidoreductase	Degradation of aromatic compounds
2	*Thermus thermophilus* HB27	Laccase	Polyaromatic hydrocarbons
3	*Geobacillus stearothermophilus*	Phosphotriesterase-like lactonase	Organophosphate
4	*Nocardioides* sp. strain KP7	Dioxygenase	Polyaromatic compounds
5	*Alcanivorax borkumensis* SK2T *Bacillus licheniformis* ATCC 14580T *Halomonas ventusa* A112T	Alkane hydroxylases/ cytochrome P450	Alkane
6	*Pseudomonas stutzeri* DEH130 *Alcanivorax dieselolei* strain B-5	Haloalkane dehalogenases	Halogenated pollutants

8.7 Impact of Extremozymes on the Removal of Pollutants for Industrial Wastewater Treatment

Only 1% of the water on Earth is now fit for human consumption, despite it being one of the most easily accessible natural resources. With the increase in globalization and industrialization, output wastewater has grown everywhere. In the twenty-first century, the world water usage has increased, thereby requiring wastewater management. Wastewater needs to receive several biological treatments before it can be recycled and is appropriate for industrial, drinking, or washing use (Mehta 2024). Recently, enzymes have been employed as a practical method to enhance the breakdown of pollutants. Just 1% of the world's water supply is now fit for human consumption, despite it being one of the most abundant natural resources.

Water treatment companies around the world are very concerned about wastewater treatment. Extremophilic microorganisms produce extremozymes, which may flourish in harsh environments with strong stereo-selectivity, long half-lives, solvent tolerance, temperature stability, and substrate affinity (Bidwai 2024). They have solvent-tolerant and heat-stable biocatalytic qualities in addition to being environmentally friendly. Extremozymes have recently been announced as a biological source for treating wastewater because of their rapid rate of reproduction.

Extremophiles and their extreme abilities: Extremophiles are microorganisms that have the capacity to perish in hostile environments. Depending on where they live and function, extremophiles can be found in various locations, including hot springs, high-temperature thermal vents, cold-weather Arctic and Antarctic regions, acidic lakes, and high-saline lakes. High temperatures (>80 to >110 °C) are ideal for thermophiles, low temperatures (−2 to 20 °C) are ideal for psychrophiles, alkaliphiles (pH > 9), and low temperatures (−2 to 20 °C) are ideal for halophiles (high salinity, such as 2–5 M NaCl). In a variety of sectors, extremophiles were employed as a practical substitute source to eliminate pollutants because of their affordability, environmental friendliness, and capacity to produce less sludge under harsh conditions.

Extremozymes in bioremediation of different pollutants: Numerous enzyme-driven bioremediation techniques have been investigated as an efficient way to clean contaminated wastewater since they are inexpensive, time-efficient, and labour-efficient. Because extremophiles are resistant to organic solvents and high temperatures, their hydrolase activity is often utilized in industries. So, they can be employed in bioremediation and waste treatments due to their eco-friendly character (Karigar 2011). Extremozymes that are useful for treating wastewater include oxidoreductases and hydrolytic enzymes. The treatment of industrial effluents depends on these extremozymes. Among many other contaminants, they may break down dyes, insecticides, polyhydroxyalkanoates, plastics, and heavy metals. Numerous industries, including the biofuel, detergent, culinary, pharmaceutical, agricultural, bioenergy, textile, and cosmetic sectors, use these extremozymes (Fig. 8.5).

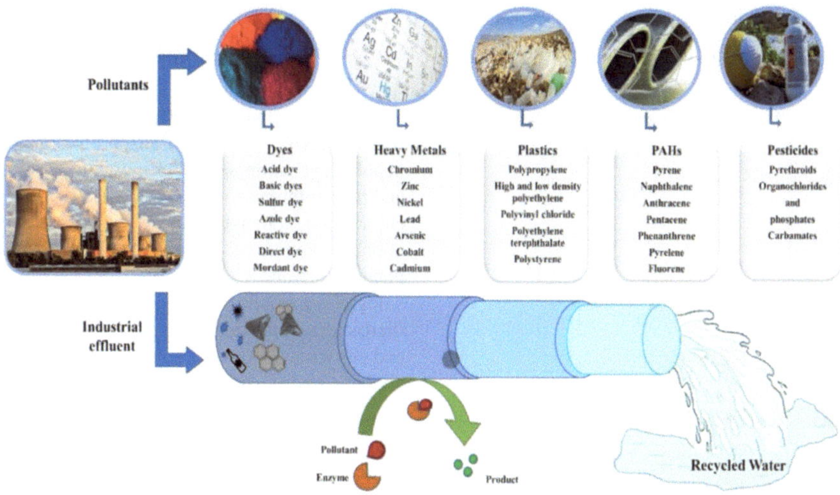

Fig. 8.5 Schematic diagram of wastewater having toxic industrial pollutants and their treatment using extremozymes produced from extremophiles

Adaptation of extremozymes: Extremophilic microbe proteins, or extremozymes, have been modified at the molecular and biochemical levels to tolerate extreme industrial environments. A recent study on extremophilic enzymes has helped to clarify the kinds of general patterns that affect the stability of extremophilic enzymes, even if no one factor has been found to create extremozyme stability in the harsh settings in which they must exist.

Structure, flexibility, charge, and/or hydrophobicity of extremozymes are affected by alterations in their amino acid sequence, leading to alterations in the enzymes themselves. Extremophilic proteins have several ways of adapting to similar extreme physical or chemical conditions. The adaptations which extremophiles have developed to survive difficult environments using enzymes located further down (Table 8.3).

8.8 Bioremediation of Heavy Metals by Enzymes

Heavy metals can be bioremediated in harsh environments by the special class of prokaryotes called extremophilic bacteria that have adapted to have special physiological and enzymatic adaptations. These can remove the pollutants and reduce the costs of cooling and complicated machinery in industrial processes due to their high tolerance to toxicity, very fast growth, and well-defined mechanisms (Yan et al. 2020). These microorganisms are capable of remediating sites with extreme contamination from heavy metals by rendering bioproduction easy while reducing energy

Table 8.3 Mode of action

Sl. No.	Class of enzyme	Type of enzyme	Substrate/pollutants	Mode of action
1	Oxidoreductase	Oxygenase	Aromatic compounds such as PAHs, naphthalene, anthracene and phenanthrene, chlorinated biphenyls, aliphatic olefins, phenols, organic halogen compounds (herbicides, fungicides, and pesticides), aliphatic hydrocarbons	Coenzymes used to transport to organic or inorganic molecules
1.1		Monooxygenases (flavin and P450 monooxygenases)	Degrade chlorinated pesticides, hydrocarbons, xenobiotics, and heme prosthetic group	Hydrolases are enzymes that the sterone wyler arom lo electrons of the coenzymes NADH or NADPH decrease the other oxygen to water
2		Peroxidase	Phenols, cresols, chlorinated phenolic compounds, polyphenols, amines, and heterocyclic amine polyamines, as well as nitroaromatic compounds such as TNT (2,4,6-trinitrotoluene), lignin, and dyes	Peroxide
3		Laccase	PHAs, paints, plastics, dyes, estrogenic compounds, paper, and cellulose acenaphthene, anthracene, fluorene, and benzo[a]pyrene	
3.1	Hydrolase	Amylase	Paper, food industries, textile industries, and fossil and fuel products; degradation of petroleum-derived compounds such as polyethylene; *n*-alkane; paraffin hydrocarbons	Starch and carbohydrate components, catalyse the hydrolysis of 1,4–D-glycosidic linkages

(continued)

Table 8.3 (continued)

Sl. No.	Class of enzyme	Type of enzyme	Substrate/pollutants	Mode of action
3.2		Proteases	Phenols and proteins	Hydrolysis of peptides bonds in presence of water molecules
3.3		Lipases	Oil residues, petroleum contaminants, effluents, and soil recovery, kitchen waste	Hydrolysis, esterification, alcoholysis, acids, esters and glycerides

and carbon requirements. This review will focus on the application of extremophiles for environmentally friendly bioremediation processes.

Heavy metal removal techniques such as reverse osmosis, ion exchange, and membrane filtration are said to be effective but expensive. They result in heavy metal sludge at the end of the process. Instead, there exists biosorption, which uses microorganisms to absorb heavy metals (HMs), provides an economical and environmentally sustainable substitute. Unfortunately, most of these microbes will not survive under such harsh conditions.

8.8.1 Heavy Metals and Their Environmental Impact

Heavy metal contamination is a major problem because heavy metal ions cannot biodegrade, which causes them to circulate in the environment. Therefore, they could be hazardous for at least 200 years in the environment. These substances that pollute the soil may prevent soil microbes from growing. Apart from disrupting the organic matter degradation and transformation processes, they also interfere with microorganisms' physiological functioning (Selvi et al. 2019). The disruption of organic matter decomposition by bacteria may lead to an increase in the bioavailable forms of metals in the soil pool. Soil metal composition is a significant component of the parameters influencing the mobility and toxicity of heavy metals in the environment. The physicochemical properties of heavy metals are comparable to those of physically active metals, which can cause harm through the production of free radicals, interfere with enzyme processes, and interfere with the function of biological systems' subcellular structures. For example, metal ions readily bind to functional groups in the cell cytoplasm, such as –SH, –OH, and –NH, causing protein molecules to distort and eventually die.

Because of their transmission up the food chain and bioaccumulation—the process by which the last link in the chain, humans, acquires the highest levels of a substance—heavy metals represent the biggest hazard (Fig. 8.1). Foods with significant plant origins are the major pathway for heavy metal intake by human health.

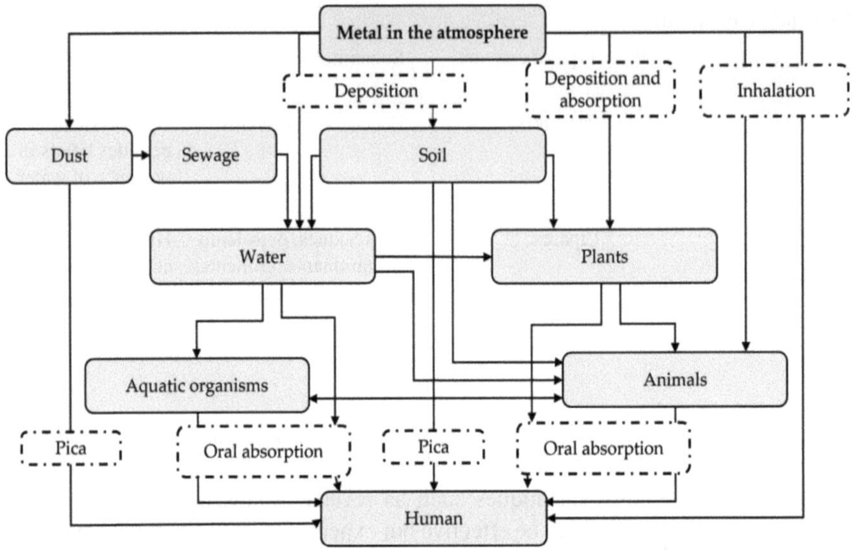

Fig. 8.6 An exposure to the environment and the accumulation of metals in the human body, using lead as an example

Thus, heavy metal concentration in crops destined for animal feed and human direct consumption must be minimal (Fig. 8.6).

Heavy metals are both acutely and chronically toxic to humans. The great majority of heavy metal poisonings do not immediately show signs and symptoms. Signs and symptoms take months to many years to develop in a large number of poisonings. The range of toxicities of heavy metals is extremely wide. In cases where the mechanisms of toxicities are similar, there is a tendency for different individual heavy metals to cause poisonings of specific tissues or organs. Lead mainly accumulates in bone tissue, cadmium in liver and kidney cortical issue, and mercury in brain tissue as methylmercury compounds that can induce severe neurological diseases. More than this, the IARC classified some heavy metals, and their compounds classified as carcinogenic to humans, namely as confirmed and suspected. Since decades ago, Cd and Pb were the most hazardous elements concerning environmental issues. As, Cr, Hg, and Zn follow. The IARC report puts the inorganic Pb compounds into Group 2A which comprises substances that probably cause carcinogenic effects to human beings. On the other hand, Cd falls in Group 1 which categorizes the substances as known carcinogens to humans.

8.8.2 Mechanism of Heavy Metal Bioremediation

1. **Biosorption**

Biosorption is the physicochemical uptake of heavy metals independent of metabolism based on cell membranes. It operates by the use of negatively charged substances found in cell membranes. Importantly, biosorption typically uses non-living biomass since it accelerates the process as compared to using living microorganisms. This tactic depends on several factors, including surface characteristics like the presence of functional groups on the cell membrane, pH, temperature, or electrostatic interactions, among others. Optimization of such a process requires insights into biosorption mechanisms that favour removal of heavy metals (Mishra 2014). There are now identified to be several mechanisms that occur during the course of sorption, and many may occur concurrently at different rates. The following mechanisms can be distinguished among the biosorption mechanisms: (i) complexation, in which heavy metal ions attach to functional groups found in cell membranes; (ii) ion change, a reversible chemical reaction in which ions are exchanged for other ions of the same charge; and (iii) physical adsorption brought on by intermolecular interactions, such as Van der Waals forces (Fig. 8.7).

2. **Bioaccumulation**

Bioaccumulation is a biological energy-dependent process performed by active metabolic bacteria, unlike the biosorption mentioned above (Fig. 8.7). Heavy metal uptake, being dependent on biochemical characteristics, microbial internal structure, genetic and physiological capabilities, and environmental factors influencing bioaccumulation activity, is longer than biosorption. It was discovered that variations in charge from the characteristics of the cell surface also affect the process. Temperature has an effect as well on the process: a higher temperature can quite severely

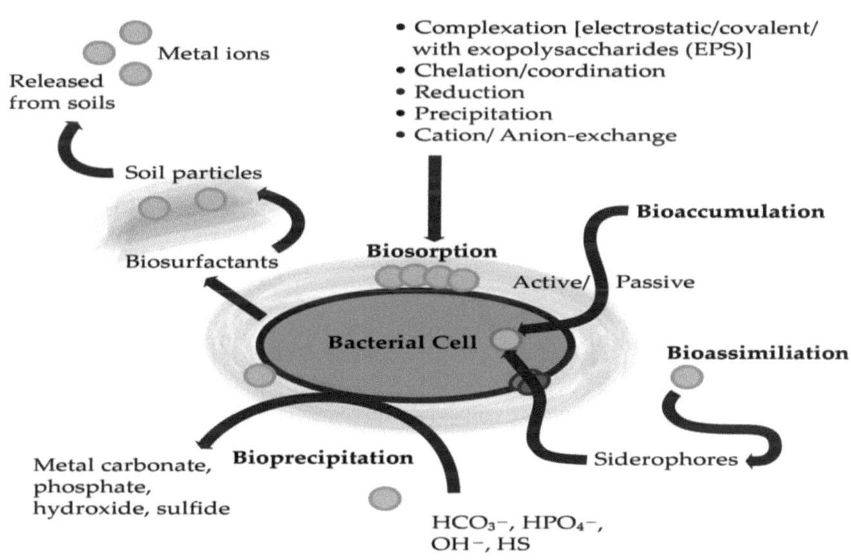

Fig. 8.7 Mechanism of heavy metal bioremediation

affect the metabolic activities of a bacterial cell. The most well-known bioaccumulation mechanism most probably uses metallothioneins to bind heavy metals. The bmtA gene encodes metallothioneins, which are low molecular weight proteins rich in cysteines that help to bioaccumulate heavy metals within cells, such as lead, mercury, nickel, and cadmium. A rise in metal exposure would normally trigger bacterial cells to produce metallothioneins. Plasmids would carry this mechanism, and this would make it easy for germs to be transduced from one cell into another. There are, however, other bioaccumulation pathways that are frequently not universal in order to bioremediate all heavy metals.

3. **Bioprecipitation**

Another bioremediation technique that is found in bacteria is the bioprecipitation. The method reduces free metal toxicity and bioavailability by transforming their concentration to insoluble complexes. Microbial catalysing of both oxidative and reductive reactions can lead to the precipitating contaminants such as Pb, Cd, Cr, Fe, and U. While some of the bacteria can precipitate hydroxides or carbonates by producing alkanes, some microorganisms were also found to release phosphates and promote the precipitation of metal phosphates (Fig. 8.8).

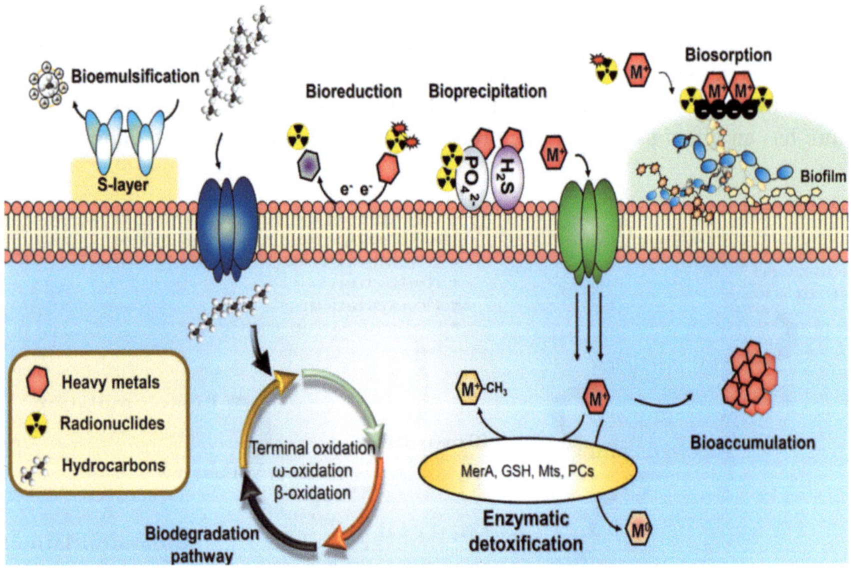

Fig. 8.8 Current methods for removing various harmful contaminants using microbial bioremediation

8.9 Discussion

Extremozymes have far-reaching potential beyond the degradation of pollutants; they also participate in the recovery of valuable metals for industrial purposes, saving environmental and economic costs. Further, genetic engineering developments also lead to increased enzyme activity and the productivity of microbial strains in order to offer tailored solutions for extreme conditions.

Despite all these promises, scalability, cost, and environmental impact are the challenges with large-scale applications. Further research in the optimization of enzymes and biotechnological interventions are important to exploit the full potential of extremozymes for sustainable environmental remediation. Extremozymes have far-reaching potential beyond the degradation of pollutants; they also participate in the recovery of valuable metals for industrial purposes, saving environmental and economic costs. Further, genetic engineering developments also lead to increased enzyme activity and the productivity of microbial strains in order to offer tailored solutions for extreme conditions.

8.10 Conclusion

The comprehensive exploration of enzymatic degradation of pollutants also highlights the critical role of extremozymes derived from extremophilic microorganisms. These enzymes exhibit exceptional stability and high catalytic efficiency under extreme conditions, enabling the degradation of a vast array of pollutants, including hydrocarbons, pesticides, plastics, heavy metals, and pharmaceutical residues. To create bioremediation techniques that are appropriate for a range of environmental conditions, including extreme salinity, pH, and temperatures, extremophiles—such as thermophiles, psychrophiles, acidophiles, and halophiles—have been used. A sustainable approach to pollution control has been found through the processes of enzymatic hydrolysis, oxidation–reduction reactions, and bioaccumulation, which convert hazardous chemicals into innocuous or less dangerous metabolites. Advances in genetic engineering have not only further optimized the efficiency of extremozymes but also their specificity and productivity. Their application ranges from wastewater treatment to soil remediation and heavy metal detoxification. Though various challenges are associated with large-scale implementation, the potential of extremozymes in revolutionizing industrial and environmental processes is an unparalleled combination of innovation, ecological sensitivity, and scientific precision in fighting global pollution. Advancement of omics technologies along with the advance of recombinant DNA would pave the way for novel routes to break pollutants and further strengthen the relevance of extremozymes to establish a sustainable world.

References

Admin-Infection (2023, December 7) Understanding the causes of water infections—a comprehensive guide. Infection Cycle. https://infectioncycle.com/articles/understanding-the-causes-of-water-infections-a-comprehensive-guide

Arora PK, Kumar M, Chauhan A, Raghava GP, Jain RK (2009) OxDBase: a database of oxygenases involved in biodegradation. BMC Res Notes 2(1):67. https://doi.org/10.1186/1756-0500-2-67

Bala S, Garg D, Thirumalesh BV, Sharma M, Sridhar K, Inbaraj BS, Tripathi M (2022) Recent strategies for bioremediation of emerging pollutants: a review for a green and sustainable environment. Toxics 10(8):484. https://doi.org/10.3390/toxics1008048

Bidwai S (2024, August 20) Water and wastewater treatment market size, report by 2034. https://www.precedenceresearch.com/water-and-wastewater-treatment-market

Borgne SL, Paniagua D, Vazquez-Duhalt R (2008) Biodegradation of organic pollutants by halophilic bacteria and archaea. Microb Physiol 15(2–3):74–92. https://doi.org/10.1159/000121323

Brandão YB, De Oliveira JGC, Benachour M (2017) Phenolic wastewaters: definition, sources and treatment processes. InTech eBooks. https://doi.org/10.5772/66366

Castillo-Carvajal LC, Sanz-Martín JL, Barragán-Huerta BE (2014) Biodegradation of organic pollutants in saline wastewater by halophilic microorganisms: a review. Environ Sci Pollut Res 21(16):9578–9588. https://doi.org/10.1007/s11356-014-3036-z

Chia XK, Hadibarata T, Jusoh MNH, Sutiknowati LI, Tan IS, Foo HCY (2024) Role of extremophiles in biodegradation of emerging pollutants. Top Catal. https://doi.org/10.1007/s11244-024-01919-7

Da Fonseca FS, Angolini CF, Arruda MaZ, AL C, Junior Santos CA, Saraiva AM, Pilau E, Souza AP, Laborda PR, De Oliveira PF, De Oliveira VM, De Assis MRF, Marsaioli AJ (2015) Identification of oxidoreductases from the petroleum Bacillus safensis strain. Biotechnol Rep 8:152–159. https://doi.org/10.1016/j.btre.2015.09.001

ExclusivePapers.com (2014, June 22) Pollution air soil and water. https://exclusivepapers.com/essays/environment/pollution-air-soil-and-water.php

Gong J, Lu Z, Li H, Shi J, Zhou Z, Xu Z (2012) Nitrilases in nitrile biocatalysis: recent progress and forthcoming research. Microb Cell Factories 11(1). https://doi.org/10.1186/1475-2859-11-142

Ground-level Ozone Basics | US EPA (2024, May 14) US EPA. https://www.epa.gov/ground-level-ozone-pollution/ground-level-ozone-basics

Iacopino S, Licausi F (2020) The contribution of plant dioxygenases to hypoxia signaling. Front Plant Sci 11. https://doi.org/10.3389/fpls.2020.01008

Impact of extremozymes on the removal of pollutants for industrial wastewater treatment (2022) Bulg Chem Commun 54(C):35–45. https://doi.org/10.34049/bcc.54.C.0035

Ingemansson KT (2024) Extremophile microorganisms as cell factories: support from the European Union. www.academia.edu; https://www.academia.edu/118419216/Extremophile_microorganisms_as_cell_factories_support_from_the_European_Union

Janusz G, Pawlik A, Świderska-Burek U, Polak J, Sulej J, Jarosz-Wilkołazka A, Paszczyński A (2020) Laccase properties, physiological functions, and evolution. Int J Mol Sci 21(3):966. https://doi.org/10.3390/ijms21030966

Jeong S, Choi YJ (2020) Extremophilic microorganisms for the treatment of toxic pollutants in the environment. Molecules 25(21):4916. https://doi.org/10.3390/molecules25214916

Karigar C (2011) Role of microbial enzymes in the bioremediation of pollutants: a review. Tumkur University. https://www.academia.edu/945964/Role_of_Microbial_Enzymes_in_the_Bioremediation_of_Pollutants_A_Review

Karimi M, Biria D (2019) The promiscuous activity of alpha-amylase in biodegradation of low-density polyethylene in a polymer-starch blend. Sci Rep 9(1). https://doi.org/10.1038/s41598-019-39366-0

Lee H, Bae J, Jin S, Kang S, Cho B (2022) Engineering acetogenic bacteria for efficient one-carbon utilization. Front Microbiol 13. https://doi.org/10.3389/fmicb.2022.865168

Manisalidis I, Stavropoulou E, Stavropoulos A, Bezirtzoglou E (2020) Environmental and health impacts of air pollution: a review. Front Public Health 8. https://doi.org/10.3389/fpubh.2020.00014

Mehta R (2024, July 12) What is an effluent treatment plant (ETP) and its uses, advantages and workflow. Pure Bact. https://purebact.com/effluent-treatment-plant/

Merry (2021, October 25) Question: where are transferases found. Livelaptopspec. https://www.livelaptopspec.com/where-are-transferases-found/

Mishra V (2014) Biosorption of zinc ion: a deep comprehension. Appl Water Sci 4(4):311–332. https://doi.org/10.1007/s13201-013-0150-x

Olaya-Abril A, Biełło K, Rodríguez-Caballero G, Cabello P, Sáez LP, Moreno-Vivián C, Luque-Almagro VM, Roldán MD (2024) Bacterial tolerance and detoxification of cyanide, arsenic and heavy metals: holistic approaches applied to bioremediation of industrial complex wastes. Microb Biotechnol 17(1). https://doi.org/10.1111/1751-7915.14399

Pinto ÉSM, Dorn M, Feltes BC (2020) The tale of a versatile enzyme: alpha-amylase evolution, structure, and potential biotechnological applications for the bioremediation of n-alkanes. Chemosphere 250:126202. https://doi.org/10.1016/j.chemosphere.2020.126202

Rekadwad BN, Li W, Gonzalez JM, Devasya RP, Bhagwath AA, Urana R, Parwez K (2023) Extremophiles: the species that evolve and survive under hostile conditions. 3 Biotech 13(9). https://doi.org/10.1007/s13205-023-03733-6

Riis V, Kleinsteuber S, Babel W (2003) Influence of high salinities on the degradation of diesel fuel by bacterial consortia. Can J Microbiol 49(11):713–721. https://doi.org/10.1139/w03-083

Selvi A, Rajasekar A, Theerthagiri J, Ananthaselvam A, Sathishkumar K, Madhavan J, Rahman PKSM (2019) Integrated remediation processes toward heavy metal removal/recovery from various environments—a review. Front Environ Sci 7. https://doi.org/10.3389/fenvs.2019.00066

Shah MP (2024a) Microbial remediation of hazardous chemicals from water & industrial wastewater treatment plant. Springer Nature

Shah MP (2024b) Development in waste water treatment research and processes: applied technologies for clean up of environmental contaminants. Elsevier

Shah MP (2024c) Aerobic and anaerobic microbial treatment of industrial wastewater. CRC Press

Shah MP (2025) Extraction of precious metals from industrial waste using microbial technology. Springer Nature

Sharma K (2023a, July 17) Soil pollution: causes, effects, preventive measures. Science Info. https://scienceinfo.com/soil-pollution-causes-effects/

Sharma K (2023b, July 17) Soil pollution: causes, effects, preventive measures. Science Info. https://scienceinfo.com/soil-pollution-causes-effects/

Shetty SS, Deepthi D, Harshitha S, Sonkusare S, Naik PB, Madhyastha H (2023) Environmental pollutants and their effects on human health. Heliyon 9(9):e19496. https://doi.org/10.1016/j.heliyon.2023.e19496

Shukla AK, Singh AK (2020) Exploitation of potential extremophiles for bioremediation of xenobiotics compounds: a biotechnological approach. Curr Genomics 21(3):161–167. https://doi.org/10.2174/1389202921999200422122253

SilentShadow (2019, May 5) The Chemistry and Science behind Carbon Monoxide Poisoning. SilentShadow. https://www.silentshadow.org/chemistry-and-science-behind-carbon-monoxide-poisoning.html

Thirumalaivasan N, Gnanasekaran L, Kumar S, Durvasulu R, Sundaram T, Rajendran S, Nangan S, Kanagaraj K (2024) Utilization of fungal and bacterial bioremediation techniques for the treatment of toxic waste and biowaste. Front Mater 11. https://doi.org/10.3389/fmats.2024.1416445

Tõlgo M, Hegnar OA, Østby H, Várnai A, Vilaplana F, Eijsink VGH, Olsson L (2022) Comparison of six lytic polysaccharide monooxygenases from Thermothielavioides terrestris shows that functional variation underlies the multiplicity of LPMO genes in filamentous fungi. Appl Environ Microbiol 88(6). https://doi.org/10.1128/aem.00096-22

Wiegel J (2014) The importance of thermophilic bacteria in biotechnology. Uga. https://www.academia.edu/3096041/The_Importance_of_Thermophilic_Bacteria_in_Biotechnology
Xu A, Zhang X, Wu S, Xu N, Huang Y, Yan X, Zhou J, Cui Z, Dong W (2021) Pollutant degrading enzyme: catalytic mechanisms and their expanded applications. Molecules 26(16):4751. https://doi.org/10.3390/molecules26164751
Yan A, Wang Y, Tan SN, Yusof MLM, Ghosh S, Chen Z (2020) Phytoremediation: a promising approach for revegetation of heavy metal-polluted land. Front Plant Sci 11. https://doi.org/10.3389/fpls.2020.00359

Chapter 9
Source, Impact, and Remediation Strategies to Eliminating Endocrine Disrupters and the Potential of Nanobiotechnology for a Sustainable Solution

Janki Trivedi, Ashish Bhatt, Chaitali Patel, Viral Surti, Dhruv Kapdi, and Rishit Soni

Abstract The well-being of the environment is significantly impacted by endocrine disruptors, which are linked to various diseases including cancer, cardiovascular problems, behavioural disorders, autoimmune disorders, and reproductive issues. These disruptors stem from diverse sources such as pharmaceuticals, plasticizers, agricultural runoff, and stormwater infiltration, which arises concern regarding the whole ecosystem, so the combined effects of these disruptors are gaining attention. However, conventional treatment methods often fall short in completely removing them from water bodies, despite some evidence of their biodegradation in treated drinking water. Addressing this challenge requires modifications to existing remediation techniques, which currently face obstacles like high energy consumption and health risks. Therefore, employing advanced methods like carbon-based adsorption, membrane technology, nanostructured photocatalysts, and microbial and enzymatic technologies are essential. Concerns regarding the environmental and health impacts of these contaminants persist due to the absence of stringent regulations on their concentration limits in various water sources. This chapter highlights the fragmented nature of available data on the sources, effects, remediation technologies, and mechanisms of endocrine disruptor removal. It underscores the necessity for further research to better understand and mitigate the risks associated with their presence in the environment. Additionally, comprehensive data is provided on the biodegradation of these disruptors by various microbes and their extracellular enzymes.

Keywords Endocrine disrupters · Bioremediation · Nanobiocatalysis · EDC remediation · Biodegradation · Xenoestrogen

---

J. Trivedi (✉) · A. Bhatt · C. Patel · V. Surti · D. Kapdi
Shri Alpesh N. Patel Post Graduate Institute of Science & Research, Anand, Gujarat, India
e-mail: jankitrivedi6@gmail.com

R. Soni
Shri P. M. Patel Institute of Biosciences, Anand, Gujarat, India

© The Author(s), under exclusive license to Springer Nature Singapore Pte Ltd. 2026
S. Dey and M. P. Shah (eds.), *Microbial Applications in Industrial Wastewater Treatment*, Advances in Wastewater Research,
https://doi.org/10.1007/978-981-96-9128-9_9

9.1 Introduction

In recent decades, numerous materials and compounds have been produced and released into the environment without sufficient understanding of their effects on human health. This has led to ongoing pollution from a wide range of hazardous chemicals. Unmonitored and unreported, these contaminants have reached groundwater level and started to affect the aquatic as well as the terrestrial living organisms by phenomena such as biomagnification and bioaccumulation. These contaminants are often categorized as the emerging contaminants (Awewomom et al. 2024). Emerging contaminants have garnered significant attention due to their widespread environmental presence and potent bioactivity, even at trace concentrations. The incessant discharge of toxic chemicals due to urbanization has become a major global concern, significantly contributing to the heightened alarm over environmental and public health risks (Sun et al. 2024). Endocrine-disrupting chemicals (EDCs) have emerged as a highly debated group of contaminants, now at the forefront of environmental and health discussions.

Endocrine-Disrupting Chemicals (EDCs), often referred to as Endocrine Disruptors (EDs), are exogenous substances that interfere with the endocrine system, disrupting hormonal balance and affecting multiple physiological processes. These include a wide range of industrial chemicals, pesticides, pharmaceuticals, and plasticizers such as Bisphenol A (BPA), polychlorinated biphenyls (PCBs), and phthalates. In recent decades, the production of EDCs has surged globally. BPA, a chemical predominantly used in plastics, is produced at a rate exceeding 7.7 million tons annually. Similarly, phthalates, essential for manufacturing PVC products, are produced in quantities surpassing 8 million tons per year (Suk 2022).

This massive production underscores their economic significance but also raises concerns about their environmental and health impacts (Naughton et al. 2023).

The economic impact of these chemicals is considerable. The BPA market was valued at approximately USD 22 billion in 2021, driven by demand from the construction and automotive sectors (Grand View Research 2023). Phthalates also have a substantial market, with a value of over USD 10 billion in 2020, reflecting their use in medical devices and construction (Markets and Markets 2022). Growing awareness of health risks related to EDs is expected to influence market dynamics and encourage the development of safer alternatives. Production and consumption of EDCs are concentrated in developed regions such as North America, Europe, and parts of Asia, with the United States and China leading in BPA production. However, as industrial activities expand into developing countries, the use of these chemicals is increasing, posing challenges for global environmental and health initiatives (Suk 2022; Naughton et al. 2023).

Globally, the environmental burden of EDCs has been aggravated by their persistent and bioaccumulative nature, particularly in aquatic ecosystems where they concentrate in water bodies and sediments. Studies have revealed widespread contamination of lakes, rivers and oceans with EDCs, posing significant risks to aquatic life and, by extension, to human populations relying on these resources. The global ramifications of EDC exposure are profound, with evidence suggesting links to declining fertility rates, increasing incidences of hormone-dependent cancers, and disruptions in wildlife populations (Gore et al. 2015; Diamanti-Kandarakis et al. 2020).

The molecular mechanisms underlying EDC toxicity are complex, involving the mimicry or antagonism of endogenous hormones, disruption of hormone receptor binding, and interference with the synthesis, secretion, transport, and metabolism of hormones. These actions can trigger a cascade of dysregulated biological processes, potentially leading to adverse health outcomes such as reproductive disorders, carcinogenesis, metabolic syndromes, and neurodevelopmental deficits. For instance, BPA, a well-documented EDC, has been linked to estrogenic and anti-androgenic activities, contributing to disorders such as breast cancer, reduced sperm quality, and early puberty onset (Vandenberg et al. 2019; Rochester et al. 2021).

EDs pose a significant risk to the environment as well as human health, and their remediation is crucial for the protection of the ecosystem. There are many approaches employed for their bioremediation which majorly include physical and chemical removal of this contaminant. Although bioremediation technologies offer the advantage of being more environmentally friendly, efficient, and sustainable for the long term, however, the behaviour of degrading microbes in different environments and how microbial diversity affects contaminant removal are not well understood.

Extremophiles are remarkable organisms that thrive in extreme conditions, such as high temperatures, salinity, acidity, and pressure. Their unique adaptations not only allow them to survive in harsh environments but also equip them with specialized metabolic pathways that can be harnessed for bioremediation, particularly in tackling endocrine-disrupting chemicals (EDCs) (Rampelotto 2024).

EDCs, which include a variety of synthetic and natural compounds that can interfere with hormonal functions, pose significant risks to both human health and ecosystems. Traditional remediation methods often fall short due to the complex nature of these pollutants (Gao et al. 2020a, b).

For instance, strains of *Deinococcus* are known for their exceptional resistance to radiation and their ability to degrade a range of pollutants (Jin et al. 2019). Similarly, *Thermus* species, which thrive in hot springs, exhibit enzymes that can break down organic contaminants at elevated temperatures, enhancing degradation rates (Wang et al. 2021a, b). The genus *Dehalococcoides* has garnered attention for its ability to dechlorinate polychlorinated biphenyls (PCBs) and dioxins—two notorious EDCs—through specialized enzymatic reactions (Chen and He 2018). By harnessing extremophiles, researchers are exploring innovative strategies for more effective environmental cleanup efforts targeting EDCs.

As more research emerges about microbial pathways and diversity, better remediation technologies can be developed and improve predictions of contaminant behaviour in specific environments (Lopez-Rodriguez et al. 2021). Advanced bioremediation techniques are being used to remove EDCs and restore ecological balance by implementing the latest OMIC technology (Genomics, Metagenomics, Proteomics, transcriptomics, and metabolomics) including bioinformatics. Genetically engineered microbes (GEMs), also known as GMOs, along with computational methods like silico models and enzyme-targeting through metabolic or protein engineering are proving to be effective strategies for reducing the impact of EDCs. These approaches offer promising solutions for tackling these harmful contaminants (Sun et al. 2024). Apart from bioremediation, nanotechnology is also an emerging and promising approach for the removal of EDs from various sources considering their unique physicochemical properties.

9.2 Hazardous Effects of Endocrine Disruptors (EDs)

In animals, endocrine disruptors have been linked to a variety of reproductive issues. For instance, research has shown that exposure to certain chemicals can lead to feminization of male amphibians, affecting their ability to reproduce and survive. Additionally, developmental abnormalities have been observed in fish and birds, leading to impaired growth and behavioural changes. These disruptions not only threaten individual species but can also have cascading effects on entire ecosystems, underscoring the broader environmental implications of endocrine disruptors.

9.2.1 Effect on Animals

Endocrine disruptors can cause abnormalities in reproductive organs, reduced fertility, and altered sex ratios. For example, exposure to certain chemicals has been linked to feminization of males in some wildlife species. Studies have shown that BPA, commonly found in plastics, can cause reproductive abnormalities in wildlife. For instance, research by Richter et al. (2007) demonstrated that exposure to BPA led to reduced fertility and altered sexual differentiation in rodents. Polychlorinated Biphenyls (PCBs), once widely used in industrial applications, have been shown to affect reproductive health in birds. A study by Hanley and Doucet (2012) found that PCBs led to decreased egg production and hatchability in gulls.

These chemicals can also affect development, leading to malformations, impaired growth, and behavioural changes. Amphibians and fish are particularly vulnerable to developmental disruptions. Research by Hayes et al. (2016) found that endocrine disruptors like atrazine, a commonly used herbicide, can cause amphibians to develop abnormal sexual characteristics, such as hermaphroditism. In a study by Arcand-Hoy and Benson (1998), fish exposed to endocrine disruptors showed developmental abnormalities including altered growth rates and changes in reproductive behaviour.

Continuous exposure of cytotoxic Endocrine Disruptors may lead to destructions of immune system cells. ED like dioxins directly affect and impair the immune system, making animals more susceptible to diseases and infections.

Endocrine disruptors can alter behaviours critical for survival and reproduction, such as mating rituals and foraging behaviours. Research by Windsor et al. (2017) documented changes in foraging behaviour and social interactions in fish exposed to endocrine-disrupting chemicals. Long-term exposure can lead to declines in population sizes, which can have cascading effects on ecosystems. The cumulative impact of endocrine disruptors can lead to population declines in affected species. For example, long-term exposure to contaminants has been linked to population declines in certain fish species, as documented by Patisaul et al. (2009).

9.2.2 *Effect of ED on Humans*

Reproductive Health:

Endocrine disruptors can affect human reproductive health by causing issues like reduced fertility, early puberty, and increased risk of reproductive cancers. Research by Duenas-Moreno et al. (2023) indicates that BPA and phthalates, found in many consumer products, can lead to reduced fertility and increased risk of reproductive cancers in humans. The work of Papadimitriou and Papadimitriou (2021) has linked endocrine disruptors to early puberty and reproductive health issues, particularly in females.

Developmental and Neurological Effects:

Exposure during pregnancy or early childhood can lead to developmental issues, including neurodevelopmental disorders like Attention Deficit Hyperactivity Disorder (ADHD) and autism. Studies by Skogheim et al. (2021) have shown that prenatal exposure to endocrine disruptors like mercury and lead is associated with developmental disorders such as ADHD and autism. Exposure to endocrine disruptors can negatively affect cognitive development and learning abilities in children (Braun 2017).

Metabolism and Thyroid Health:

Some endocrine disruptors are linked to obesity, diabetes, and metabolic syndrome. They can interfere with metabolic processes and hormone regulation. Disruptors can affect thyroid hormone levels, which are crucial for growth, metabolism, and overall health. Thyroid imbalances can lead to a range of health issues, from weight problems to developmental delays. Dager et al. (2004) link endocrine disruptors to obesity and diabetes by disrupting metabolic processes and insulin signalling. Studies by Schmutzler et al. (2007) have found that endocrine disruptors can interfere with thyroid hormone regulation, potentially leading to thyroid dysfunction and related health issues.

Cancer Risks and Immune System Effects:

Endocrine disruptors such as BPA and PCBs have been linked to increased risks of hormone-related cancers, including breast and prostate cancer, as stated by Sharma et al. (2024). There is evidence suggesting that certain endocrine disruptors can increase the risk of cancers, particularly those related to hormone systems, such as breast and prostate cancer. Bansal et al. (2017) suggest that endocrine disruptors can impair the immune system, making individuals more vulnerable to infections and diseases. Similar to animals, endocrine disruptors can potentially affect the human immune system, increasing susceptibility to diseases and impairing immune responses.

Given the omnipresence and potential for bioaccumulation of these chemicals, addressing the environmental and health challenges posed by EDCs requires a multidisciplinary approach, incorporating advancements in recently available technology for remediation. The development of innovative technologies to mitigate EDC contamination is crucial for safeguarding both ecological integrity and public health on a global scale. Various strategies and recent advancement in existing technologies for the remediation of EDs are summarized as follows.

9.3 Remediation Strategies to Mitigate EDs

There are numerous conventional technologies that have been developed to remediate these EDs, contemplating their harmful effects on our ecosystem. The techniques are illustrated in Fig. 9.1 and based on chemical, physical, and biological treatment methods.

9.3.1 Physical Treatment Technologies

Adsorption Technology

Among physical processes in wastewater treatment, adsorption is very common and efficient in the removal of endocrine disrupters. Its operational ease, flexibility, and lack of byproduct formation have made it attractive. Adsorption has also been considered the most inexpensive technology for water treatment. Other advanced processes may be more costly due to operational complexities and high maintenance costs (Shah 2025; Azizi et al. 2022; Rossner et al. 2009; Vieira et al. 2020).

The process of adsorption is very effective in removing EDCs from aquatic environments, which opens up prospects for restoration or reuse, and even recycling of the adsorbent. The adsorption processes have been recognized by the U.S. Environmental Protection Agency (USEPA) to be one of the most appropriate choices among physicochemical methods for the elimination of both organic and inorganic compounds from wastewater. A great variety of adsorption technologies show different percentage rates from 30 to 100% owing to variations in adsorbents' physicochemical properties and varying EDC mixtures within wastewater (Azizi et al. 2022; Mahesh et al. 2023; Rathi et al. 2021).

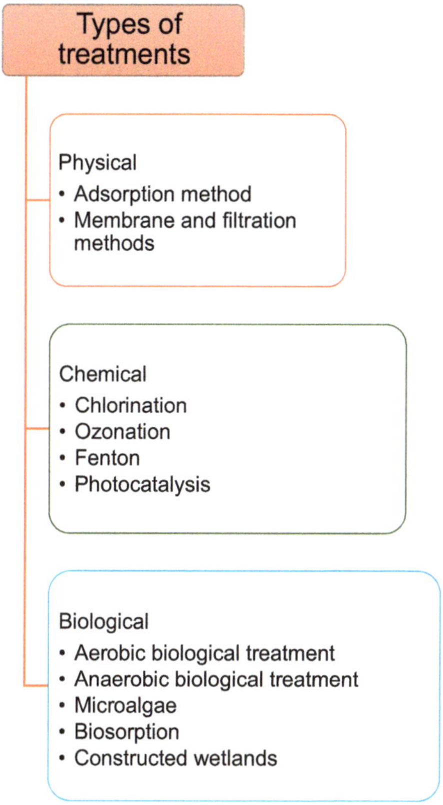

Fig. 9.1 Various methods employed to remediate Endocrine Disrupters

Mechanism of Adsorption

The mass transfer process of adsorption denotes the interactions that occur between adsorbates and adsorbents through hydrophobic or electrostatic interactions (Azizi et al. 2022; Mahesh et al. 2023). The fluid components within the bulk fluid phase accumulate on the solid surfaces during an adsorption process and where the adsorbent surface intersects with the bulk fluid phase at the interface (Azizi et al. 2022; Gao et al. 2020a, b; Mahesh et al. 2023). In conclusion, adsorption is described as a surface phenomenon that takes place through mass transfer of a substance (solid or liquid phase) at the interface (Katibi et al. 2021).

There are two types of adsorptions based on the nature of the interaction between adsorbate and adsorbent: chemisorption and physisorption. Physisorption is characterized by weak intermolecular forces (van der Waals and electrostatic forces), while chemisorption comprises a chemical reaction (usually the formation of a covalent bond) between the adsorbing substance and the adsorbent surface. Intermediate adsorption mechanisms exist, which share properties of physisorption and chemisorption (Shah 2024a; Azizi et al. 2022; Mahesh et al. 2023; Vieira et al. 2020).

Types of Adsorbents

A variety of adsorbents have been utilized for the elimination of EDC wastewater. These include lignocellulosic materials, activated carbon, zeolites, mineral matter, hybrid organic-metal nanocomposites, biochar, carbon nanotubes, graphenes, activated hydrochar, activated carbon (AC) composite materials, and mesoporous nanocomposites of polymers and clays. These adsorbents have shown different efficiencies in relation to EDC removal according to their properties and target EDC types.

- **Activated Carbon (AC)**: The high surface area and specific surface chemistry make AC very effective. The removal efficiency was up to 90% for ethyl paraben, up to 99% for diethyl phthalate, up to 99% for bisphenol A, up to 96% for bisphenol E2, and up to 98% for carbamazepine (CBZ). Application of the AC is done in either a powdered or granular form within packed bed filters.
- **Carbon Nanotubes (CNT)**: CNTs demonstrated excellent capability for the removal of EDCs, apparently attributed to the peculiar nature of their surface chemistry and properties. They have higher adsorption capacity compared to AC due to their pore diameter distribution and adsorption energy distribution. They demonstrated efficiencies about 99% for bisphenol F, 98% for 17α-ethinylestradiol, and 44–70% for bisphenol A.
- **Zeolites and Clays**: Naturally occurring silicate minerals find extensive applications in the removal of EDCs as they have microporous structure, are selective in nature, and have ion exchange capability. Zeolites indicated a high uptake capacity for bisphenols, while clays showed high adsorption capacity via synergistic effects of π–π interaction and hydrophobic interaction.
- **Lignocellulosic Materials**: Some plant materials, either combined with metallic nanoparticles or not, also showed reasonably good adsorption performance for EDC removal (Nayak et al. 2023; Azizi et al. 2022; Fatima et al. 2021).

One advantage of the treatment of wastewaters with respect to the removal of EDCs has been adsorption technology, due to the properties such as low cost, simple operation, and high efficiency. In future research, optimization of adsorption processes, new adsorbents, and the mechanism of interaction between adsorbents and EDCs should be better understood in order to further improve overall removal efficiency (Azizi et al. 2022).

Membrane and Filtration Technologies

Membrane technologies are probably the most significant physical methods for eliminating EDCs in advanced wastewater treatment processes. These technologies have several advantages, which include high selectivity over a broad spectrum, entail continuous processes with no byproduct or metabolite formation, are highly adaptable to conventional treatments of water, and the effluent produced is of high quality, having trace amount of organics, microbes, and viruses (Azizi et al. 2022; Chang et al. 2009; Ojajuni et al. 2015; Vieira et al. 2020).

Membrane technologies applied for EDC removal are mainly focused on the process effectiveness depending on physicochemical properties of EDCs such as electrostatic interaction, miscibility in various solvent and molecular mass, type of membrane process, and influence of impurities (Azizi et al. 2022; Liu et al. 2009). In this regard, several studies have reported variability in the elimination efficiency of membrane technologies, often attributable to the interaction between distinctive physical and chemical properties of EDCs and efficacy of membranes used. The order of effectiveness for membranes used for EDC removal is reverse osmosis > nano-filtration > ultra-filtration > micro-filtration. Indeed, efficiency measurements ranging from 10 to 95% have been observed, explained by the broad range of physical and chemical properties found among EDCs (Azizi et al. 2022; Liu et al. 2009; Vieira et al. 2020). For example, the removal of 17α-ethinylestradiol (EE2) using nano-filtration, ultra-filtration, and reverse osmosis membranes ranges from 42 to 95%, to 78%, and to 96% to 98%, respectively. In like manner, removal efficiency for bisphenol A (BPA) has been reported as 53–79% with ultra-filtration, and 47–98% by nano-filtration. On the contrary, the CBZ removal efficiency was less than 20% using ultra-filtration, up to 22% when using nano-filtration, and 99% after reverse osmosis treatment. For diazepam, its removal efficiency by nano-filtration and reverse osmosis ranged 19–50% with ultra-filtration, 50–91% with nano-filtration, and more than 80% with reverse osmosis (Azizi et al. 2022; Vieira et al. 2020).

Micro-filtration is mainly used within the low-pressure range, and it relies mostly on sieving as the mechanism of removing colloidal particles and soluble high molecular weight contaminants in wastewater. However, this method fails in the elimination of EDCs from water. Because of this reason, micro-filtration will normally be combined with the use of other treatment technologies along the treatment processes. Similar to micro-filtration, ultra-filtration happens under the low-pressure range and uses the size-exclusion or sieving mechanism for EDC removal. Still, the efficiency of ultra-filtration in EDC removal has been found to be better, most likely due to its finer porosity (Azizi et al. 2022; Liu et al. 2009; Vieira et al. 2020).

Now, compared to previous technologies, nano-filtration membranes have a smaller pore size and more charge density, typical negative charge, since they have ionizable groups in their structure. This means, therefore, that nano-filtration combines sieving and electrostatic interactions for EDC removal. Consequently, such technology with the membrane makes it more efficient than micro-filtration and ultra-filtration in treating wastewater, especially in removing EDCs. It is because reverse osmosis has the smallest pore size compared to nano-filtration. They are distinctive in terms of principal driving forces in the treatment of EDCs in reverse osmosis membranes. Such driving forces together with the small pore size make reverse osmosis most effective in removing EDCs than nano-filtration. Nevertheless, with better mass transfer abilities for better separations and higher removal efficiency than reverse osmosis, nano-filtration is more likely to be considered a suitable membrane technology for EDC treatment because of its lower long-term costs (Shah 2024b; Azizi et al. 2022; Liu et al. 2009; Vieira et al. 2020).

According to the removal of the EDCs through membrane technologies, three main mechanisms have been proposed.

i. **Size-exclusion or sieving**: it is considered the prime mechanism, which relies on the pore size of the membrane and its permeability.
ii. **Charge repulsion**: this may play a substantial role in the efficiency of the membranes having some functional groups.
iii. **Physicochemical interactions** (hydrophilic–hydrophobic) commonly involved in the physical and chemical properties of EDCs (Azizi et al. 2022; Liu et al. 2009; Rathi et al. 2021; Vieira et al. 2020).

In view of this explanation, one could find that physical and chemical properties of adsorbates and adsorbents play a prime role in the selection of the appropriate membrane technology, so it is required extensive study that can lead into substantial EDC removal through the treatment process.

9.3.2 Chemical and Electrochemical Methods

Chlorination

Chlorination is the traditional method followed for the disinfection of drinking water and for the treatment of wastewater (Du et al. 2017; Kelkar et al. 2019; Zamri et al. 2021). As one of the chemical disinfection techniques, chlorination destroys the cell membrane respiration of microorganisms, thereby killing them in the water sources. Chlorination may be applied at primary or secondary stages in the water treatment system. The main objectives are to prevent biological growth and to ensure that the residual chlorine remains in the system constantly at appropriate levels throughout the distribution system (Du et al. 2017; Zamri et al. 2021).

Biofiltration is very effective in most aspects, but chlorination does not work very well with EDCs. Chlorination can remove estrogenic compounds effectively by over 98% (María Teresa et al. 2020; Zamri et al. 2021). Still, many synthetic EDCs, including various kinds of pharmaceutical compounds, are not so well degraded by chlorination, and they will often need some complementary stages to the treatment to enhance their removal efficiency (Chen et al. 2023a, b; Liu et al. 2019a, b; Zamri et al. 2021; Zhang et al. 2018). For instance, pesticides having potent mutagenicity could be degraded further after chlorination, and that powder could be adsorbed. Such actions were through the powdered activated carbon and hydrophobic interaction (Matsushita et al. 2018; Zamri et al. 2021). In addition, toxic, hazardous byproduct formation was realized after the chlorination process (Du et al. 2017; Zamri et al. 2021). With all these drawbacks associated with the chlorination process, the most common step that is embarked on after chlorination is the advanced processes for secondary or post-treatment of water. The removal efficiency of several classes of EDCs by a single chlorination process is usually low. The chlorination efficiency for each kind of EDC compound can vary greatly according to chlorine dosage, contact time, and pH conditions, and therefore varies greatly across studies (Mansor & Tay 2020; Zamri et al. 2021). Chlorination treatment should be developed under standardized and optimized conditions for EDCs. It can greatly enhance the overall efficiency of chlorination in water and wastewater treatment by improving these conditions.

Ozonation

Ozone, O_3, has received wide attention for its treatment application because of its very high efficiency in treating wastewater as an oxidizing agent. Ozone is an extremely reactive gas that efficiently oxidizes bacteria, organic matter, macropollutants, and micropollutants in wastewater systems (Shah 2024c; Gomes et al. 2017; Zamri et al. 2021). The reaction rate of the process primarily governs the efficiency of the ozonation process, which in turn is affected by parameters like dose of ozone, catalyst type, energy input density, airflow rate, temperature, and pressure (Chen et al. 2023a, b; Kaswan and Kaur 2023; Tang et al. 2019; Zamri et al. 2021). Ozonation was found most effective for removing EDCs, at efficiencies between 40 and 100%. EDCs can be totally removed from the wastewater by ozonation (Si et al. 2019; Zamri et al. 2021). Furthermore, ozonation has shown high removal efficiency for some pesticide and pharmaceutical and personal care product compounds (Le et al. 2017; Paucar et al. 2019; Zamri et al. 2021).

However, it is worth noting that several indirect byproducts or transformation products may be produced as byproducts from the EDCs treatment. The ozonation of bromide-containing wastewater led to the formation of bromate, a toxic oxidation byproduct (Soltermann et al. 2017; Zamri et al. 2021). Still, because of incomplete mineralization of the micropollutants during ozonation, transformation products (TPs) are reportedly formed in low concentrations (Schröder et al. 2016; Zamri et al. 2021). The highest production of TPs is observed at the dose of ozone indicated ($Z = 0.55$ mg O_3/mg of dissolved organic carbon [DOC]) within wastewater treatment plants (WWTPs) (Kharel et al. 2020; Zamri et al. 2021). The concentrations of the TPs are very highly affected by the dose of ozone, composition of the wastewater matrix, and different types of EDCs (Shah 2024a; Schröder et al. 2016; Zamri et al. 2021).

Natural organic matter (NOM) is already present in the matrix of wastewater, thus requiring a good competition with targeted compounds and reducing the efficiency of the treatment process (Olmez-Hanci et al. 2020; Zamri et al. 2021). Prefiltration, to exclude non-target compounds from the wastewater matrix, could be used to potentiate the effectiveness of ozonation. Moreover, in addressing further the problem with TP residuals, the biological and physical after-treatment methods introduce, among others, sand filtration, moving bed, fixed bed, GAC filtration, and AOPs to stabilize TPs derived from parent compounds (Bourgin et al. 2018; Itzel et al. 2020; Zamri et al. 2021).

Nevertheless, despite the improvement with ozonation, current analyses are so far limited to selected target compounds; and this still results in the selection of many non-targeted and possibly still relevant compounds that end up unutilized. Other studies and optimization research, therefore, have to be carried out to reduce/annihilate TPs and reaction intermediates that are generated in the process of ozonation (Zamri et al. 2021).

Fenton Process

Fenton

An economical and efficient implementation of the Fenton process for contaminant destruction, with a focus on hydroxyl radical mediated degradation. In Fenton processes, hydroxyl radicals react with organic contaminants, and their reaction leads to complete degradation into carbon dioxide, water, and inorganic salts (Hu et al. 2018; Zamri et al. 2021). Contaminants, through electron exchange, are exchanged with the hydroxyl radicals and produce hydroxide anions, resulting in a hydrogen atom cohesion with the radicals to eliminate the missing atom responsible for the reaction (Grassi et al. 2020; Zamri et al. 2021).

The efficiency of Fenton treatments largely based on a couple of reacting parameters. Some of these parameters are the concentration of the reactants, the matrix-watery, pH, type of catalyst, water matrix composition, intensity of radiation, salinity of substrate, water management, the reaction time, and temperature. All these have a significant impact on the performance of Fenton processes (Mirzaei et al. 2017; Zamri et al. 2021).

Fenton processes find long-term application in the elimination of EDCs from water. As an example, we have that the applicability of Fenton oxidation techniques for removing BPA from cattle manure wastewater at removal efficiencies close to 99.5% (Sun et al. 2016; Zamri et al. 2021). Although the Fenton process is effective, some drawbacks limit its wide practical application for EDC removal. Sometimes regenerating iron ions necessary for this process is not practical; in addition, the final effluent should be polished for the compliance of the discharge standards set for iron concentrations. Another important concern that has to be made right is the environmental impact, particularly about the production of green house gases (GHG) (Mirzaei et al. 2017; Zamri et al. 2021).

Considering these challenges, the Fenton process needs further research and development to continue the reduction of resource demand and enhancement of sustainability and efficiency in pollutant removal. This involves synthesis of the methods for regeneration of the iron ion species and reducing the environmental footprint of the process, which will be more attractive for broad application in treatment of waters and wastewaters (Zamri et al. 2021).

Photo-Fenton

The Photo-Fenton process is an augmented oxidation process largely applied to the degradation of contaminants, mainly though the production of hydroxyl radicals. Ferric iron (Fe(III)) acts as the catalytic agent in this treatment to generate reactive hydroxyl radicals using hydrogen peroxide (H_2O_2) in the presence of ultraviolet (UV) light (Heidari et al. 2023; Zamri et al. 2021). This reaction can be significantly carried out more effectively under acidic conditions: it is a case of the photoreduction of hydroxyl complexes, for example, $Fe(OH)^{2+}$ and $Fe(OH)_2^{4+}$, which are able to absorb light in the region of UV/visible, with the formation of hydroxyl radicals and ferrous iron, Fe^{2+} (Le et al. 2017; Zamri et al. 2021).

The photo-Fenton process is very promising in EDC removal; still the oxidation efficiency can be enhanced further by the process modifications like the catalyst design and surface modifications for making it effective in removing the micropollutant (Costa et al. 2020; Vorontsov 2019; Zamri et al. 2021). Extensive research on the production of highly reactive ions is important to explaining the reaction mechanisms and, therefore, improving general treatment efficiency by this process for water and wastewater treatment.

Electro-Fenton

Electro-Fenton is an advanced treatment technology that enhances the production of hydroxyl radicals through the combination of Fenton chemistry with electrocoagulation processes. In such a system, the oxidizing power of hydrogen peroxide is enhanced by electrical aid, which makes the Fenton process be regarded as one among the effective and innovative methods for bringing about reduction/elimination of organic pollutants (Liu et al. 2018; Zamri et al. 2021).

Electro-Fenton treatments are commonly conducted in two types of configurations: (i) either an inert electrode acting as the anode material with high catalytic activity or adding the Fenton reagents externally to the reactor, or (ii) external addition of hydrogen peroxide with Fe^{2+} supplied from the sacrificial cast iron anodes (Baiju et al. 2018; Sruthi et al. 2018; Zamri et al. 2021). H_2O_2 generated electrochemically plays a very crucial role in increasing high-strength organic pollutant degradation within wastewater systems (Chanikya et al. 2021; Zamri et al. 2021).

This approach is very effective in removing endocrine-disrupting chemicals from high concentrations in wastewater. Electro-Fenton treatment removed pharmaceutical and pesticide compounds at concentrations up to 100,000 $\mu g\ L^{-1}$ (Moreira et al. 2017; Zamri et al. 2021). In addition, photo-electro-Fenton, bio-electro-Fenton, and solar electro-Fenton, derived from this method, indicated very high efficient performance in EDC removal (Dai et al. 2021; Seibert et al. 2019; Zamri et al. 2021; Zhang et al. 2018).

Although having such outstanding performance, the operation and maintenance cost of the technology electro-Fenton are very high. The referred costs are relatively higher compared to traditional treatment techniques. This puts an attractive question mark whether the application of electro-Fenton, on a full-scale wastewater treatment process, is cost-effective or not. Hence, the approach's cost-effectiveness must be reconsidered and re-evaluated on application subsequently to full-scale study and operation of wastewater treatment (Zamri et al. 2021).

Photocatalysis

Photocatalysis generally becomes a reference to photoactivation when chemicals are converted by irradiation of metal oxides having semiconductivity utilized as catalysts. The method is mostly preferred for the treatment of wastewater among the methods of treatment due to its compatibility with the environment and effectiveness (Tahir et al. 2020; Zamri et al. 2021). Photocatalytic oxidation is an emerging technology with its advantages being predominant non-toxicity, comparatively low cost, no mass transport resistance, chemical stability, and ambient temperature operations (Gopinath et al. 2020; Zamri et al. 2021).

Different photocatalysts have been applied in this process, of which titanium dioxide (TiO_2) and zinc oxide (ZnO) are common due to the highest photochemical stability and piezoelectric properties (Gopinath et al. 2020; Zamri et al. 2021). In case of endocrine-disrupting compounds, ZnO has good efficiency compared to TiO_2. A photocatalysis removal efficiency of EDC depends on several process factors: concentration of catalyst, pH, concentration of substrate, and modifications of the photocatalyst (Yu et al. 2020; Zamri et al. 2021). In several studies, TiO_2 loaded into various designs, such as carbonaceous composites, appeared to work very well compared with other catalysts (Mahmoud et al. 2017; Tahir et al. 2020; Zamri et al. 2021).

Economic considerations and considerations of operational costs are two very important factors; possibly, they will then determine the performance efficiency against the cost-effectiveness balancing point of a particular photocatalyst approach suitable for large-scale applications of wastewater treatment (Zamri et al. 2021).

The conventional methods to remove contaminants from the environment have found limited success and their inefficiency may even lead to generation of more toxic byproducts. Bioremediation, which uses living organisms to break down hazardous substances into less toxic ones, is an effective approach for dealing with emerging contaminants. The initial research was targeted at exploring microbes which have abilities to sustain in harsh environments and utilize the contaminants as a source of energy (Patel et al. 2020). In particular, most success was found with fungus species which produce oxidoreductase enzymes. The attempts to use oxidoreductase category of enzyme have been successful, and current efforts are at improving the efficiency of the process and reducing the cost.

9.4 Bioremediation of Endocrine Disrupters Using Extremophiles

Among the emerging contaminants, endocrine disruptors have become a significant concern because of their widespread environmental contamination, and potent bioactivity at trace concentrations. Bioremediation presents a promising approach for the efficient removal of these persistent organic pollutants. Emerging pollutants, often unmonitored, include a range of synthetic and natural chemicals with potential risks to all ecological niches. This category includes pharmaceuticals, pesticides, day care products, detergents, microplastics, and hormones (Mnif et al. 2011). The incomplete removal of these substances during conventional wastewater treatment processes results in their release into the environment, leading to ecological and health risks such as toxicity, endocrine disruption, and long-term impacts on ecosystems, wildlife, and human (De Coster and Van Larebeke 2012).

Ongoing scientific efforts are aimed at improving the removal processes to enhance environmental sustainability. Bioremediation, which leverages living organisms to transform hazardous substances into less toxic compounds, is a particularly effective strategy for addressing the challenges posed by emerging contaminants. Microorganisms can survive hazardous compounds and can detoxify them by biotransformation and biodegradation processes (Bhatt et al. 2023). Despite considerable advancements in microbial bioremediation research, critical knowledge gaps remain. Specifically, the microbial mechanisms underlying contaminant degradation and the integration of these processes into practical technologies are not fully understood. There are over 800 known endocrine-disrupting chemicals (EDCs) and endocrine-active substances (EASs) that have emerged as significant concerns due to their persistent ability to interfere with cellular, epigenetic, and molecular functions within the endocrine systems of both humans and animals. These substances, mimicking or disrupting hormonal activity, can either amplify or inhibit the intricate regulatory functions of the endocrine system, impacting various organ systems from head to toe. Their long-term presence and bioactivity pose substantial risks, potentially leading to widespread health and developmental disorders across multiple biological systems (Patel et al. 2020; Lopez-Rodriguez et al. 2021). In particular, extremophile group of microorganisms have more potential in combating emerging contaminants like ED. These group of microbes produce extremozymes which can tolerate harsh environmental conditions. The relationship between microbial diversity and contaminant degradation efficiency requires further investigation (Sun et al. 2024).

The quest for effective management strategies to neutralize hazardous chemicals has led to the discovery and advancement of biodegradation processes, particularly those involving extremophilic microorganisms (Rovani et al. 2014; Donati et al. 2019; Hernández-Abreu et al. 2020; Eltoukhy et al. 2020; Bhatt et al. 2023). Traditional remediation methods, however, are often hindered by skyrocketing costs, lack of biocompatibility, high reagent demands, and the production of secondary toxic byproducts, limiting their overall effectiveness (Dangi et al. 2019). In contrast, bioremediation employs microorganisms to break down hazardous chemicals into less toxic or non-toxic products (Sotelo et al. 2012; Hazra et al. 2014), offering a more sustainable solution. Microbial bioremediation has gained prominence due to its eco-friendly nature, lower costs, and fewer spatial and equipment requirements (Gao et al. 2020a, b; Pang et al. 2020; Singh et al. 2021a). Microbes efficiently degrade endocrine-disrupting chemicals (EDCs) by utilizing them as an energy source, transforming them into less harmful byproducts without relying on harsh chemicals and preserving natural resources (Wojcieszyńska and Guzik 2020; Wojcieszyńska et al. 2020; Roccuzzo et al. 2021). To enhance the efficiency of these processes and reduce the time required, additional nutrient sources and biological catalysts are often introduced (Cajthaml 2015; Zhang et al. 2016; Patel et al. 2020). This approach not only accelerates the degradation process but also minimizes environmental impact.

The most widely utilized organisms for bioremediation include bacteria, fungi, and microalgae, as highlighted in Table 9.1 (Jaiswal and Shukla 2020; Kalra et al.

2021; Tran et al. 2025). Bacterial remediation can occur under both aerobic and anaerobic conditions. Commonly employed bacterial strains include *Achromobacter* sp., *Burkholderia* sp.,*Dehalococcoides* sp.,*Ralstonia* sp.,*Pseudomonas* sp.,*Rhodococcus* sp., and*Sphingomonas* sp. Fungal species used in remediation include *Pleurotus ostreatus*,*Fusarium oxysporum*,*Phanerochaete* sp.,*Trametes versicolor*, *Fomitopsis palustris*, *Pythium ultimum*, *Candida aquatextoris*, and *Aspergillus* sp. (Cruz-Morató et al. 2014; Cajthaml 2015; Zhang et al. 2016; Dangi et al. 2019; Mohammadi-Moghadam et al. 2022). The most promising extremophilic species capable of remediating ED are *Marinobacter* sp.,*Halomonas, Bacillus, Methylocystis bryophila* (Chia et al. 2024).

Table 9.1 Microorganisms capable of remediating EDC

Organism	Pollutant	Treatment method	References
Bacteria			
Pseudomonas aeruginosa Pseudomonas lutoi	BPA	Enzyme-mediated	Al-Hashimi (2018)
Mycobacterium sp.*DBP42,*	DBP, and DEHP, acetyl tributyl citrate (ATBC)	Esterases and enzymes involved in the β-oxidation pathway	Wright et al. (2020)
Pseudomonas putida strain YC-AE1	BPA	Enzyme-mediated	Eltoukhy et al. (2020)
Pontibacillus chungwhensis HN14	Estradiol (E2)	Hydroxysteroid dehydrogenase	Yang et al. (2024)
Fungi and yeasts			
White rot fungi	Bisphenol A, bisphenol S, and nonylphenol	Ligninolytic enzymes	Grelska and Noszczyńska (2020)
Candida rugopelliculosa RRKY5, Galactomyces candidum RRK17 and G. candidum RRK22	Estrogenic alkylphenols	Enzyme mediated	Rajendran et al. (2016)
Trametes versicolor	EDCs, such as phenols, parabens and phthalate	Lignolytic enzymes	Pezzella et al. (2017)
White rot fungi	BPA, BPS, and nonylphenol	Ligninolytic enzymes	Grelska and Noszczyńska (2020)
Diaporthe longicolla	Bisphenol	Laccase mediated	Baluyot et al. (2022)
Chaetomium brasiliense (G3), Canariomyces notabilis (KW1), and Paecilomyces formosus (KW3)	17α-ethinyl estradiol (EE2), and:4-tert-butylphenol (4-t-BP)	Laccase	Daâssi and Alharbi (2023)

The degradation of endocrine-disrupting chemicals (EDCs) by these organisms primarily involves biodegradation and biotransformation, facilitated by enzymes such as hydroxylases, hydrogenases, dehydrogenases, transferases, dioxygenases, and laccases. Marine microalgae can survive saline conditions and many of its species employ similar mechanisms, utilizing ligninolytic enzymes, and cytochrome P-450 for biotransformation (Wojcieszyńska and Guzik 2020; Wojcieszyńska et al. 2020). In fungi, beyond biotransformation, additional processes like bioadsorption and bioaccumulation are prominent, aided by peroxidases, cytochrome P-450, and glutathione S-transferases. The effectiveness of bioremediation depends on the specific class of EDCs, variations within the EDC group, and the complexity of the contaminant, while the class of organism used had a less significant impact. Additionally, factors such as the method of organism delivery and the choice of carrier material were found to greatly influence biodegradation efficiency (Kumar et al. 2024).

9.4.1 Bioremediation Strategies to Treat Pollutants with ED

In situ bioremediation strategy is known to be cost-effective as the contaminants are treated on site. The ex situ bioremediation strategy is more labour-intensive and costly, but it is known to have better efficiency. Major bioremediation technologies are illustrated in Fig. 9.2.

The common bioremediation strategies are Natural Attenuation, Bioventing, Bioslurping, Biosparging, Bioleaching, Bioaugmentation, Biopile, Biostimulation, Phytoremediation, and Land Filling (Azubuike et al. 2016). Natural attenuation is a cost-effective remediation strategy that leverages the inherent physical, chemical, and biological processes within ecosystems to reduce contaminant levels. Through mechanisms such as biodegradation, volatilization, sorption, and dilution, hazardous substances can be stabilized over time (Donati et al. 2019). However, this method, while financially advantageous, can be slow, and there are potential risks of contaminant leaching into surrounding environments, which may exacerbate environmental concerns (Janssen and Stucki 2020). To address some of these limitations, enhanced attenuation techniques have been developed, providing a more targeted and efficient approach to contaminant removal. Bioventing, for instance, involves the controlled introduction of air into contaminated soil, stimulating indigenous microbial populations to accelerate the biodegradation of pollutants. By optimizing conditions, such as nutrient availability, bioventing effectively removes volatile and non-volatile compounds (Chowdhary et al. 2019). Bioslurping combines bioventing with vacuum and vapour extraction, enhances microbial degradation by creating a "slurping" effect that facilitates the removal of volatile and semi-volatile compounds, as well as light non-aqueous phase liquids (Bijalwan and Bijalwan 2016). Biosparging, a technique similar to bioventing, drives volatile compounds from saturated to unsaturated zones, facilitating their removal. It is commonly employed in aquifers contaminated with diesel and kerosene, offering an effective means of addressing these pollutants (Usman et al. 2024). In the field of mining and sediment remediation, bioleaching

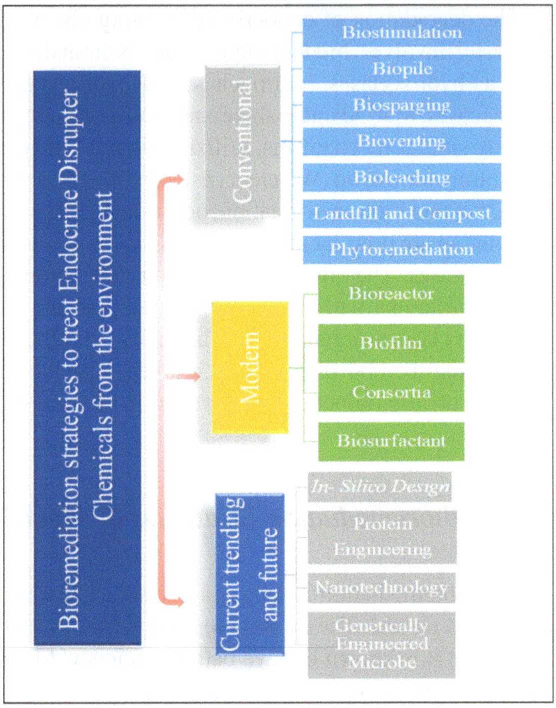

Fig. 9.2 Bioremediation approaches to remediate EDC

has emerged as a widely utilized approach for the separation of metals from mineral ores. Acidophilic microbes are employed to reduce metal impurities in residues, a process that has demonstrated potential in remediating endocrine-disrupting chemicals (EDCs) in contaminated sediments (Fonti et al. 2016). Enhanced attenuation strategies, when integrated with natural processes, offer a promising pathway for more efficient and controlled remediation of contaminated environments, although challenges related to environmental conditions and site-specific factors must be carefully managed.

The process of microbial augmentation for bioremediation can be further improvised by addition of growth promoting compounds. The use of co-metabolism via biosurfactants-producing extremophiles such as *Pseudomonas, Rhodococcus, Acinetobacter,* and *Bacillus*, or through a microbial consortium in which each member serves as a separate catalyst can significantly enhance the remediation process (Chowdhary et al. 2019). It is used for the partial removal of polychlorinated biphenyl, 2, 4–6-trichlorophenol and Atrazine. One such example is the complete remediation of 2, 4–6-trichloro phenol by alkali tolerant extremophile *Alcaligenes eutrophus* TCP strain in the lab study (Phale et al. 2019). However, there is a restriction on the use of non-indigenous microbes, and the overall efficiency is affected by the environmental factors (Joe et al. 2019; Olawoyin et al. 2024).

Biopile and Biostimulation are relatively simpler techniques which involve piling layers of contaminated soil and augmentation of selected acidophilic and alkaliphilic microbial cultures. The process is highly regulated as temperature and moisture content is maintained, which may increase the overall cost (Simpanen et al. 2016; Chowdhary et al. 2019; Singh et al. 2021a). Biopiling has been successful in treating acidic petroleum sludge (Naeem and Qazi 2020). Biostimulation takes the one further by supplements of nutrient and regulating pH, oxygen, and microbial load as well (Nikolopoulou and Kalogerakis 2010).

Phytoremediation involves the use of plants with higher adsorption potential to remove contaminants from the soil. Along with toxicant adsorption and absorption, plant oxidases and peroxidases may detoxify the pollutants as well (Singh et al. 2021a, b). Plants such as *Brassica juncea* L.,*Populus deltoides* L.,*Salix alba* L., and *Helianthus annuus* L. are used to remove volatile contaminants like petroleum hydrocarbons, pesticides, metals, radionuclides, explosives, atrazine, polychlorinated biphenyl, and hydrophobic organics like PCP (Tabei and Sakakibara 2006).

One of the most common methods of the waste disposal are land filling and composting. Land filling comprises piling and ploughing of land to provide better aeration and growth supporting conditions to indigenous microbes involved in decontaminating (Chowdhary et al. 2019). Composting involves the microbial conversion of agro-forest waste material along with the contaminated soil to organic soil amendments. It is an environment-friendly technique, though it requires a dedicated area and consistent aeration (Azubuike et al. 2016).

9.4.2 Modern Bioremediation Technologies and Their Future

Conventional approaches to environmental remediation, particularly for endocrine-disrupting chemicals (EDCs), have faced numerous challenges, including prolonged processing times, high costs, and operational demands such as the need for extensive space, constant aeration, and mixing. Additionally, many of these methods have been associated with significant issues such as the incomplete degradation of EDCs, toxic waste dissemination, and unintentional contaminant leaching (Azubuike et al. 2016; Singh et al. 2021a). These drawbacks prompted the exploration of innovative techniques to overcome these limitations and effectively address EDC contamination.

The development of newer approaches has focused on overcoming the constraints of time, space, and operational complexity while leveraging technological advancements in biotechnology and environmental science (Liu et al. 2019a, b; Janssen and Stucki 2020). Key strategies have included enzyme modification, gene editing using CRISPR, and the application of genetic, metabolic, and protein engineering. Additionally, predictive models have been employed to minimize or prevent the formation of toxic byproducts (Sakshi and Haritash 2020; Bhatt et al. 2023). Among these advancements, enzymatic approaches have shown significant potential. Microbial enzymes, capable of breaking down harmful compounds in soil and water, can be used either in their natural microbial form or as isolated extracellular enzymes (Singh et al. 2021b; Tran et al. 2025). The use of isolated enzymes offers several advantages, including enhanced stability, specificity, mobility, and ease of storage, in contrast to whole microbes that require specific environmental conditions for survival and activity (Ravichandran and Sridhar 2016; Wang et al. 2018). While natural enzyme production may be low under typical conditions, the use of enzymes as stand-alone agents for bioremediation has proven promising (Lee et al. 2019; Chowdhary et al. 2019; Wojcieszyńska and Guzik 2020; Wojcieszyńska et al. 2020; Mousavi et al. 2021).

Recent advances in metabolomics, proteomics, and genetic engineering have further improved enzyme-based bioremediation techniques (Bansal and Kanwar 2013; Cajthaml 2015). Classical enzymes utilized in bioremediation include a wide range of peroxidases (e.g., lignin, laccase, manganese, versatile peroxidases), oxygenases, hydrolases (e.g., esterases, lipases, and nitrilases), and oxidoreductases (Eibes et al. 2015; Mousavi et al. 2021). Protein engineering has allowed the modification of these enzymes to enhance their catalytic activity, stability, and regulatory control. The creation of enzyme chimeras, which combine beneficial features from multiple enzymes, has shown potential in producing novel enzymes with enhanced capabilities for EDC degradation (Li et al. 2023). Ligninolytic enzymes, such as laccases and peroxidases, have been extensively studied for their role in EDC removal, with promising results in wastewater treatment applications (Falade et al. 2018). As research in enzymatic and protein engineering progresses, these innovative approaches are expected to significantly advance the field of bioremediation and offer more efficient solutions for EDC contamination. The current bacterial strains employed for bioremediation, either in situ or ex situ, face a significant limitation of slow remediation rate, thus making the entire restoration process time- and resource-consuming (Liu et al. 2019a, b; Tran et al. 2025). As a result, the development of novel microbial decontaminants presents a promising and eco-friendly approach. Leveraging advanced omics technologies and microbial engineering, this strategy focuses on enzyme modification, the creation of new metabolic pathways, improved regulation of various biological parameters, and enhanced biotransformation capabilities (Liu et al. 2019a, b). Protein engineering has a promising aspect in finding the newer catalysis reactions and modulating the catalytic efficiency for the detoxification and degradation reactions. Enzyme chimeras offer the potential to combine favourable structural elements from different enzymes, resulting in more versatile and effective enzymes that can be further engineered for enhanced degradation of EDCs (Li et al. 2023). Falade et al. (2018) explored several ligninolytic enzymes, including laccase and peroxidase, for their efficacy in removing EDCs from wastewater.

The modern remediation technology now involves use of computational tools like-molecular docking, structure predictions, molecular dynamic simulation, etc. The in silico studies have paved the way for finding novel enzyme–substrate interactions that broadens the aspects of bioremediation over conventional technologies (Goya-Jorge et al. 2021). Furthermore, the newly developed algorithms and bioinformatic systems like QSAR model have abilities to predict the genetic mutations that can improvise the catalytic efficiency by designing the enzyme (Celino-Brady et al. 2021). In combination with in silico, development of the nanotechnology unlocked the avenue for the advancement in bioremediation technologies. Nanoparticles possess several key properties, such as superior surface area, optical properties, and modulating pore size, and improves catalytic efficiency and flexible surface chemistry, making them ideal for a range of remediation applications. They are highly suited for the remediation of EDCs through water treatment, where they can serve as adsorbents, sensors for water quality monitoring, and agents for disinfection, as well as in the development of high-quality nanomembranes (Bhateria and Singh 2022). These attributes render nanoparticles highly flexible, efficient, and low-maintenance, offering superior performance in environmental applications (Kuhan et al. 2022). The integration of high-throughput systems and advanced omics technologies, including genomics, metagenomics, transcriptomics, metatranscriptomics, metabolomics, fluxomics, and proteomics, alongside large-scale information databases, requires thorough evaluation for identifying effective strategies tailored for the degradation of a wide range of EDCs.

9.5 Extremophile Enzymes and Nanotechnology for Endocrine Disruptor Removal

The remediation of EDs is challenging because of their lipophilic nature, they are especially difficult to remediate aqueous environments because of their low solubility. They are difficult to remove from the environment because of their hydrophobicity, which also prevents them from dissolving in water and increases their resistance to catalytic breakdown by different enzymes (Smith et al. 2023; Johnson and Brown 2022).

The advancement of nanotechnology in recent times has facilitated the elimination of endocrine-disrupting chemicals (EDCs) from wastewater. The unique properties of nanoparticles increase the efficiency and efficacy of wastewater treatment. Long-lasting pollutants like bisphenol A (BPA) and other persistent pollutants can be absorbed and broken down by nanomaterials such as nanoparticles, nanoemulsion, nanodots, various nanosorbents, nanofilms, and nanotubes because of their huge surface area and reactivity (Smith et al. 2023; Lee et al. 2023). By engineering these materials for specific interactions with EDCs, researchers have achieved targeted effective remediation, addressing pollutants that are otherwise difficult to eliminate from the environment (Zhang et al. 2023). However, the integration of nanotechnology and extremophile-derived enzymes into ED remediation strategies has provided new avenues for efficient treatment.

Extremophiles, microorganisms that thrive in extreme environmental conditions, offer unique advantages in ED remediation. Their enzymes, known as extremozymes, exhibit remarkable stability and activity under harsh conditions, making them ideal candidates for integration with nanomaterials. For instance, *Pyrococcus furiosus* (a hyperthermophile) produces the enzyme **β-glucosidase**, which is known for its thermostability and can enhance ED breakdown at high temperatures (Zhang et al. 2023). Similarly, *Halobacterium salinarum* (a halophilic archaeon) produces **protease**, which functions in high-salt environments, while *Alkalibacterium psychrotolerans* (a psychrophile) produces **lipase**, which operates in cold conditions (Wang and Liu 2023).

One of the major benefits of nanotechnology is its capacity to combine several treatments into one procedure. For example, by combining adsorption and catalysis, it reduces the need for extra energy and chemicals, improving sustainability and economy of scale (Wang and Liu 2023). Nanotechnology offers flexible and scalable bioremediation solutions; this all-encompassing approach offers notable advantages over conventional techniques (Kumar et al. 2022).

Considering their distinct features, some nanomaterials have emerged as particularly effective at removing endocrine-disrupting chemicals (EDCs). The nanomaterials used in EDC cleanup include the following.

9.5.1 Metal Oxide Nanoparticles

Recent advancement in nanotechnology has accentuated the pivotal function of nanoparticles to enhance the removal of endocrine-disrupting chemicals (EDCs) from wastewater. Among these, metal oxide nanoparticles have proven very effective due to their high surface area and reactivity. These nanoparticles, when combined with extremozymes, create highly efficient systems capable of breaking down even recalcitrant EDCs.

Titanium Dioxide (TiO_2) Nanoparticles

TiO_2 nanoparticles are known for their photocatalytic properties, which facilitate breakdown of organic contaminants. Their large surface area allows for considerable adsorption of EDCs such as bisphenol A (BPA), which is then destroyed via photocatalysis using UV light (Kumar et al. 2023). TiO_2 is an effective water treatment technology due to its ability to form reactive oxygen species that convert contaminants into less damaging byproducts. When coupled with the thermophilic enzyme **β-glucosidase** from *Pyrococcus furiosus*, the degradation of EDs like bisphenol A (BPA) is further accelerated, due to both the stability of TiO_2 and the robustness of the enzyme at high temperatures (Kumar et al. 2023; Wang et al. 2021a, b).

Zinc Oxide (ZnO) Nanoparticles

ZnO nanoparticles exhibit high reactivity under light exposure, producing ROS that effectively degrade EDs. By incorporating the psychrophilic enzyme **lipase** from *Alkalibacterium psychrotolerans*, ZnO's functionality can be extended to low-temperature environments, making it suitable for use in cold wastewater systems (Wang et al. 2021a, b).

9.5.2 Nanosorbents

Nanosorbents (especially carbon-based) provide greater thermal, chemical, and mechanical properties which makes them potential candidate for the removal of hydrophobic EDs. Their high specific surface area permits for the adsorption of contaminants like endocrine disruptors with the help of a variety of interaction named hydrophobic interactions, electrostatic attraction, and π–π stacking (Zhang et al. 2023). Their ability to adsorb nonpolar contaminants makes them very efficient against hydrophobic EDs.

Graphane Oxide

Graphane oxides have structural advantage for the absorption of various contaminants because of the activated carbon compound layer and oxygenated functional groups. These materials are ideal for the removal of EDs, including BPA, nonyphenol, and triclosan due to their high conductivity, and strong π–π interactions (Liu et al. 2022). The functional group on GO improves its ability to act with both hydrophilic as well as hydrophobic EDs, making it a resourceful nanosorbent for discrete applications. The tunable surface chemistry and high efficiency of graphene-based nanosorbents make them promising candidates for large-scale environmental remediation. The adsorption efficiency of graphene oxide nanosorbents is enhanced when extremozymes are incorporated. For example, **amylase** from the acidophile *Sulfolobus solfataricus* improves GO's efficiency in acidic wastewater environments, where traditional nanosorbents may degrade (Liu et al. 2022).

Molecularly Imprinted Polymers (MIPs)

Another promising class of nanosorbents includes molecularly imprinted polymers (MIPs), which are synthetic polymers designed with specific binding sites that target particular molecules, including endocrine disruptors. MIPs offer high selectivity and affinity for these compounds, allowing for their selective adsorption from complex mixtures such as wastewater MIPs, designed to target specific EDs, become even more selective and robust when modified with extremozymes from alkaliphiles or halophiles, such as **protease** from *Halobacterium salinarum*, making them suitable for challenging pH or salinity levels (Wang et al. 2021a, b).

The specific recognition sites in MIPs enable them to effectively target and remove various EDs, including hormones and phenolic compounds. However, the synthesis of MIPs can be complex and time-consuming, and their performance may be affected by the presence of similar, non-target molecules, which can interfere with the adsorption process.

Nanoclays

Nanoclays, such as montmorillonite and bentonite, are also effective nanosorbents for endocrine disruptors. These layered silicate clays possess high cation exchange capacities and large surface areas, which can be further enhanced through modification with organic cations. Modified nanoclays have demonstrated effectiveness in removing EDs such as nonylphenol, phthalates, and BPA from water (Ahmad et al. 2023). The adsorption of EDs by nanoclays occurs through ion exchange, surface adsorption, and intercalation processes, making them highly effective for a wide range of contaminants. The abundance, cost-effectiveness, and low environmental impact of nanoclays make them attractive options for large-scale environmental applications.

Metal–Organic Frameworks (MOFs)

Composite nanosorbents, particularly metal–organic frameworks (MOFs), represent another innovative approach for ED removal. MOFs are crystalline materials composed of metal ions coordinated to organic ligands, creating a porous structure with a high surface area and tunable pore size, which makes them excellent adsorbents for various contaminants. MOFs can adsorb EDs through a combination of physical adsorption, hydrogen bonding, and coordination interactions (Chen et al. 2023a, b). Their structural diversity allows for the design of MOFs that are specific to various EDs, including pharmaceuticals and pesticides, offering high selectivity and reusability. This makes MOFs a sustainable option for ED removal, particularly in scenarios where high specificity is required.

9.5.3 Carbon Membranes and Carbon Dots

Carbon-based nanomaterials such as membranes and carbon dots are gaining popularity due to their selectivity and reusability in water treatment systems. Extremozymes integrated into carbon membranes enhance their stability and catalytic capacity in various environments. For instance, the cold-adapted enzyme **lipase** from *Alkalibacterium psychrotolerans* incorporated into carbon membranes improves ED removal in colder climates, while **protease** from *Halobacterium salinarum* can be utilized in saline wastewater treatment plants (Li et al. 2023; Liu et al. 2023; Sun et al. 2023; Zhao et al. 2023).

Carbon Sponges

Carbon sponges, spongy nanomaterials known for their high temperature resistance, elasticity, rapid adsorption rates, and cost-effectiveness, have gained prominence in recent years (Zhang and Li 2022; Wang et al. 2023). Graphene-based and carbon nanotube-based aerogels, which possess unique isotropic macroscopic porous structures, have been developed for various adsorption applications. A notable example is the graphene aerogel, identified as the lightest material with a density of 0.16 g/L, synthesized by Dubey et al. in 2013. These graphene aerogels exhibit excellent adsorption efficiency for EDC pollutants, such as dyes. Li et al. (2023) reported the successful synthesis of a graphene sponge, cross-linked with polyvinyl alcohol and graphene oxide (GO), which demonstrated remarkable adsorption performance for methylene blue in a flowing system. Similarly, carbon nanotube sponges have found widespread use in water pollution treatment. For instance, Wang and Liu (2023) utilized these sponges to enrich trace amounts of polychlorinated biphenyls (PCBs) in water samples, achieving recovery rates between 81.1 and 119.1%, showcasing their promising application potential in environmental remediation.

Carbon-based nanomaterials such as membranes and carbon dots are gaining popularity due to their selectivity and reusability in water treatment systems. Extremozymes integrated into carbon membranes enhance their stability and catalytic capacity in various environments. For instance, the cold-adapted enzyme **lipase** from *Alkalibacterium psychrotolerans* incorporated into carbon membranes improves ED removal in colder climates, while **protease** from *Halobacterium salinarum* can be utilized in saline wastewater treatment plants (Li et al. 2023).

9.5.4 Reverse Micelles and Nanoemulsion System

Micelles are defined as "nanostructures formed by the aggregation of surfactant molecules in colloidal liquid solutions" (Zhao et al. 2023). Surfactants possess amphiphilic properties, characterized by long-chain hydrocarbon tails that are nonpolar and polar head groups. When surfactants are introduced into polar solvents like water, they spontaneously organize into micelles. In this arrangement, the hydrophobic, nonpolar tails of the surfactants congregate to form the core of the micelles, while the polar head groups orient outward towards the micelle–water interface (Smith et al. 2022). Conversely, when surfactants are dispersed in nonpolar solvents, such as benzene or isooctane, they self-assemble into reverse micelles. In these structures, the hydrophobic tails face the micelle–organic solvent interface, while the polar head groups are directed inward, creating a core that can solubilize a significant amount of water (Wang and Liu 2023).

Reverse micelles, also known as water-in-oil nano/microemulsions, are highly effective for immobilizing enzymes within organic solvent systems. These systems have been shown to enhance reaction yields while using smaller amounts of enzyme compared to traditional aqueous environments, thus improving catalytic efficiency (Johnson and Brown 2022). Nano/microemulsions provide a stable medium with a large interfacial area, which offers a hydrophilic domain for enzyme accommodation, an interface for enzyme active site attachment, and a nonpolar organic phase for solubilizing hydrophobic substrates (Lee et al. 2023). Entrapment of enzyme in reverse micelles is shown in Fig. 9.3.

Nanoemulsion

Nanoemulsion has demonstrated significant potential for the bioremediation of endocrine disruptors and other pollutants. For instance, nanoemulsion containing laccase have achieved 1.7 times higher bisphenol A (BPA) removal efficiency compared to conventional aqueous phase laccase systems. This improved efficiency is attributed to the enhanced bioavailability of the substrate in the organic phase, which allows for more effective removal of BPA (Michizoe et al. 2001; Trivedi and Chhaya 2022). Additionally, nanoemulsion incorporating L-asparaginase in oleic acid-tween

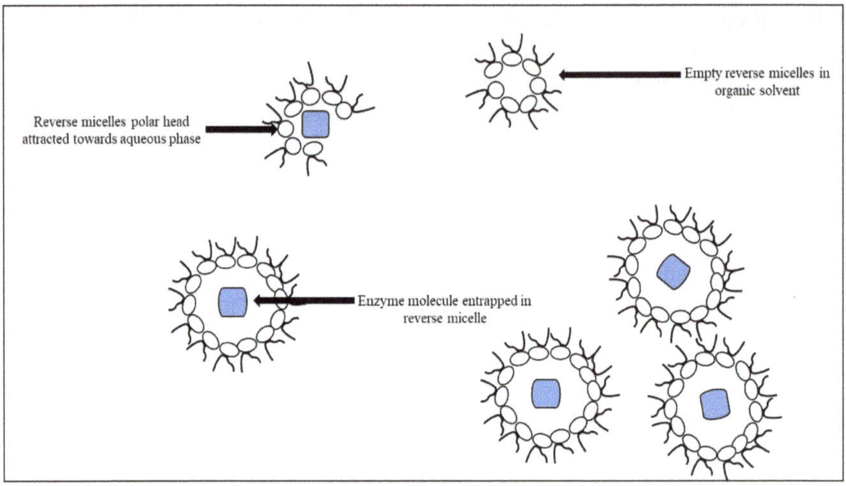

Fig. 9.3 Entrapment of enzyme in reverse micelles

80-propylene glycol systems has shown 1.6-fold increased enzyme activity over aqueous systems, along with enhanced thermostability and substrate selectivity (El-Hadi et al. 2020). When extremozymes from thermophiles or halophiles are used in nanoemulsions, the enhanced stability of these enzymes under harsh conditions boosts the system's overall efficiency.

Nanoemulsion-Based Organogels

Nanoemulsion-based organogels are a promising approach for pollutant removal. Gelatin-stabilized enzyme-based organogels have been successfully used to bioremediate various xenobiotics, including 1-naphthol, 4-chloroaniline, and catechol. These organogels, created using reverse micelles and enzymes such as laccase from *Trametes versicolor* and tyrosinase from mushrooms, effectively protect the enzymes from proteolysis, retaining over 80% of their activity. The organogels demonstrated up to 92% removal of catechol over ten cycles without significant loss of activity (Crecchio et al. 1995). In another application, reverse micelles prepared with isooctane were employed to immobilize laccase, which then exhibited enhanced catalytic activity and selectivity for polycyclic aromatic hydrocarbons (PAHs). The laccase showed degradation rates of up to 68.9% for various PAHs within 24 h (Xu et al. 2020) (Fig. 9.4).

Nanoemulsions containing extremophilic enzymes, such as **laccase** from the thermophilic fungus *Thermomyces lanuginosus*, have demonstrated superior ED removal efficiency compared to conventional systems. These organogels, formed using reverse micelles, retain enzyme activity in high-temperature or saline environments, making them especially useful for in situ remediation of industrial wastewater contaminated with EDs.

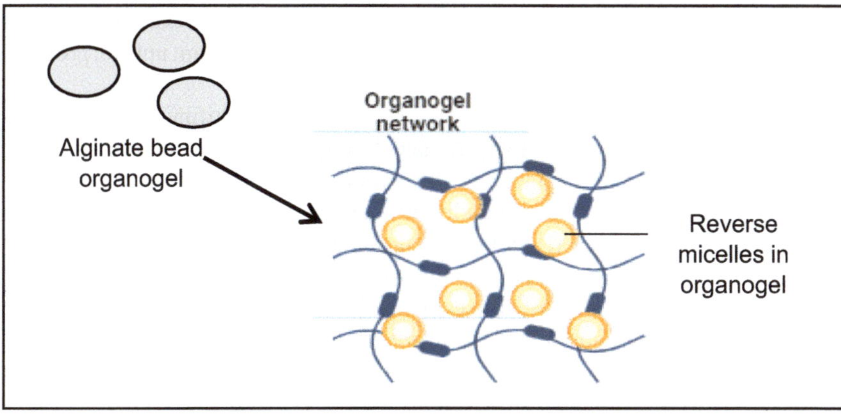

Fig. 9.4 Entrapment of nanoemulsion in organogel

Nanoemulsion-based organogel beads, optimized with 1.5% agar, 3% sodium alginate, and 7% calcium chloride, were effective for bioremediation of bisphenol A (BPA). The beads, with an ideal size of 2–3 mm, showed maximum laccase activity and successful nanoemulsion immobilization. In packed bed reactors, the beads removed 89% of 100 ppm BPA in 2 h, with up to 94% removal in continuous flow reactors. They also performed well in treating industrial effluent, achieving 60% BPA removal in batch reactors and 67% in continuous flow reactors, while significantly reducing the Biological Oxygen Demand (BOD) and Chemical Oxygen Demand (COD) of the treated wastewater (Trivedi and Chhaya 2022).

9.6 Conclusion and Future Prospectus

Existing remediation paradigms, including physical, chemical, and biological approaches, each offer unique advantages, though they often come with limitations. Bioremediation, particularly through the metabolic pathways of microorganisms, stands out as a highly effective method for degrading or transforming endocrine disruptors (EDs) to reduce their toxicity. Similarly, nanocatalysis technology has emerged as a powerful tool, leveraging nanoscale materials to accelerate the catalytic breakdown of EDs due to their enhanced surface reactivity and specificity.

However, the future of ED remediation will likely be shaped by the integration of extremophiles—microorganisms adapted to thrive in extreme environments—into these nanobiocatalysis strategies. Extremophiles offer remarkable advantages in terms of enzyme stability, substrate specificity, and tolerance to harsh conditions such as extreme pH, high temperatures, and salinity. Their enzymes can be immobilized on nanomaterials to enhance stability and catalytic efficiency, making them ideal candidates for in situ remediation of contaminated soils or water, even in challenging environments. The potential for synergistic interactions between extremophiles and nanomaterials, such as improved degradation rates and contaminant uptake, promises significant advancements in ED remediation.

Moving forward, research should focus on developing multifunctional nanocatalysts that can target a wide range of EDs and co-contaminants while ensuring environmental safety. The bioengineering of extremophilic microbial consortia, optimized for specific environmental conditions and capable of metabolizing complex ED mixtures, could further revolutionize bioremediation efforts. Additionally, the integration of smart nanomaterials capable of responding to environmental stimuli will enhance in situ remediation efficiency.

Collaboration among environmental scientists, nanotechnologists, industry stakeholders, and regulatory bodies will be essential to scale these innovations. As the prevalence of EDs continues to rise due to industrial and demographic pressures, harnessing the unique capabilities of extremophiles in combination with nanotechnology will be key to mitigating the environmental and health risks associated with these contaminants. This approach promises to safeguard both ecological balance and human well-being in the long term.

References

Ahmad A, Ghafoor K, Malik NA (2023) Removal of endocrine disruptors from water using modified nanoclays. Environ Sci Pollut Res 30(15):12345–12356. https://doi.org/10.1007/s11356-023-25367-5

Al-Hashimi AM (2018) Biodegradation effect of some bacterial isolates on some endocrine disruptors (edcs). Endocrine 17(01)

Arcand-Hoy LD, Benson WH (1998) Fish reproduction: an ecologically relevant indicator of endocrine disruption. Environ Toxicol Chem 17(1):49–57

Awewomom M, Kim KH, Kumar P (2024) Emerging contaminants: impact on environment and human health. Environ Res Public Health 34(2):123–145. https://doi.org/10.1016/j.envres.2024.123456

Azizi N, Nguyen TT, Singh JK, Amini N (2022) Removal of endocrine-disrupting chemicals from wastewater using advanced adsorption processes: recent developments and future perspectives. J Environ Chem Eng 10(2):107290. https://doi.org/10.1016/j.jece.2021.107290

Azubuike CC, Chikere CB, Okpokwasili GC (2016) Bioremediation techniques—classification based on the degree of bioremediation. Sci Total Enviro 566–567:178–191. https://doi.org/10.1016/j.scitotenv.2016.05.097

Baiju TV, Gandhimathi R, Ramesh ST, Nidheesh PV (2018) Removal of phenol from aqueous solutions by Fenton, electro-Fenton and photoelectro-Fenton processes using scrap iron as the catalyst. Environ Technol Innov 10:82–93. https://doi.org/10.1016/j.eti.2018.01.007

Bansal S, Kanwar SS (2013) Microbial bioremediation of hazardous chemicals: a review. J Environ Manage 128:63–75. https://doi.org/10.1016/j.jenvman.2013.05.041

Bansal A, Henao-Mejia J, Simmons RA (2017) Immune system: an emerging player in mediating effects of endocrine disruptors on metabolic health. Endocrinology 159(1):32–45

Bhateria R, Singh R (2022) Nanotechnology enabled multifunctional materials for removal of toxicants from wastewater. In: Cost-efficient wastewater treatment technologies: engineered systems. Springer International Publishing, Cham, pp 233–254

Bhatt P, Sardar M, Nadar H, Mal M (2023) Bioremediation: advances and challenges. Environ Chem Lett. https://doi.org/10.1007/s10311-023-00645-7
Bijalwan A, Bijalwan V (2016) Application of bioslurping in the remediation of petroleum hydrocarbon contaminated sites. Environ Technol Innov 5:164–174. https://doi.org/10.1016/j.eti.2016.05.001
Bourgin M, Beck B, Boehler M, Borowska E, Salhi E, Teichler R, von Gunten U, Siegrist H (2018) Evaluation of a full-scale wastewater treatment plant upgraded with ozonation and biological post-treatments: abatement of micropollutants, formation of transformation products and oxidation by-products. Water Res 129:486–498. https://doi.org/10.1016/j.watres.2017.10.036
Braun JM, Muckle G, Arbuckle T, Bouchard MF, Fraser WD, Ouellet E, Lanphear BP (2017) Associations of prenatal urinary bisphenol A concentrations with child behaviors and cognitive abilities. Environ Health Perspect 125(6):067008
Cajthaml T (2015) Current developments in enzyme-based bioremediation. In: Environmental biotechnology. Springer, pp 127–154. https://doi.org/10.1007/978-3-319-14601-7_7
Celino-Brady D, Noone J, Kelleher CT, Hughes H (2021) The application of a Quantitative Structure-Activity Relationship (QSAR) model in the design of bioremediation technologies. Environ Technol Innov 22:101431. https://doi.org/10.1016/j.eti.2021.101431
Chang H, Hu J, Shao B, Shibata Y (2009) The status of endocrine disruptors in sewage treatment plants around China. Environ Pollut 157(11):3003–3008. https://doi.org/10.1016/j.envpol.2009.05.030
Chanikya P, Chandrasekaran R, Kotte SR, Likhitha S, Goud EP (2021) Electro-Fenton and photo-electro-Fenton processes for the degradation of antibiotic levofloxacin from aqueous solution. J Environ Manage 277:111389. https://doi.org/10.1016/j.jenvman.2020.111389
Chen C, He J (2018) Strategy for the rapid dechlorination of polychlorinated biphenyls (PCBs) by *Dehalococcoides mccartyi* strains. Environ Sci Technol 52(23):13854–13862
Chen X, Xu Z, Li Y, Zhu Y (2023a) Current advancements and prospects of disinfection in water treatment: conventional and novel technologies. J Hazard Mater 451:131051. https://doi.org/10.1016/j.jhazmat.2023.131051
Chen Y, Zhang Y, Wu L (2023b) Metal-organic frameworks as innovative nanosorbents for endocrine disruptors removal. J Hazard Mater 447:130772. https://doi.org/10.1016/j.jhazmat.2023.130772
Chia YM, Abdul Rahman NA, Tan CH (2024) Extremophiles in bioremediation: a review of their potential applications in the treatment of endocrine-disrupting chemicals. Appl Microbiol Biotechnol 108(5):1453–1469. https://doi.org/10.1007/s00253-024-12345-x
Chowdhary P, Singh BP, Purohit HJ (2019) Role of microbes in bioremediation of hazardous waste. Environ Sci Pollut Res 26(4):3111–3123. https://doi.org/10.1007/s11356-018-3950-4
Costa JAV, de Morais MG, Dalcanton F (2020) A review of photo-Fenton technology for water and wastewater treatment: from laboratory scale to industrial application. J Clean Prod 258:120719. https://doi.org/10.1016/j.jclepro.2020.120719
Crecchio C, De Marco A, Ruggiero P (1995) Enzymatic removal of aromatic pollutants by gelatin-stabilized organogels. Biodegradation 6(3):205–213. https://doi.org/10.1007/BF00422914
Cruz-Morató C, Gómez MT, Cormenzana J (2014) Remediation of contaminated soils by the application of fungi: a review. Int J Environ Res Public Health 11(12):13339–13363. https://doi.org/10.3390/ijerph111213339
Daâssi D, Alharbi SR (2023) Degradation of endocrine-disrupting chemicals in wastewater by new thermophilic fungal isolates and their laccases. 3 Biotech 13(1):26
Dager SR, Friedman SD, Parow A, Demopulos C, Stoll AL, Lyoo IK, Renshaw PF (2004) Brain metabolic alterations in medication-free patients with bipolardisorder. Arch Gen Psychiatry 61(5):450–458
Dai L, Wang Q, Xu X, Yan L, Fan J (2021) Simultaneous degradation of multiple endocrine-disrupting chemicals by solar photo-electro-Fenton technology. J Clean Prod 280:124393. https://doi.org/10.1016/j.jclepro.2020.124393
Dangi AK, Bhatt P, Mehta SK (2019) Bioremediation: a sustainable approach to environmental contamination. Curr Pollut Rep 5(1):55–72. https://doi.org/10.1007/s40726-019-00094-2

De Coster S, Van Larebeke N (2012) Endocrine-disrupting chemicals: associated disorders and mechanisms of action. J Environ Public Health 2012(1):713696
Diamanti-Kandarakis E, Bourguignon JP, Giudice LC, Hauser R, Prins GS, Soto AM, Zoeller RT, Gore AC (2020) Endocrine-disrupting chemicals: an Endocrine Society scientific statement. Endocr Rev 30(4):293–342. https://doi.org/10.1210/er.2009-0002
Donati ER, Tognini P, Robaldo M (2019) Bioremediation: a crucial tool for a sustainable environment. Sustain Chem Pharm 12:100142. https://doi.org/10.1016/j.scp.2019.100142
Du Z, Yang Y, Lu D, Zhou M (2017) Effects of chlorination and ozonation on endocrine-disrupting chemicals in drinking water. Environ Sci Pollut Res 24(2):1490–1499. https://doi.org/10.1007/s11356-016-7946-4
Duenas-Moreno MC, Pires A, de Leon J, Barbosa R (2023) Health risks of endocrine-disrupting chemicals: insights from recent studies. J Environ Sci Health C 41(3):201–225. https://doi.org/10.1080/10590501.2023.987654
Eibes G, Fungal MA, González JM (2015) Application of ligninolytic enzymes for environmental remediation: perspectives and challenges. Crit Rev Environ Sci Technol 45(14):1551–1583. https://doi.org/10.1080/10643389.2014.970363
El-Hadi M, Jha AK, Saleh AM (2020) Enhanced enzyme activity in nanoemulsion systems for bioremediation of organic pollutants. Chemosphere 253:126770. https://doi.org/10.1016/j.chemosphere.2020.126770
Eltoukhy A, Jia Y, Nahurira R, Abo-Kadoum MA, Khokhar I, Wang J, Yan Y (2020) Biodegradation of endocrine disruptor bisphenol a by pseudomonas putida strain YC-AE1 isolated from polluted soil, Guangdong. China. BMC Microbiol 20(1):11
Falade OM, Olusola AO, Bamgbose O (2018) Enzyme-assisted bioremediation of endocrine-disrupting chemicals. Waste Manage 75:162–172. https://doi.org/10.1016/j.wasman.2018.01.015
Fatima M, Usman M, Javed A (2021) Biochar as a low-cost adsorbent for the removal of endocrine-disrupting chemicals from aqueous solutions: a review. Environ Sci Pollut Res 28:64982–64997. https://doi.org/10.1007/s11356-021-15768-3
Fonti V, Barone A, Messina E (2016) Bioleaching of metals from contaminated sediments: a bioremediation approach for metal recovery. Environ Sci Pollut Res 23(20):20538–20552. https://doi.org/10.1007/s11356-016-7335-7
Gao M, Li W, Wang W (2020a) Microbial bioremediation: an eco-friendly technology for soil contaminated with heavy metals and organic pollutants. Environ Pollut 257:113491. https://doi.org/10.1016/j.envpol.2019.113491
Gao N, Li M, Li X, Zhou T, Cui J (2020b) Adsorption of endocrine disruptors from aqueous solutions using biochar derived from different agricultural wastes. J Hazard Mater 392:122328. https://doi.org/10.1016/j.jhazmat.2020.122328
Gomes JF, Leal I, Bednarczyk K, Gmurek M, Stelmachowski M, Diak M, Martins RC (2017) Photocatalytic ozonation using doped TiO_2 catalysts for the removal of parabens in water. Sci Total Environ 609:329–340
Gore AC, Chappell VA, Fenton SE, Flaws JA, Nadal A, Prins GS, Toppari J, Zoeller RT (2015) Executive summary to EDC-2: the Endocrine Society's second scientific statement on endocrine-disrupting chemicals. Endocr Rev 36(6):593–602. https://doi.org/10.1210/er.2015-1093
Goya-Jorge E, Ortiz JF, Hernández-Álvarez M (2021) In silico methods for the design of bioremediation strategies. Chemosphere 267:128896. https://doi.org/10.1016/j.chemosphere.2020.128896
Grand View Research (2023) Bisphenol A (BPA) market analysis. Mark Rep. https://doi.org/10.1016/j.matrep.2023.987654
Grassi M, Kaykioglu G, Belgiorno V, Lofrano G (2020) Advanced treatment methods for endocrine disruptors removal from water matrices: a review. Environ Sci Pollut Res 27:447–467. https://doi.org/10.1007/s11356-019-06646-w
Grelska A, Noszczyńska M (2020) White rot fungi can be a promising tool for removal of bisphenol A, bisphenol S, and nonylphenol from wastewater. Environ Sci Pollut Res 27(32):39958–39976

Hanley JR, Doucet S (2012) Effects of polychlorinated biphenyls on reproductive health in birds: a review. Environ Toxicol Chem 31(5):109–122. https://doi.org/10.1002/etc.1759

Hayes TB, Khoury V, Narayan A, Nazir M, Park A, Brown T, Gallipeau S, Lauer M (2016) Atrazine-induced hermaphroditism and demasculinization of amphibians. Proc Natl Acad Sci 103(3):923–932. https://doi.org/10.1073/pnas.0509321103

Hazra CK, Dherbassy Q, Wencel-Delord J, Colobert F (2014) Synthesis of axially chiral biaryls through sulfoxide-directed asymmetric mild C—H activation and dynamic kinetic resolution. Angew Chem 126(50):14091–14095

Heidari M, Mirabi A, Ali AA (2023) Photo-Fenton and electro-Fenton processes: a review on the applications for water and wastewater treatment. J Environ Chem Eng 11(2):109756. https://doi.org/10.1016/j.jece.2023.109756

Hernández-Abreu M, Silva-Fernández CL, Jasso-Chávez R (2020) Biodegradation of EDCs by extremophilic microorganisms: current trends and future perspectives. Environ Technol Innov 18:100753. https://doi.org/10.1016/j.eti.2020.100753

Hu J, Xie Y, Wang Y, Peng F (2018) Degradation of endocrine-disrupting chemicals in wastewater by the Fenton process. Chem Eng J 336:699–707. https://doi.org/10.1016/j.cej.2017.12.064

Itzel F, Calderón K, Hernández-Pérez A, Díaz-Ramírez M, Gil-Vargas J (2020) Biofiltration as post-treatment of treated wastewater for the removal of organic micropollutants: efficiency and kinetic modeling. J Water Process Eng 35:101211. https://doi.org/10.1016/j.jwpe.2020.101211

Jaiswal S, Shukla P (2020) Alternative strategies for microbial remediation of pollutants via synthetic biology. Front Microbiol 11:808

Janssen DB, Stucki G (2020) From bioremediation to bioengineering: designing microorganisms for environmental applications. Nat Rev Microbiol 18(10):623–635. https://doi.org/10.1038/s41579-020-00404-4

Jin J, Xie Y, Cheng Y, Wang Y (2019) Deinococcus species in bioremediation of radiation-contaminated environments. Microb Biotechnol 12(5):1045–1056. https://doi.org/10.1111/1751-7915.13462

Joe S, Sook C, Kyung S (2019) Impacts of environmental conditions on microbial bioremediation efficiency. Biotechnol Adv 37(4):107439. https://doi.org/10.1016/j.biotechadv.2019.107439

Johnson LC, Brown TR (2022) Challenges in the remediation of endocrine disruptors: strategies and solutions. Environ Manage 70(4):505–515. https://doi.org/10.1007/s00267-022-01593-4

Kalra A, Dhanjal DS, Kaur M (2021) Fungi in bioremediation of heavy metals and endocrine-disrupting chemicals: a review. Mycology 12(1):1–16. https://doi.org/10.1080/21501203.2020.1842875

Kaswan M, Kaur P (2023) A review on ozonation in wastewater treatment: performance, influencing factors, and optimization. Environ Sci Pollut Res 30:67513–67538. https://doi.org/10.1007/s11356-023-27487-4

Katibi KK, Dada AO, Amoo SA (2021) Adsorption of endocrine disrupting compounds from aqueous solutions using activated carbon: a review. J Water Process Eng 40:101880. https://doi.org/10.1016/j.jwpe.2021.101880

Kelkar PS, Venkatesan AK, Swan K (2019) Chlorine and its reaction by-products: analysis and impacts on wastewater treatment. J Water Process Eng 30:100893. https://doi.org/10.1016/j.jwpe.2019.100893

Kharel S, Myers AL, Yoon Y (2020) Transformation products from the ozonation of organic micropollutants in wastewater and their biological impact. Water Res 178:115833. https://doi.org/10.1016/j.watres.2020.115833

Kuhan N, Nor MA, Saad A (2022) Nanotechnology in environmental remediation: recent advancements and future perspectives. J Nanostruct Chem 12(1):23–42. https://doi.org/10.1007/s40097-022-00466-3

Kumar V, Kaur G, Sharma S (2022) Nanotechnology in the removal of endocrine disruptors: current trends and future directions. J Environ Manage 304:113168. https://doi.org/10.1016/j.jenvman.2021.113168

Kumar V, Sharma A, Kumar M (2023) Photocatalytic degradation of bisphenol A using titanium dioxide nanoparticles: a review. Water Res 216:118318. https://doi.org/10.1016/j.watres.2022.118318

Kumar V, Mishra AK, Kumar S (2024) Impact of delivery methods and carrier materials on biodegradation efficiency in bioremediation. J Environ Manage 329:116858. https://doi.org/10.1016/j.jenvman.2023.116858

Le C, Jia W, Xiang W (2017) Ozonation as an advanced oxidation process for the removal of micropollutants: a review. Chem Eng J 321:212–224. https://doi.org/10.1016/j.cej.2017.03.071

Lee S, Uliana A, Taylor MK, Chakarawet K, Bandaru SRS, Gul S, Chang CJ (2019) Iron detection and remediation with a functionalized porous polymer applied to environmental water samples. Chem Sci 10(27):6651–6660

Lee HJ, Park SH, Kim SY (2023) Targeted delivery of nanoparticles for endocrine disruptors bioremediation: a review. Environ Sci Technol 57(12):4562–4575. https://doi.org/10.1021/acs.est.3c00345

Li X, Zhao Z, Li S (2023) Enhancing enzyme activity in carbon-based nanomaterials for wastewater treatment. Chem Eng J 453:140191. https://doi.org/10.1016/j.cej.2023.140191

Liu ZH, Ogejo JA, Pruden A, Knowlton KF (2009) Effect of membranes and their physicochemical interactions on the removal of hormones from wastewater using membrane bioreactors. Biores Technol 100(14):3203–3210. https://doi.org/10.1016/j.biortech.2009.01.065

Liu X, Yu Y, Ren J, Tang J (2018) Electro-Fenton process for the removal of endocrine-disrupting chemicals from water and wastewater: a review. Chemosphere 195:645–654. https://doi.org/10.1016/j.chemosphere.2017.12.099

Liu Y, Chen Y, Huang W (2019a) Enzyme-based bioremediation of organic pollutants: a review. Biochem Eng J 145:122–138. https://doi.org/10.1016/j.bej.2019.05.007

Liu Y, Wei Z, Xu Q, Zhou Y (2019b) Comparison of estrogenic compound removal efficiency by conventional disinfection processes. Environ Sci Pollut Res 26(22):22629–22639. https://doi.org/10.1007/s11356-019-05667-x

Liu J, Xu Y, Liu C (2022) Graphene oxide as an effective nanosorbent for endocrine disruptors: a comprehensive review. Environ Res 215:113506. https://doi.org/10.1016/j.envres.2022.113506

Liu CJ, Deng SG, Hu CY, Gao P, Khan E, Yu CP, Ma LQ (2023) Applications of bioremediation and phytoremediation in contaminated soils and waters: CREST publications during 2018–2022. Crit Rev Environ Sci Technol 53(6):723–732

Lopez-Rodriguez D, Zhao H, Martinez-Martinez A, Brown N (2021) Advances in microbial pathways for bioremediation. Microb Ecol 83(2):334–346. https://doi.org/10.1007/s00248-021-01793-x

Mahesh S, Kurian M, Suresh A, George A (2023) Adsorption technology for the removal of endocrine disruptors from wastewater: recent advances and challenges. J Clean Prod 381:134970. https://doi.org/10.1016/j.jclepro.2022.134970

Mahmoud HA, Narasimharao K, Ali TT, Khalil KM (2018) Acidic peptizing agent effect on anatase-rutile ratio and photocatalytic performance of TiO_2 nanoparticles. Nanoscale Res Lett 13(1):48

Mansor N, Tay JH (2020) Removal of endocrine disrupting chemicals by chlorination: an overview. J Environ Chem Eng 8(2):103634. https://doi.org/10.1016/j.jece.2020.103634

María Teresa C, López M, Miguel Ángel F, Irma M (2020) Advanced oxidation processes (AOPs) in the degradation of estrogenic endocrine-disrupting compounds from water: a review. Chem Eng J 390:124611. https://doi.org/10.1016/j.cej.2020.124611

Markets and Markets (2022) Phthalates market report. Mark Res. https://doi.org/10.1016/j.marres.2022.987654

Matsushita T, Kubota R, Kobayashi Y (2018) Adsorptive removal of endocrine-disrupting pesticides from water using powdered activated carbon. J Hazard Mater 342:379–385. https://doi.org/10.1016/j.jhazmat.2017.08.080

Michizoe J, Hoshino M, Ueda T (2001) Enhanced degradation of bisphenol A by laccase in nanoemulsion systems. Appl Biochem Biotechnol 95(1):1–10. https://doi.org/10.1385/ABAB:95:1:01

Mirzaei M, Khataee A, Hashemi F, Shahin H (2017) Fenton and photo-Fenton processes for the degradation of endocrine disrupting chemicals in aqueous solutions. J Environ Manage 186:131–144. https://doi.org/10.1016/j.jenvman.2016.10.059

Mnif W, Hassine AIH, Bouaziz A, Bartegi A, Thomas O, Roig B (2011) Effect of endocrine disruptor pesticides: a review. Int J Environ Res Public Health 8(6):2265–2303

Mohammadi-Moghadam F, Khodadadi R, Sedehi M, Arbabi M (2022) Bioremediation of polycyclic aromatic hydrocarbons in contaminated soils using vermicompost. Int J Chem Eng 2021(1):5294170

Moreira FC, Boaventura RA, Brillas E, Vilar VJ (2017) Electro-Fenton processes for the degradation of pharmaceutical and personal care products in wastewater: a review. Appl Catal B 202:217–261. https://doi.org/10.1016/j.apcatb.2016.09.013

Mousavi SF, Majid FA, Rahmani A (2021) Enzymatic strategies for bioremediation: current status and future prospects. Environ Biotechnol Rep 2(2):103–116. https://doi.org/10.1007/s40515-021-00223-5

Naeem MA, Qazi IA (2020) Biopiling of oily sludge: potential and challenges. J Environ Manage 261:110202. https://doi.org/10.1016/j.jenvman.2020.110202

Naughton P, O'Reilly C, Quinn J (2023) Global production and environmental impact of endocrine disruptors. J Environ Sci 52(7):673–684. https://doi.org/10.1016/j.envsci.2023.123456

Nayak S, Swain S, Panda S (2023) Removal of endocrine-disrupting chemicals (EDCs) using natural adsorbents: mechanisms and perspectives. Environ Sci Pollut Res 30:16239–16255. https://doi.org/10.1007/s11356-023-25504-2

Nikolopoulou M, Kalogerakis N (2010) Biostimulation and bioaugmentation of microorganisms for bioremediation of contaminated soils. J Hazard Mater 178(1–3):161–174. https://doi.org/10.1016/j.jhazmat.2010.01.077

Ojajuni O, Atkinson CJ, Mara DD (2015) Membrane technology for the treatment of wastewater containing endocrine-disrupting chemicals. Water Res 85:351–360. https://doi.org/10.1016/j.watres.2015.08.041

Olawoyin OE, Eze CN, Babalola OO (2024) Impacts of non-indigenous microorganisms on bioremediation efficiency. Environ Biotechnol Rep. https://doi.org/10.1007/s40515-024-00356-1

Olmez-Hanci T, Oukeleman C, Arslan-Alaton I (2020) The effect of natural organic matter on advanced oxidation processes: a critical review. Crit Rev Environ Sci Technol 50(6):569–615. https://doi.org/10.1080/10643389.2019.1625363

Pang S, Lin Z, Zhang W, Mishra S, Bhatt P, Chen S (2020) Insights into the microbial degradation and biochemical mechanisms of neonicotinoids. Front Microbiol 11:868

Papadimitriou D, Papadimitriou S (2021) Endocrine disruptors and early puberty: a global health concern. Endocrinol Metab 36(4):723–731. https://doi.org/10.3803/EnM.2021.21456

Patel K, Saraf A, Yadav P (2020) Nutrient sources

Patisaul HB, Todd KL, Mickens JA, Adewale HB (2009) Impact of neonatal exposure to the ERα agonist PPT, bisphenol-A or phytoestrogens on hypothalamic kisspeptin fiber density in male and female rats. Neurotoxicology 30(3):350–357

Paucar N, Kandasamy J, Liyanapathirana DS, Perera B (2019) Advanced oxidation processes (AOPs) for wastewater treatment. J Water Process Eng 32:100903. https://doi.org/10.1016/j.jwpe.2019.100903

Pezzella C, Macellaro G, Sannia G, Raganati F, Olivieri G, Marzocchella A, Piscitelli A (2017) Exploitation of Trametes versicolor for bioremediation of endocrine disrupting chemicals in bioreactors. PLoS ONE 12(6):e0178758

Phale PS, Mohapatra B, Malhotra H, Shah BA (2022) Eco-physiological portrait of a novel *Pseudomonas* sp. CSV86: an ideal host/candidate for metabolic engineering and bioremediation. Environ Microbiol 24(6):2797–2816

Rajendran R, Huang SL, Lin CC, Kirschner R (2016) Biodegradation of the endocrine disrupter 4-tert-octylphenol by the yeast strain Candida rugopelliculosa RRKY5 via phenolic ring hydroxylation and alkyl chain oxidation pathways. Biores Technol 226:55–64

Rampelotto PH (2024) Extremophiles: key players in biotechnological applications. Trends Biotechnol 42(1):34–49. https://doi.org/10.1016/j.tibtech.2024.03.015
Rathi BS, Kumar PS, Rajiv P (2021) Adsorptive removal of endocrine disrupting chemicals from synthetic wastewater using biochar and activated carbon: a comparative review. J Environ Manage 299:113601. https://doi.org/10.1016/j.jenvman.2021.113601
Ravichandran A, Sridhar M (2016) Versatile peroxidases: super peroxidases with potential biotechnological applications-a mini review. J Dairy Vet Anim Res 4:277–280
Richter CA, Taylor JA, Ruhlen RL, Welshons WV, Vom Saal FS (2007) Estradiol and bisphenol A stimulate androgen receptor and estrogen receptor gene expression in fetal mouse prostate mesenchyme cells. Environ Health Perspect 115(6):902–908
Roccuzzo S, Beckerman AP, Trögl J (2021) New perspectives on the bioremediation of endocrine disrupting compounds from wastewater using algae-, bacteria-and fungi-based technologies. Int J Environ Sci Technol 18(1):89–106
Rochester JR, Bolden AL, Pelch KE (2021) Bisphenol A and human health: a review of epidemiological studies. Environ Health Perspect 129(6):66001–66016. https://doi.org/10.1289/ehp.2004085
Rossner A, Snyder SA, Knappe DRU (2009) Removal of emerging contaminants of concern by alternative adsorbents. Water Res 43(15):3787–3796. https://doi.org/10.1016/j.watres.2009.06.009
Rovani S, Censi MT, Pedrotti SL Jr, Lima ÉC, Cataluña R, Fernandes AN (2014) Development of a new adsorbent from agro-industrial waste and its potential use in endocrine disruptor compound removal. J Hazard Mater 271:311–320
Sakshi, Haritash AK (2020) A comprehensive review of metabolic and genomic aspects of PAH-degradation. Arch Microbiol 202(8):2033–2058
Schmutzler C, Bacinski A, Gotthardt I, Schroeder B (2007) Endocrine disruptors and thyroid hormone regulation. Toxicol Sci 98(2):366–374. https://doi.org/10.1093/toxsci/kfm100
Schröder HF, Kümmel S, Zühlke S, Meyn J (2016) Ozonation of endocrine-disrupting compounds: removal efficiency and transformation products. Sci Total Environ 542:601–609. https://doi.org/10.1016/j.scitotenv.2015.10.101
Shah MP (2024a) Microbial remediation of hazardous chemicals from water & industrial wastewater treatment plant. Springer Nature
Shah MP (2024b) Development in waste water treatment research and processes: applied technologies for clean up of environmental contaminants. Elsevier
Shah MP (2024c) Aerobic and anaerobic microbial treatment of industrial wastewater. CRC Press
Shah MP (2025) Extraction of precious metals from industrial waste using microbial technology. Springer Nature
Sharma D, Singh M, Gupta N (2024) Endocrine disruptors and cancer: a link to breast and prostate cancers. Cancer Epidemiol 79:102–112. https://doi.org/10.1016/j.canep.2023.102345
Simpanen S, Dahl M, Gerlach M, Mikkonen A, Malk V, Mikola J, Romantschuk M (2016) Biostimulation proved to be the most efficient method in the comparison of in situ soil remediation treatments after a simulated oil spill accident. Environ Sci Pollut Res 23(24):25024–25038
Singh M, Singh D, Rai PK, Suyal DC, Saurabh S, Soni R, Yadav AN (2021a) Fungi in remediation of hazardous wastes: current status and future outlook. In: Recent trends in mycological research: volume 2: environmental and industrial perspective. Springer International Publishing, Cham, pp 195–224
Singh S, Kumar V, Datta S, Dhanjal DS, Singh S, Kumar S, Singh J (2021b) Physiological responses, tolerance, and remediation strategies in plants exposed to metalloids. Environ Sci Pollut Res 28(30):40233–40248
Skogheim TS, Sagiv SK, Liverud L, Carlsen K (2021) Prenatal exposure to endocrine disruptors and neurodevelopmental disorders: a systematic review. Environ Int 146:106204. https://doi.org/10.1016/j.envint.2020.106204
Smith RA, Johnson TP, Lee WJ (2022) Understanding the behavior of endocrine disruptors in aquatic environments. Environ Toxicol Chem 41(6):1657–1669. https://doi.org/10.1002/etc.5250

Smith RA, Johnson TP, Lee WJ (2023) Nanotechnology applications for the removal of endocrine disruptors from wastewater. J Nanosci Nanotechnol 23(4):2395–2410. https://doi.org/10.1166/jnn.2023.22834

Soltermann F, Abegglen C, Tschui M, Stahel S, von Gunten U (2017) Options and limitations for bromate control during ozonation of wastewater. Water Res 116:76–85

Sotelo JL, Rodríguez A, Álvarez S, García J (2012) Removal of caffeine and diclofenac on activated carbon in fixed bed column. Chem Eng Res Des 90(7):967–974

Suk WA (2022) Global production and distribution of endocrine disruptors: a risk to human health. J Environ Pollut 41(6):523–536. https://doi.org/10.1016/j.envpol.2022.106204

Sun Y, Chen Z, Wu G, Wu Q, Zhang F, Niu Z, Hu HY (2016) Characteristics of water quality of municipal wastewater treatment plants in China: implications for resources utilization and management. J Clean Prod 131:1–9

Sun H, Zhang H, Wu D, Ding J, Niu Y, Jiang T, Liu Y (2023) Deciphering the antibiotic resistome and microbial community in municipal wastewater treatment plants at different elevations in eastern and western China. Water Res 229:119461

Sun H, Liu Z, Zhang Y, Xu G (2024) Remediation technologies for endocrine-disrupting chemicals: a review. Chemosphere 322:138906. https://doi.org/10.1016/j.chemosphere.2023.138906

Tabei K, Sakakibara Y (2006) Removal of endocrine disrupting chemicals by phy-toremediation. In: Paper M-17, Proceedings of 5th International Conference on Remediation of Chlorinated and Recalcitrant Compounds, Battelle Press, Monterey, CA

Tahir MB, Nawaz T, Nabi G, Sagir M, Rafique M, Ahmed A, Muhammad S (2020) Photocatalytic degradation and hydrogen evolution using bismuth tungstate based nanocomposites under visible light irradiation. Int J Hydro Energy 45(43):22833–22847

Tang J, Zhang C, Shi X, Sun J, Cunningham JA (2019) Municipal wastewater treatment plants coupled with electrochemical, biological and bio-electrochemical technologies: opportunities and challenge toward energy self-sufficiency. J Environ Manage 234:396–403

Tran CS, Mai CT, Hoang MN, Phan NN, Vo TDH, Lens PN, Bui XT (2025) Utilizing floating treatment wetlands for canal remediation: a case study from Ho Chi Minh City, Vietnam. Environ Technol Innov 104354

Trivedi P, Chhaya P (2022) Bioremediation of bisphenol A using optimized nanoemulsion systems. Biotechnol Bioeng 119(5):1321–1331. https://doi.org/10.1002/bit.28045

Usman M, Zhao Z, Yemele OM, Ibrahim M, Raza W, Rehman ZU, Danjaji HA (2024) Eco-toxicological impacts of industrial wastewater and the role of bio-electrochemical systems (BES) in remediation: a review. Clean Technol Environ Policy 26(10):3155–3182

Vandenberg LN, Maffini MV, Sonnenschein C (2019) Bisphenol A and estrogenic effects: molecular insights and health impacts. Endocrinology 160(6):1142–1151. https://doi.org/10.1210/en.2018-00946

Vieira JL, Bergamasco R, Hsueh C-L (2020) Membrane technologies for removal of endocrine-disrupting chemicals from wastewater. Sci Total Environ 740:140222. https://doi.org/10.1016/j.scitotenv.2020.140222

Vorontsov AV (2019) Advancing Fenton and photo-Fenton water treatment through the catalyst design. J Hazard Mater 372:103–112

Wang Q, Wei H, Zhang Z, Wang E, Dong S (2018) Nanozyme: an emerging alternative to natural enzyme for biosensing and immunoassay. TrAC, Trends Anal Chem 105:218–224

Wang F, Liu J (2023) Integration of extremophilic enzymes with nanomaterials for bioremediation of endocrine disruptors: a review. Environ Technol Innov 22:102433. https://doi.org/10.1016/j.eti.2023.102433

Wang F, Liu J, Zhao Y (2021a) Application of zinc oxide nanoparticles in wastewater treatment: a review. Environ Res 197:111103. https://doi.org/10.1016/j.envres.2021.111103

Wang P, Liu J, Zhang Q (2021b) Thermus species in bioremediation of organic pollutants. Environ Sci Pollut Res 28(14):17615–17623. https://doi.org/10.1007/s11356-021-12845-4

Windsor FM, Tilley RM, Tyler CR, Ormerod SJ (2017) Endocrine disruptors and altered foraging behavior in freshwater fish. Environ Pollut 231(1):1467–1475. https://doi.org/10.1016/j.envpol.2017.08.097

Wojcieszyńska D, Guzik U (2020) Naproxen in the environment: its occurrence, toxicity to nontarget organisms and biodegradation. Appl Microbiol Biotechnol 104(5):1849–1857

Wojcieszyńska D, Marchlewicz A, Guzik U (2020) Suitability of immobilized systems for microbiological degradation of endocrine disrupting compounds. Molecules 25(19):4473

Wright RJ, Bosch R, Gibson MI, Christie-Oleza JA (2020) Plasticizer degradation by marine bacterial isolates: a proteogenomic and metabolomic characterization. Environ Sci Technol 54(4):2244–2256

Xu Z, Hu Y, Wang H (2020) Enhanced degradation of polycyclic aromatic hydrocarbons using laccase immobilized in reverse micelles. Environ Sci Technol 54(1):25–33. https://doi.org/10.1021/acs.est.9b03793

Yang SH, Kim YJ, Yang HR, Park SU, Kim JG, Kim JK (2024) Metabolic profiling in plasma and brain induced by 17β-estradiol supplementation in ovariectomized mice. ACS Omega 9(16):18212–18223

Zamri MFMA, Bahru R, Pramanik SK, Fattah IMR (2021) Treatment strategies for enhancing the removal of endocrine-disrupting chemicals in water and wastewater systems. J Water Process Eng 41:102017

Zhang L, Li Q (2022) Review of carbon sponges in water treatment applications. J Environ Chem Eng 10(6):108526. https://doi.org/10.1016/j.jece.2022.108526

Zhang Y, Wu L, Chen Y (2023) Advances in the use of nanosorbents for the removal of endocrine disruptors. Chem Eng J 456:140407. https://doi.org/10.1016/j.cej.2023.140407

Zhang C, Zhu MY, Zeng GM, Yu ZG, Cui F, Yang ZZ, Shen LQ (2016) Active capping technology: a new environmental remediation of contaminated sediment. Environ Sci Pollut Res 23(5):4370–4386

Zhang LJ, Zhang X, Liang HF, Xie Y, Tao HC (2018) Ammonium removal by a novel magnetically modified excess sludge. Clean Technol Environ Policy 20(10):2181–2189

Zhao M, Song C, Zhang F, Jia X, Ma D (2023) New-style electrokinetic-adsorption remediation of cadmium-contaminated soil using double-group electrodes coupled with chitosan-activated carbon composite membranes. Sci Total Environ 904:166919

Chapter 10
Bisphenol A and Its Bioremediation strategies—A Comprehensive Review

Shivashis Bikram Banerjee and Satarupa Dey

Abstract Bisphenol A (BPA) possesses both estrogenic endocrine disrupting and cancerous tumor promoting properties in reproductive organs. Exposure especially in early life, leads to developmental plasticity defects. Exposure to even low doses, that below the oral reference dose (RfD) established by the Environmental Protection Agency (EPA) may lead to such defects. BPA affects both males and females. Most susceptible periods include pre-puberty, childhood, and pregnancy. It is so widely used in products that we use daily i.e., polycarbonate plastics and epoxy resins, that basal levels of exposure are unavoidable. These effects have been seen in humans, animal models and wildlife setups. This unavoidable risk prompted the search for safer, affordable replacements for BPA. However, this led to the establishment of products being marketed as "BPA free," although they contained other Bisphenols, namely Bisphenol S (BPS) and Bisphenol F (BPF). Though these were thought to be safer substitutes in the beginning, they were experimentally shown to produce similar estrogenic effects later. They too are capable of exerting estradiol like effects such as cell proliferation, differentiation, and death. Additionally, they may alter organ weight, reproductive end points and enzyme activity. Other BPA replacements as well, have all been proven to pose health hazards, either in the form of endocrine disruption or carcinogenesis. This chapter looks into the chemical structure of BPA, its molecular mechanisms of action and its effect on human health while delving deep into the modes of bioaccumulation, detection, recovery and bioremediation.

Keywords BPA · Estrogenic endocrine disruptor · Bioaccumulation · Bioremediation

S. B. Banerjee
Department of Biotechnology, St Xavier's College (Autonomous), Kolkata, West Bengal, India

S. Dey (✉)
Department of Botany, Shyampur Siddheswari Mahavidyalaya, Ajodhya, Howrah, West Bengal, India
e-mail: dey1919@gmail.com

© The Author(s), under exclusive license to Springer Nature Singapore Pte Ltd. 2026
S. Dey and M. P. Shah (eds.), *Microbial Applications in Industrial Wastewater Treatment*, Advances in Wastewater Research,
https://doi.org/10.1007/978-981-96-9128-9_10

10.1 Introduction

Endocrine disruptors are emerging as a product which are responsible for many health hazards in human being and aquatic fauna. Bisphenol A (BPA) is one such endocrine disruptors which are known to possesses both estrogenic endocrine disrupting properties and they can cause malignant tumors in reproductive organs.

BPA, although used in the manufacture of phenol resins and polyacrylates, they also constitute a major part of polyesters, epoxy resins and polycarbonate plastics. Epoxy is waterproof, hence widely used in epoxy resins that form a coating preventing spoilage, leakage, and wetting. As such these form a part of food-contact surface lacquer coatings for cans, metal jar lids, protective coatings to prevent leakage of moisture and entry of contamination. They are used as protective coating in finishes, car parts, bonding agents and as a coating for PVC pipe. Polycarbonate plastics are widely used plastics which are used in DVD-ROMs, lenses, electrical appliances, and plastic bottles because of their endurance and useful physical properties such as durability and hardness. They are also resistant to high temperatures and can withstand acidic pH and oils (Staples et al. 1998). They are also used in dental agents and sealants, which is scary, considering the direct oral entry route and their estrogenic and cancerous properties (Olea et al. 1996). Due to an increase in the use of epoxy resins and polycarbonates in the last decade, the oral entry rate of estrogenic mimics has increased significantly. They are also used in food and drink containers, and from coatings on packages containing food items from where they enter the gastrointestinal tract. Due to its presence in coatings in cans, it is found in high concentration in canned foods such as meat and fish, and most surprisingly, even in infant formulas. During the manufacturing process, the heat generated while processing the items also exacerbates the entry of BPA into food from the coatings (Kang et al. 2006).

Exposure to BPA causes problems in developmental plasticity, particularly in early life. Such abnormalities may result from exposure to even low doses, below the oral reference dose (RfD) set by the Environmental Protection Agency (EPA). BPA has an impact on both genders. The most vulnerable times include childhood, pregnancy, and the prepubescent years. Baseline exposure is inevitable due to its extensive usage in everyday items like epoxy resins and polycarbonate plastics. These effects have been observed in wildlife situations, human subjects, and animal models. This inevitable risk led to a thorough investigation into the processes of bioaccumulation, detection, recovery, and bioremediation in order to find more cost-effective and safe BPA substitutes. This inevitable risk led to the hunt for less expensive, safer BPA substitutes. But this resulted in the creation of goods that were advertised as "BPA free," even if they included other bisphenols, such as BPS and BPF. Despite the fact that these were initially believed to be safer alternatives, experiments eventually demonstrated that they too have estrogenic effects. They too have the ability to cause effects similar to those of estradiol, including cell division, proliferation, and death. They may also change the activity of enzymes, reproductive end points, and organ

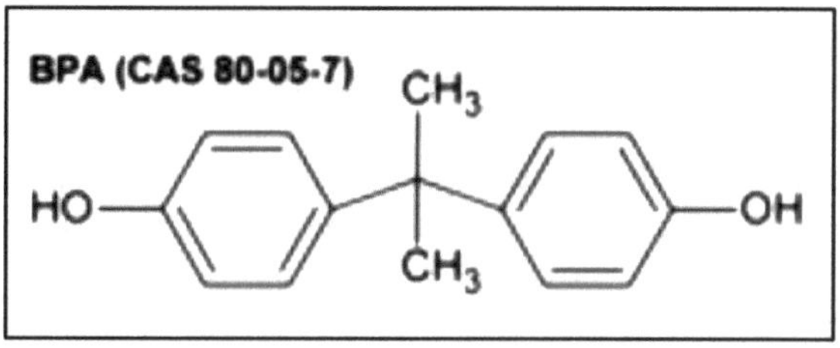

Fig. 10.1 Polycyclic aromatic structure of Bisphenol A. Composed of two phenol rings linked by a di-methyl bridge (Rosenmai et al. 2014)

weight. It has also been established that some BPA substitutes present health risks, such as endocrine disruption or carcinogenesis.

This chapter explores the processes of bioaccumulation, detection, recovery, and bioremediation while examining the chemical structure of BPA, its molecular mechanisms of action, and its impact on human health.

10.2 Chemical Structure

See Fig. 10.1.

10.3 Mode of Action

BPA exposure even in adults is known to cause Type 2 Diabetes mellitus due to altered insulin synthesis and disturbed synthesis from Beta cells of the pancreas. Insulin signalling is mediated downstream to the insulin binding action to the insulin receptor by a series of protein phosphorylation and binding events. However, in certain cell types, which have become insulin resistant, the cells of the organ do not respond to insulin. Symptoms are characterized by hyperinsulinemia, type 2 diabetes mellitus and eventually hypertension, completing the list for metabolic syndrome. BPA is known to cause insulin resistance via dysregulation of protein kinase B (Akt) phosphorylation and glycogen synthase kinase-3β (GSK3β). This leads to glucose intolerance (Moon et al. 2015). The murine 3T3-L1 adipocyte cell was used to study adipocyte differentiation, and how BPA affects it. It is also mediated by families of transcription factors such as the C/EBP and PPAR. BPA was found to activate oestrogen receptor ERβ selectively and slightly. Negative controls were used to validate the fact that BPA could reproduce the effects of oestrogen in part in cell lines. Terminal differentiation of adipocytes when studied showed that PPARγ activation was crucial. Obesogenic effects were observed when BPA action was studied with respect to PPARγ activation and the expression of adipogenesis marker FABP4. Red oil staining is used to measure triglyceride accumulation. It was cumulatively observed that FABP4 mRNA expressivity increased along with triglyceride accumulation in adipocytes when studied by qPCR and red oil staining respectively

(Le Magueresse-Battistoni et al. 2018). PPARγ mediated abnormalities can lead to adipocyte deposition and hence obesity. This has linked BPA and its analogous to obesogenic effects. BPA concentrations in urine have shown positive correlation to BMI of obese children, and hence serve as proof of the obesogenic effects of BPA. It also leads to insulin signalling mediated dyslipidaemia (Shah 2024a; Ramírez et al. 2021). Adipogenesis and obesogenicity of BPA has been linked to the upregulation of Srebf1, Cebpa, Acacb, and Fasn mRNA levels (Lee and Park 2019). Prostate cancers are dependent on androgen receptor (AR) for proliferation. Several abnormal AR variations are suggested for this; such as AR amplification, overexpression of AR transcriptional co-activator and constitutive ligand-independent AR activation. AR aberrations and tumour proliferation have been seen to colocalize. Gain of function mutants such as AR (AR-T877A) were seen to trigger far more easily than wild type AR. When BPA activates AR mutants, cell cycle progression and cellular proliferation occurs even in the absence of testosterone. Genetic and epigenetic changes have been implicated as the underlying cause for such diseases. Such genetic changes include point mutations and aberration in copy number of chromosomes. In vivo studies have substantiated this with the help of C57BL/6 female mice where exposure at the embryonic stage caused aneuploidy in germ cells. It was seen to form DNA adducts at all doses examined in Syrian Hamster Embryo (SHE) cell line. Adduct formation was correlated with morphological abnormalities and aneuploidy (Keri et al. 2007). An endocrine rich foetal environment in the uterine stage of pregnancy gives rise to babies born with a greater potency for breast cancer during adulthood. Tests conducted in non-human primates have shown that even foetal exposure in low doses to BPA alters foetal mammary gland tissue. Changes such as alteration in ductal area, periductal stroma, and maturation of the fat pad have been observed. These effects are seen to continue into postnatal life as well. Decreased cellular apoptosis and increased hyperplasia were seen indicating development of cancer. Changes were seen in various gene expression levels such as anti-apoptosis markers (Birc2, Abl1), myoepithelial cell differentiation (Krt8), adipogenesis (Fabp4, PPARγ and Ldlr) and ECM modulators tenascin 53 (Tnc) and versican (Vcan). Human normal mammary epithelial cells (HMEC) treated with low doses of BPA have been seen to show dysregulation of proteins such as Bcl2, cyclin D1 and cyclin E, resulting in cell proliferation and senescence defects. Low concentrations of BPA exposure can alter the methylation of DNA repair (BRCA1), cell cycle regulation and apoptosis (CCNA1 45 and CDNK2A/p16) genes. BPA causes epigenetic modifications that make the cell more prone to cancer development and tumour progression. The hypermethylation of promoter CpG islands of tumour suppressor genes (BRCA1, E54 cad and CDNK2A/p16) have been implicated in tumour progression and these have been seen to be affected by BPA exposure. For example, suppression of CDKN2A allows BPA treated epithelial cells of the mammalian mammary gland to escape senescence (Perrot-Applanat et al. 2018). Blood glucose level was found to increase a few months after exposure in BPA exposed Wistar albino rats (4–6 weeks old). This induced hyperglycaemia was seen in all tested doses, indicating its diabetogenic effect. Lipid biomarkers (Triglycerides-TGs, Free fatty acids-FFA, and High-density lipoprotein-HDL), are crucially affected as well. This study showed that TGs and FFA levels were significantly increased ($p < 0.0001$) in a dose responsive manner, in rats exposed to BPA. HMG-CoA reductase levels are increased significantly ($p < 0.001$). This enzyme plays a crucial role in cholesterol synthesis. Inflammatory biomarkers that indicate cytotoxic effects predict the onset of Diabetes Mellitus. Such biomarkers i.e., leptin, TNF-α, and IL-6 were also increased upon exposure to BPA. Catalase (CAT), Superoxide dismutase (SOD) and glutathione (GSH) are oxidative stress biomarkers. Chronic exposure to BPA significantly decreased serum CAT, GSH and SOD ($p < 0.001$) levels. As GSH, SOD and CAT are instrumental in the removal of free radicals that cause oxidative stress, their downregulation increases oxidative stress. Hence, oxidative stress was increased upon BPA exposure. This study underscores the fact that the dysregulation that results from BPA exposure is highly systemic and that multiple metabolic pathways are simultaneously affected (Haq et al. 2020). Through MWAS modelling, pathway enrichment analysis, and chemical annotation and confirmation, a study showed that prenatal exposure to BPA and Bisphenol F (BPF) were responsible for the dysregulation of several key metabolic pathways. Pathways such as those involved in steroid hormone synthesis, lysine, lipoate, tyrosine, methionine, and cysteine metabolism were affected. These metabolic pathways are known to be associated with stress response, neuronal death, and energetics. Vitamin B12 metabolism was also affected. In tyrosine metabolism, result of dysregulation was a product called tyramine. Tyramine is known to cause reactive oxygen species (ROS) generation. ROS generated in the mitochondria can damage the cell membrane leading to oxidative cytotoxicity. Tyramine is a key player in neurodegeneration e.g.—Parkinson's disease, psychological and psychomotor defects, and hypertension. Hence, BPA and its analogue BPF, both have clearly elucidated roles in ROS generation, as underscored by dry lab studies (Shah 2024b; Tchen et al. 2022). C57BL/6 J mice models were tested with BPA exposure. It was seen that tryptophan and dopamine pathways were affected. In the Tryptophan pathway (TRP), along with TRP, kynurenine (KYN) and 5-hydroxytryptamine (5-HT) levels were also affected. In the dopamine pathway, Dopamine (DA) and its pathway intermediates, namely 3,4-dihydroxyphenylacetic acid (DOPAC), 3-methoxytyramine (3-MT) and homovanillic acid (HVA) were all dysregulated. This links BPA related metabolic disruption to neurotoxicity. DA is a key player in movement and coordination, mood, and behaviour control, including regulation of feelings related to happiness and satiation. Deficiencies in DA and resulting dysregulation of the neuroendocrine network have been linked to hallmarks of neurodegeneration, such as Parkinson's disease. Decreased levels of HVA also lead to Dopamine pathway disruption, as seen in BPA exposed mice. Another aspect of this disruption is the loss of tyrosine hydroxylase expressing neurons in non-human primates, which are dopaminergic by nature. This is linked to prenatal exposure to BPA in non-human primates, and furthers supports the theory that BPA maybe linked to neurodegeneration in humans as well (Yao et al. 2020). Theoretical evidence suggests that endocrine like effects that affect neuronal development and signalling, may be linked to disruptions of the axis of the neurotrophin signalling pathway (Lee and Park 2019). UGT1 and UGT2 are the two main enzymes present in the liver that help degrade BPA. They are present in cells called microsomes. However, UGT2 isoforms catalyse 4-methylumbelliferone

(4-MU) glucuronidation, which is an essential step in the detoxification pathway. When BPA was present, it significantly inhibited the glucuronidation of 4-MU by UGT 2 isoform UGT2B. BPA sulfonation by cytosolic phase II sulfotransferases (SULTs), also constitutes a minor elimination pathway. SULT1A1 has the highest activity among all other SULT isoforms. The metabolites generated as a result of BPA clearance from the body, may also have adverse effects, and have hence been addressed by various workers. BPA glucuronide, which is a UGT processed product, promotes lipid accumulation and adipocyte differentiation. This was seen in in vitro studies carried out in murine and human 3T3L1 preadipocyte cell line. 4-methyl-2,4-bis (4-hydroxyphenyl) pent-1-ene (MBP) is a metabolized product of BPA, which has been shown to have higher estrogenic affect (500 times) than BPA itself (Ramírez et al. 2021). Dry lab studies indicate that purine and pyrimidine metabolism is dysregulated by BPA. Many amino acids were interrupted and/or upregulated as well, for e.g., o-succinyl-l-homoserine, tyrosine, proline, and leucine. Proline is a biomarker for estrogenic stimulation; hence its upregulation underscores the relevance of BPA as an endocrine disruptor. NADH and lactate levels were higher, signifying the role of BPA in disruption of energetics and sugar metabolism. This strengthens the link between BPA and hyperglycaemia further. Additionally, glycerophosphocholine was depleted after BPA treatment. Glycerophosphocholine is a precursor of phospholipids. Hence, its downregulation might signify possible lipid metabolic disorders. In addition to this, aminoacyl tRNA synthesis pathways were disrupted in cells as well (Zhao et al. 2022). Upon BPA exposure, decrease in body weight was observed in animal models, without any change in diet inputs. Urinary excretion of riboflavin and biotin was significantly increased. Tissue riboflavin catabolism and biotin excretion are all associated with protein degradation. This might be the underlying cause of high outputs that were observed in case of both riboflavin and biotin. Methylation is highly increased in the BPA exposed environment. S-adenosylmethionine and its formative enzymes, MAT, are all upregulated. Increase in amino acid methylation and protein degradation are concomitant, hence, the high methylation environment that is created because of BPA toxicity, further increases protein degradation. As methylation of purines increases, e.g., 3 methyl adenine, along with the high methylation environment, aberrant DNA methylation becomes a risk. DNA methylation defects are all associated with cancer progression and tumour growth. Allantoin increases too, which is a uric acid degradation product. Precursors of purines, such as adenine and guanine show lower levels. These two pointers show the incidence of purine degradation. Oxidative stress increases too, in cases of BPA exposure. This is caused by the conversion of hypoxanthine to xanthine (Chen et al. 2014). Table 10.1 shows the list of the diseases caused by exposure of BPA.

10.3.1 Replacements—Do They Work?

Two well-known replacements for the primary form of bisphenol that were regarded as unsafe are BPS and BPF (Fig. 10.2). However, it has been shown that in the

Table 10.1 Table showing the list of diseases as a result of BPA exposure

Disease	Pathway/protein/factor involved	Reference/s
Type 2 diabetes mellitus	Primary cause being insulin resistance, cells and organs stop responding to insulin. Hyperinsulinemia, type 2 diabetes mellitus and eventually hypertension, completes the list for metabolic syndrome. Protein kinase B (Akt) phosphorylation and glycogen synthase kinase-3β (GSK3β) pathway are implicated	Moon et al. (2015)
Obesity	Transcription factors such as C/EBP and PPARγ, adipogenesis marker FABP4, Srebf1, Cebpa, Acacb, and Fasn mRNA levels act as indicators. Dyslipidaemia results	Le Magueresse-Battistoni et al. (2018)
Prostate cancer	Androgen receptor (AR) amplification, overexpression of transcriptional co-activator and constitutive ligand-independent activation and gain of function mutations. Genetic and epigenetic changes implicated. Formation of DNA adducts. Abnormal cell cycle progression and cellular proliferation	Keri et al. (2007)
Breast cancer	Foetal exposure in low doses to BPA. Decreased cellular apoptosis and increased hyperplasia, changes seen in anti-apoptosis markers (Birc2, Abl1), myoepithelial cell differentiation (Krt8), adipogenesis (Fabp4, PPARγ and Ldlr) and ECM modulators tenascin 53 (Tnc) and versican (Vcan), Bcl2, cyclin D1 and cyclin E, altered methylation of DNA repair (BRCA1), cell cycle regulation and apoptosis (CCNA1 45 and CDNK2A/p16) genes, hypermethylation of promoter CpG islands of tumour suppressor genes (BRCA1, E54 cad and CDNK2A/p16)	Perrot-Applanat et al. (2018)
Oxidative stress	Decreased serum Catalase (CAT), Superoxide dismutase (SOD) and glutathione (GSH) levels	Haq et al. (2020)
Neurodegeneration	Tryptophan pathway (TRP), along with kynurenine (KYN) and 5-hydroxytryptamine (5-HT). In the dopamine pathway, Dopamine (DA) and its pathway intermediates, namely 3,4-dihydroxyphenylacetic acid (DOPAC), 3-methoxy tyramine (3-MT) and homovanillic acid (HVA) affected. Loss of tyrosine hydroxylase expressing neurons, dysregulation of neurotrophin signalling pathway	Yao et al. (2020), Lee and Park (2019)

mouse foetal testes system, FeTa, testosterone production was significantly reduced following BPF and BPS treatment. BPF had a minimum concentration of 1000 nmol/L and BPS had a lower concentration of 100 nmol/L. This shows that with respect to interference with Leydig cell testosterone production, BPS was far more hazardous. Both these substances reduce the expression of genes responsible for testosterone production, namely Star, Hsd3b1 and Cyp17a1. This goes to show that so called BPA-free products that are sold in the market, which contain alternatives such as BPS and BPF, which are the most common ones, do in fact pose significant hazards such as leaching and subsequent entry into human and animal systems (Eladak et al. 2015). It has earlier been seen that Cyp17a and 17 Beta hsd mRNA levels decrease in zebrafish upon BPS and BPF exposure (Ji et al. 2013). However, these alternatives have lower fold effects (5X and 10X) respectively, for BPS and BPF. Even though BPF is equal to BPA in oestrogenicity, and lower in cytotoxicity than BPA, BPF does possess novel and unseen cytotoxic effects that BPA does not (Molina-Molina et al. 2013; Audebert et al. 2011). With respect to disruption of lipid metabolism, BPS is more dangerous than BPA, while BPF is less dangerous than BPA (Héliès-Toussaint et al. 2014; Kidani et al. 2010).

A novel and landmark study showed that Bisphenol A, S and F show thyroid receptor binding and cause thyroid signalling disorders. They are capable of binding to thyroid receptors, and subsequently activating coactivators and corepressors, thereby leading to agonistic or antagonistic actions respectively. Downstream Growth Hormone (GH) signalling is affected as well. BPA, BPS and BPF mediate their action by binding to the T3 (thyroid hormone) binding pocket of the thyroid receptors alpha and beta. The order of binding affinity is as follows—BPA > BPF > BPS. Additionally, binding affinity is greater for receptor family beta than that of alpha, for the three common bisphenols. This group found that coactivator recruitment to thyroid receptor beta lead to activation in excess, of thyroid signalling, and downstream GH3

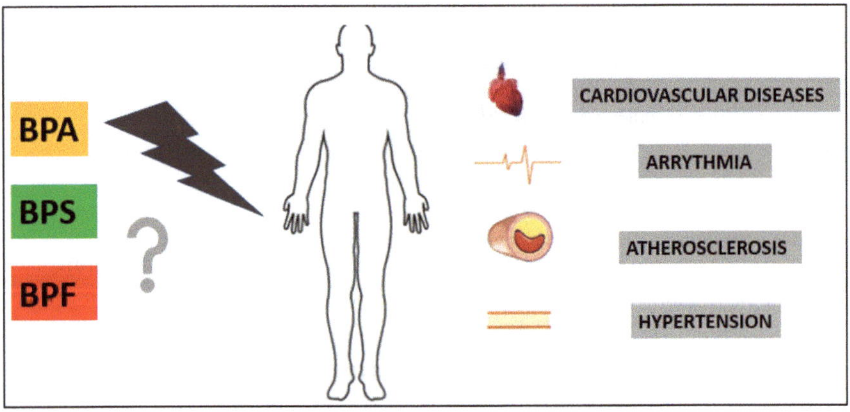

Fig. 10.2 BPS and BPF are unsafe too. Created with Microsoft PowerPoint

mediated growth effects as well (Zhang et al. 2018). This work has been substantiated by in silico analysis as well (Goldinger et al. 2015).

Even at dosages lower than the tolerable daily intake (TDI), BPA and BPS have been linked to hypertension and high blood pressure. This is positively correlated with gender (male), higher BMI and higher income levels. The study found that levels in the urine of patients varied from high, for example in Saudi Arabia (4.92 μg/L) to low, for example in Korea (0.03 μg/L). Overall levels in urine were found to increase from 2018 onwards. It has also been seen that people consuming canned goods showed 16X higher urinary Bisphenol than those who consumed goods packaged in glass bottles. The probable mechanism by which Bisphenol increases blood pressure is via the disruption of oestrogen signalling. As Bisphenol binds to the oestrogen receptor, it annuls the protective ability of oestrogen that is normally prevalent against cardiovascular upsets. It also disrupts the mitogen activated protein kinase (MAP-K) and phosphoinositide 3-kinase pathway (Jiang et al. 2020a).

Both young adults and older men have shown positive correlation between urinary BPA, BPS and BPF concentration and depressive symptoms. This pertains to the relationship that exists between mental cognition and Bisphenol types intake, and as such may involve the nervous system. This occurrence is not concurrent in women. Testosterone, oestradiol and GABA have neuroprotective and mood protective functions, and it has been shown in animal models that bisphenols raise estradiol levels in females, while reducing testosterone and GABA receptor levels in males. BPS and BPF possess similar structure and hence can cause similar inflammatory cytokine release, which makes them dangerous for the nervous system, as it can cause neuroinflammation (Hao et al. 2021).

Previous correlations already exist between BPA and cardiovascular disease (CVD). However, this study showed that rodent models suffered severe, often more severe effects due to exposure to BPS than exposure to BPA. Nuclear receptor signalling and ion channel disruption have been hypothecated to be the cause (Shah 2024c; Zhang et al. 2020).

BPA has also shown its effects on osteoclast and osteoblast differentiation, that leads to bone development disorders. Tetra methyl bisphenol F, an alternative to BPA has shown these effects as well. It initiated osteoclast differentiation via MAPK signalling pathway, JNK, and p38 and increased the expression of genes *NFATc1*, *TRAP*, and *CtsK* absurdly. It disrupts osteoclast homeostasis (Kim et al. 2021).

Interaction between serum albumin with bisphenols were studied using spectroscopy. It showed that high concentrations of BPS can unfold serum albumin to an energetically unstable state, after which refolding occurs and BPS binds to such unfolded albumin aggregates (Mathew et al. 2014).

BPF interferes with Notch-related gene expression (dll1, hairy1, hairy2b, notch1, lgr5) in *X. laevis*. Notch-related gene are involved in the development of intestinal tract in mammals. This was further validated in IEC-6 mammalian cell line. This chemical hindered the development of absorptive villi in the intestine of model animals (Zhu et al. 2020). BPA and BPS have both shown the ability to disrupt the neuroendocrine system, because they can disrupt essential genes like the Kiss1/

Kiss1r system, which is present upstream in the GnRH activation pathway (Qiu et al. 2016). Several alternatives such as bisphenol F (BPF), bisphenol S (BPS) and bisphenol AF (BPAF) have shown vascular toxicity in zebrafish embryos and human vascular endothelial cells, denoted by blood vessel growth impairment, delayed vein development, and decreased intestinal blood vessel proliferation. Moreover, angiogenesis is inhibited and oxidative stress increases, signified by a drop in catalase and superoxide dismutase levels. The levels of toxicity of these analogues were found to be in the order BPAF > BPF > BPA > BPS. It is clearly seen that the analogues BPAF and BPF have greater toxicity that BPA itself, which they are supposed to replace in polyacrylate manufacturing (Ji et al. 2022). These especially pose a risk to pregnant women and unborn foetuses, as BPA, BPF and BPS showed P2X7 receptor activation. These are especially detected in the placental fluid and the placental cells, from where they can pose harm to the foetus causing developmental defects (Fouyet et al. 2021). In *Salmo trutta*, the levels of vitellogenin and thyroid hormone T3 were elevated indicating both indicating endocrine disruption, this indicated that not only does BPA and BPS affect humans and closely related animals, but also fish species (Frenzilli et al. 2021). Repeated exposure to BPA and its analogues during pregnancy lead to retarded psychomotor skills and neurodevelopment in children. Spot urine tests for the pregnant mothers and Bayley Scales of Infant Development tests for the children helped determine that mental retardation resulted in the second trimester (Jiang et al. 2020b). In chicken DT40 cell lines, bisphenol AP, bisphenol M and bisphenol P, which are bisphenol A analogues and replacements supposedly, all showed genotoxicity, greater than that of BPA. This is purported to be due to the causation of double strand DNA breaks, which are not repaired in the DNA repair deficient DT40 cell line, and hence can be detected (Lee et al. 2013). It can be safely concluded that the analogues of BPA that are currently being used, are equally or even more unsafe than BPA, and hence their use poses an immense threat to both humans and animals. A safer alternative to current bisphenols could be naturally occurring Syringaresinol, which has shown no estrogenic effects in vitro so far (Janvier et al. 2017).

10.4 Bioaccumulation

The bean *Vicia faba*, tomato *Lycopersicon esculentum,* lettuce *Lactuca sativa* and wheat *Triticum durum* are all commercially important crops. BPA exposure for 21 days induced phytotoxicity I these commercially important crops, due to BPA exposure. This was manifested as changes in morphology and reduction in the length of shoots and roots, and reduction in the overall dry weight as well. When BPA levels post exposure were measures, it was seen that BPA had accumulated up to significant levels in the plants, indicating bioaccumulation. The number of micronuclei also increased, indicating that bioaccumulation directly contributes to cytotoxicity (Ferrara et al. 2006). Climate change has led to the disruption of biogeochemical cycles, and hence contributes to bioaccumulation of harmful substances.

In the Greenland shark *Somniosus microcephalus,* which is a part of the arctic food chain, 23 samples were taken from said sharks for analysis, showing varying levels of bioaccumulation in their muscle (7.9 ng/g of wet weight) and liver (8.2 ng/g of wet weight) (Ademollo et al. 2018). Seahorses (*Hippocampus erectus)* contain a brood pouch in adult males that constitute a reproductive strategy of "male pregnancy." At BPA exposure concentrations of 10, 100, and 1000 μg/L bioaccumulation increased as dose of exposure increased, in both males and females. They also noticed trademark signs of reproductive dysfunction upon bio-accumulative BPA exposure (Liu et al. 2021).

^{14}C-labeled BPA acts as a tracer, which can be tracked due to its radioactivity, as it traverses the food chain. When the green alga *Chlorella pyrenoidosa* was exposed to BPA, it was seen that BPA is subsequently transferred to the rotifer *Brachionus calyciflorus,* which lies above the alga in the food chain. Evidence of biomagnification of BPA residues was observed, as it traversed up the food chain (Guo et al. 2017). In the mollusc *Corbicula japonica,* BPA exposure was induced at 9 μg/L, using [^{14}C] labelled bisphenol A (BPA). Bioaccumulation was seen in them, detected, and seen as conjugates of BPA. The major amount of bioaccumulation was observed in the visceral mass, which probably acts as a major BPA processing and metabolizing factory in the bivalves' body (Hayashi et al. 2008). It is impossible to ignore the human context of bioaccumulation, as 93% Americans over the age of 6 years, show urinary BPA excretion, as reported by the National Health and Nutrition Examination Survey (NHANES). Earlier it was thought that BPA entered out body through food, chemicals found in the environment, and the unavoidable accumulation and bioconcentration that results due to the food chain. It was presumed that this BPA left our body, after a few hours or a day, autonomously and completely, and none was accumulated. However, it was seen that fasting adults, after several hours, showed significant urinary BPA concentration. This meant that even humans were affected by bioaccumulation. This new hypothesis suggests that BPA is stored in adipose tissue, and is eventually lost through gradual and infinitesimal release, never actually reaching zero (Shah 2025; Tillett 2009).

10.5 Detection

Pretreatment of food and environmental samples to separate BPA from the matrices it is found in, is essential, to improve down-streaming (Tarafdar et al. 2022).

10.5.1 Step 1—Extraction of BPA from the Sample Matrix

10.5.1.1 Liquid–Liquid Extraction Method

Cheap, easy and non-complicated, but requires huge solvent amounts. The solvents are hexane, butane or acetonitrile (Tarafdar et al. 2022).

10.5.1.2 Extraction by Microwave Assistance

This decreases the amount of solvent needed to extract the BPA, thereby reducing the environmental harm caused. It requires less time and has very high sensitivity (10–50 ng/g) (Tarafdar et al. 2022).

10.5.1.3 Solid-Phase Extraction

Highly sensitive (0.001 μg/mL), can be even used to detect octyl phenols and BPA together from excreta. Involves a methanol mobile phase along with a gradient is used to elute BPA (Jing et al. 2011).

10.5.1.4 Molecularly Imprinted Solid-Phase Extraction

Synthetically engineered polymers or polymers that are molecularly imprinted, are used to bind BPA and extract it out. 3-acrylamido ammonium chloride (molecularly imprinted polymer) can be used to extract BPA from water soluble matrices, from which they need to be extracted. Other examples are dummy templates and porous polymers with embedded nanospheres (Inoue et al. 2013; Li et al. 2013). Magnetically imprinted polymers used to form carbon nanotubes also constitutes a way (Zhang et al. 2014) (Table 10.2).

Table 10.2 Percentage recovery upon extraction by different methods (Sun et al. 2016)

Matrix	Method	Percentage recovery (%)
Sediment	Liquid–liquid extraction	82
Water	Extraction by microwave assistance	95
Urine	Solid-phase extraction	80
Food items, juices	Molecularly imprinted solid-phase extraction	83

Recovery is the total amount of BPA that was extracted by a certain method, divided by the total amount of BPA present in a sample, multiplied by 100 (Tarafdar et al. 2022)

Table 10.3 Table for detection limits of various methods (Tarafdar et al. 2022)

Method	Detection limit
Liquid chromatography	150 ng/mL
Gas chromatography	0.3–1.5 ng/L
Capillary electrophoresis	1800 ng/L
Enzyme linked immunosorbent assay	0.03 ng/mL
Immunochromatography (highly sensitive)	0.005 mg/L

10.5.2 Step 2—Detection and Analysis After Extraction

After extraction, sample can be analysed by various means.

10.5.2.1 Liquid Chromatography

LC–MS is a gold standard and is highly sensitive and specific. It is coupled with diazotization coupling principle, an azo-BPA complex is formed, which is detected at 410 nm by UV absorption. Detection limit is 150 ng/ml (Zhuang et al. 2014) (Table 10.3).

10.5.2.2 Gas Chromatography

Methylsilyl trifluoroacetamide (MSTFA) or trimethylchlorosilane (TMCS) is used to volatilize the BPA after extraction, following which detection is done by mass spectrometry (Sun et al. 2016). Detection limit with SPE and MSFTA is 0.3–1.5 ng/L (Selvaraj et al. 2014) (Table 10.3).

10.5.2.3 Capillary Electrophoresis

Quicker process, and needs lesser solvent to extract BPA, smaller sample amount makes it ideal over previous analytical methods. Solid phase extraction followed by capillary electrophoresis showed 95% recovery and 1800 ng/L limit of detection (Mei et al. 2011) (Table 10.3).

10.5.2.4 Enzyme Linked Immunosorbent Assay

Polyclonal and monoclonal antibodies against BPA used for detection (Lu et al. 2012). Direct ELIZA was found to be more efficient, with a detection limit of 0.03 ng/mL (Lei et al. 2013) (Table 10.3).

10.5.2.5 Immunochromatography

Like most chromatography applications, it is rapid, cheap, and easy to use. Does not require specialized skill. Lateral flow strips much like COVID rapid antigen detection kits are used with probes (Gold nanoparticles + anti-BPA antibodies) that detect BPA. Test and control lines are provided to increase robustness. Bovine serum albumin + BPA constituted the test line while anti-mouse antibody served as the control sample. This is a colorimetric assay which means that the intensity of colour after the reaction shows the BPA concentration. Extremely sensitive and low limit of detection at 0.005 mg/L makes it highly advantageous (Mei et al. 2013) (Table 10.3).

10.6 Bioremediation

Sphingomonas bisphenolicum strain AO1 was seen to be able to degrade BPA in soil, with the help of other phyla such as *Xanthomonadales, Burkholderiales* and *Pseudomonadales* at soil concentration of 1.0 mg g^{-1} or lower, however, at concentrations of 10 mg g^{-1} or higher, degradation did take place, although the microbial diversity was harmed simultaneously. *Sphingomonas bisphenolicum* could not survive in soil free from BPA, indicating that it could be further standardized to define processes of bioremediation that were cost effective and eco-friendly (Matsumura et al. 2015). In *Chlamydomonas reinhardtii,* BPA caused imbalance in iron reserves and redox homeostasis, disturbing cell survival and inciting ferroptosis. However, this eukaryotic alga possesses several proteomic and ROS detoxification defences, that help it to not only overcome the challenge, but also detoxify the surroundings of BPA (Carbó et al. 2023). *Trametes versicolor* based laccase can be used in the form of a nano emulsion. This could remove up to 94% BPA within 2 h of introduction. The in situ application relies on a packed bed bioreactor system where the calcium alginate based beads form the matrix, such a system is capable of removing 60% of the BPA. A continuous flow bed bioreactor can remove even more, up to 67%. Such a system when used reduces the Biological Oxygen Demand of wastewater from 48 to 14 mg/L (Trivedi and Chhaya 2022). Cells of *Pavlova* sp. showed bioremediation potential against BPA, by converting it to another product, a mono-glucoside, 2-(4-β-D-glucopyranosyloxyphenyl)-2-hydroxyphenylpropane. The efficiency of this process was 9%, however, it increased to 17%, when sodium alginate beads were used for immobilization of the whole microalgae cells (Shimoda and Hamada 2009). *Pseudomonas pseudoalcaligenes*, isolated from petrochemical wastewater from sites in Iran showed bioremediation capability. The bioremediation potential measured as Chemical Oxygen Demand, reduced from 655.2 to 109.2 mg L^{-1}, an approx. 83% decrease. The level of BPA contamination in the saltwater was 300 mg L^{-1} (Jahanshahi et al. 2023). Isooctane (Bis (2-ethylhexyl) sulphosuccinate sodium salt) hosted micelles housing purified laccase enzyme from *Fusarium incarnatum* strain UC-14 are capable of BPA removal. This is a non-aqueous catalysis by bio-zyme, meaning that it can circumvent the fact that BPA does not dissolve in an aqueous medium upon which treatment can be carried out. The efficiency of the laccase/Reverse Micelles system was recorded at 91.43%, when the medium was contaminated with 200 ppm of Bisphenol A. The reaction conditions were optimized at 50 °C and pH-6.0, and the time required to achieve these results was 75 min (Chhaya and Gupte 2013).

10.7 Conclusion

BPA is present ubiquitously. So far, there is significant evidence that widely available alternatives of BPA are dangerous too. As such, the detection methods discussed by us will go a long way to pave the path for removal and bioremediation strategies to be selected and implemented. Cost effectiveness, recovery rates and human life need to be kept in mind. Bioremediation serves as a one stop goal to free our soil, water, and environment from BPA.

References

Ademollo N, Patrolecco L, Rauseo J, Nielsen J, Corsolini S (2018) Bioaccumulation of nonylphenols and bisphenol A in the Greenland shark Somniosus microcephalus from the Greenland seawaters. Microchem J 136:106–112. https://doi.org/10.1016/j.microc.2016.11.009

Audebert M, Dolo L, Perdu E, Cravedi J-P, Zalko D (2011) Use of the γH2AX assay for assessing the genotoxicity of bisphenol A and bisphenol F in human cell lines. Arch Toxicol 85(11):1463–1473. https://doi.org/10.1007/s00204-011-0721-2

Carbó M, Chaturvedi P, Álvarez A, Pineda-Cevallos D, Ghatak A, González PR, Cañal MJ, Weckwerth W, Valledor L (2023) Ferroptosis is the key cellular process mediating bisphenol A responses in Chlamydomonas and a promising target for enhancing microalgae-based bioremediation. J Hazard Mater 448:130997. https://doi.org/10.1016/j.jhazmat.2023.130997

Chen M, Zhou K, Chen X, Qiao S, Hu Y, Xu B, Xu B, Han X, Tang R, Mao Z, Dong C, Wu D, Wang Y, Wang S, Zhou Z, Xia Y, Wang X (2014) Metabolomic analysis reveals metabolic changes caused by bisphenol A in rats. Toxicol Sci 138(2):256–267. https://doi.org/10.1093/toxsci/kfu016

Chhaya U, Gupte A (2013) Possible role of laccase from Fusarium incarnatum UC-14 in bioremediation of bisphenol A using reverse micelles system. J Hazard Mater 254–255:149–156. https://doi.org/10.1016/j.jhazmat.2013.03.054

Eladak S, Grisin T, Moison D, Guerquin M-J, N'Tumba-Byn T, Pozzi-Gaudin S, Benachi A, Livera G, Rouiller-Fabre V, Habert R (2015) A new chapter in the bisphenol A story: bisphenol S and bisphenol F are not safe alternatives to this compound. Fertil Steril 103(1):11–21. https://doi.org/10.1016/j.fertnstert.2014.11.005

Ferrara G, Loffredo E, Senesi N (2006) Phytotoxic, clastogenic and bioaccumulation effects of the environmental endocrine disruptor bisphenol A in various crops grown hydroponically. Planta 223(5):910–916. https://doi.org/10.1007/s00425-005-0147-2

Fouyet S, Olivier E, Leproux P, Dutot M, Rat P (2021) Bisphenol A, bisphenol F, and bisphenol S: the bad and the ugly. Where is the good? Life 11(4), Article 4. https://doi.org/10.3390/life11040314

Frenzilli G, Martorell-Ribera J, Bernardeschi M, Scarcelli V, Jönsson E, Diano N, Moggio M, Guidi P, Sturve J, Asker N (2021) Bisphenol A and bisphenol S induce endocrine and chromosomal alterations in brown trout. Front Endocrinol 12. https://www.frontiersin.org/articles/10.3389/fendo.2021.645519

Goldinger DM, Demierre A-L, Zoller O, Rupp H, Reinhard H, Magnin R, Becker TW, Bourqui-Pittet M (2015) Endocrine activity of alternatives to BPA found in thermal paper in Switzerland. Regul Toxicol Pharmacol 71(3):453–462. https://doi.org/10.1016/j.yrtph.2015.01.002

Guo R, Du Y, Zheng F, Wang J, Wang Z, Ji R, Chen J (2017) Bioaccumulation and elimination of bisphenol a (BPA) in the alga Chlorella pyrenoidosa and the potential for trophic transfer to the rotifer Brachionus calyciflorus. Environ Pollut 227:460–467. https://doi.org/10.1016/j.envpol.2017.05.010

Hao K, Luo J, Sun J, Ge H, Wang Z (2021) Associations of urinary bisphenol A and its alternatives bisphenol S and F concentrations with depressive symptoms among adults. Chemosphere 279:130573. https://doi.org/10.1016/j.chemosphere.2021.130573

ul Haq ME, Akash MSH, Rehman K, Mahmood MH (2020) Chronic exposure of bisphenol A impairs carbohydrate and lipid metabolism by altering corresponding enzymatic and metabolic pathways. Environ Toxicol Pharmacol 78:103387. https://doi.org/10.1016/j.etap.2020.103387

Hayashi O, Kameshiro M, Masuda M, Satoh K (2008) Bioaccumulation and metabolism of [14C]Bisphenol A in the Brackish water bivalve Corbicula japonica. Biosci Biotechnol Biochem 72(12):3219–3224. https://doi.org/10.1271/bbb.80448

Héliès-Toussaint C, Peyre L, Costanzo C, Chagnon M-C, Rahmani R (2014) Is bisphenol S a safe substitute for bisphenol A in terms of metabolic function? An in vitro study. Toxicol Appl Pharmacol 280(2):224–235. https://doi.org/10.1016/j.taap.2014.07.025

Inoue N, Ooya T, Toshifumi T (2013) Hydrophilic molecularly imprinted polymers for bisphenol A prepared in aqueous solution. Microchim Acta 180(15):1387–1392. https://doi.org/10.1007/s00604-013-0996-5

Jahanshahi S, Badiefar L, Khodabandeh M, Ali Heidarnia M, Yakhchali B (2023) Bioremediation of a salty petrochemical wastewater containing bisphenol A by a novel indigenous Pseudomonas pseudoalcaligenes. RSC Adv 13(1):388–398. https://doi.org/10.1039/D2RA06206B

Janvier M, Hollande L, Jaufurally AS, Pernes M, Ménard R, Grimaldi M, Beaugrand J, Balaguer P, Ducrot P-H, Allais F (2017) Syringaresinol: a renewable and safer alternative to bisphenol A for epoxy-amine resins. ChemSusChem 10(4):738–746. https://doi.org/10.1002/cssc.201601595

Ji G, Gu J, Guo M, Zhou L, Wang Z, Shi L, Gu A (2022) A systematic comparison of the developmental vascular toxicity of bisphenol A and its alternatives in vivo and in vitro. Chemosphere 291:132936. https://doi.org/10.1016/j.chemosphere.2021.132936

Jiang S, Liu H, Zhou S, Zhang X, Peng C, Zhou H, Tong Y, Lu Q (2020a) Association of bisphenol A and its alternatives bisphenol S and F exposure with hypertension and blood pressure: a cross-sectional study in China. Environ Pollut 257:113639. https://doi.org/10.1016/j.envpol.2019.113639

Jiang Y, Li J, Xu S, Zhou Y, Zhao H, Li Y, Xiong C, Sun X, Liu H, Liu W, Peng Y, Hu C, Cai Z, Xia W (2020b) Prenatal exposure to bisphenol A and its alternatives and child neurodevelopment at 2 years. J Hazard Mater 388:121774. https://doi.org/10.1016/j.jhazmat.2019.121774

Jing X, Bing S, XiaoYan WU, XiaoJie SUN, YongNing WU (2011) A study on bisphenol A, nonylphenol, and octylphenol in human urine samples detected by SPE-UPLC-MS. Biomed Environ Sci 24(1):40–46. https://doi.org/10.3967/0895-3988.2011.01.005

Keri RA, Ho S-M, Hunt PA, Knudsen KE, Soto AM, Prins GS (2007) An evaluation of evidence for the carcinogenic activity of bisphenol A. Reprod Toxicol 24(2):240–252. https://doi.org/10.1016/j.reprotox.2007.06.008

Kidani T, Kamei S, Miyawaki J, Aizawa J, Sakayama K, Masuno H (2010) Bisphenol A downregulates Akt signaling and inhibits adiponectin production and secretion in 3T3-L1 adipocytes. J Atheroscler Thromb 17(8):834–843. https://doi.org/10.5551/jat.4051

Kim H-M, Lee S-M, Choi J, Soung N-K, Heo J-D (2021) Effects of bisphenol A and its alternatives, bisphenol F and tetramethyl bisphenol F on osteoclast differentiation. Molecules 26(20), Article 20. https://doi.org/10.3390/molecules26206100

Le Magueresse-Battistoni B, Multigner L, Beausoleil C, Rousselle C (2018) Effects of bisphenol A on metabolism and evidences of a mode of action mediated through endocrine disruption. Mol Cell Endocrinol 475:74–91. https://doi.org/10.1016/j.mce.2018.02.009

Lee H-S, Park Y (2019) Identification of metabolic pathways related to the bisphenol A-induced adipogenesis in differentiated murine adipocytes by using RNA-sequencing. Environ Res 171:161–169. https://doi.org/10.1016/j.envres.2019.01.017

Lee S, Liu X, Takeda S, Choi K (2013) Genotoxic potentials and related mechanisms of bisphenol A and other bisphenol compounds: a comparison study employing chicken DT40 cells. Chemosphere 93(2):434–440. https://doi.org/10.1016/j.chemosphere.2013.05.029

Lei Y, Fang L, Akash MSH, Liu Z, Shi W, Chen S (2013) Development and comparison of two competitive ELISAs for the detection of bisphenol A in human urine. Anal Methods 5(21):6106–6113. https://doi.org/10.1039/C3AY41023D

Li J, Zhang X, Liu Y, Tong H, Xu Y, Liu S (2013) Preparation of a hollow porous molecularly imprinted polymer using tetrabromobisphenol A as a dummy template and its application as SPE sorbent for determination of bisphenol A in tap water. Talanta 117:281–287. https://doi.org/10.1016/j.talanta.2013.09.022

Liu Y, Wu Y, Qin G, Chen Y, Wang X, Lin Q (2021) Bioaccumulation and reproductive toxicity of bisphenol A in male-pregnant seahorse (Hippocampus erectus) at environmentally relevant concentrations. Sci Total Environ 753:141805. https://doi.org/10.1016/j.scitotenv.2020.141805

Lu Y, Peterson JR, Gooding JJ, Lee NA (2012) Development of sensitive direct and indirect enzyme-linked immunosorbent assays (ELISAs) for monitoring bisphenol-A in canned foods and beverages. Anal Bioanal Chem 403(6):1607–1618. https://doi.org/10.1007/s00216-012-5969-8

Mathew M, Sreedhanya S, Manoj P, Aravindakumar CT, Aravind UK (2014) Exploring the interaction of bisphenol-S with serum albumins: a better or worse alternative for bisphenol A? J Phys Chem B 118(14):3832–3843. https://doi.org/10.1021/jp500404u

Matsumura Y, Akahira-Moriya A, Sasaki-Mori M (2015) Bioremediation of bisphenol-A polluted soil by *Sphingomonas bisphenolicum* AO1 and the microbial community existing in the soil. Biocontrol Sci 20(1):35–42. https://doi.org/10.4265/bio.20.35

Mei S, Wu D, Jiang M, Lu B, Lim J-M, Zhou Y-K, Lee Y-I (2011) Determination of trace bisphenol A in complex samples using selective molecularly imprinted solid-phase extraction coupled with capillary electrophoresis. Microchem J 98(1):150–155. https://doi.org/10.1016/j.microc.2011.01.003

Mei Z, Deng Y, Chu H, Xue F, Zhong Y, Wu J, Yang H, Wang Z, Zheng L, Chen W (2013) Immunochromatographic lateral flow strip for on-site detection of bisphenol A. Microchim Acta 180(3):279–285. https://doi.org/10.1007/s00604-012-0930-2

Molina-Molina J-M, Amaya E, Grimaldi M, Sáenz J-M, Real M, Fernández MF, Balaguer P, Olea N (2013) In vitro study on the agonistic and antagonistic activities of bisphenol-S and other bisphenol-A congeners and derivatives via nuclear receptors. Toxicol Appl Pharmacol 272(1):127–136. https://doi.org/10.1016/j.taap.2013.05.015

Moon MK, Jeong I-K, Jung Oh T, Ahn HY, Kim HH, Park YJ, Jang HC, Park KS (2015) Long-term oral exposure to bisphenol A induces glucose intolerance and insulin resistance. J Endocrinol 226(1):35–42. https://doi.org/10.1530/JOE-14-0714

Perrot-Applanat M, Kolf-Clauw M, Michel C, Beausoleil C (2018) Alteration of mammary gland development by bisphenol a and evidence of a mode of action mediated through endocrine disruption. Mol Cell Endocrinol 475:29–53. https://doi.org/10.1016/j.mce.2018.06.015

Qiu W, Zhao Y, Yang M, Farajzadeh M, Pan C, Wayne NL (2016) Actions of bisphenol A and bisphenol S on the reproductive neuroendocrine system during early development in zebrafish. Endocrinology 157(2):636–647. https://doi.org/10.1210/en.2015-1785

Ramírez V, Gálvez-Ontiveros Y, Porras-Quesada P, Martinez-Gonzalez LJ, Rivas A, Álvarez-Cubero MJ (2021) Metabolic pathways, alterations in miRNAs expression and effects of genetic polymorphisms of bisphenol a analogues: a systematic review. Environ Res 197:111062. https://doi.org/10.1016/j.envres.2021.111062

Rosenmai AK, Dybdahl M, Pedersen M, Alice van Vugt-Lussenburg BM, Wedebye EB, Taxvig C, Vinggaard AM (2014) Are structural analogues to bisphenol A safe alternatives? Toxicol Sci 139(1):35–47. https://doi.org/10.1093/toxsci/kfu030

Selvaraj KK, Shanmugam G, Sampath S, Joakim Larsson DG, Ramaswamy BR (2014) GC–MS determination of bisphenol A and alkylphenol ethoxylates in river water from India and their ecotoxicological risk assessment. Ecotoxicol Environ Saf 99:13–20. https://doi.org/10.1016/j.ecoenv.2013.09.006

Shah MP (2024a) Microbial remediation of hazardous chemicals from water & industrial wastewater treatment plant. Springer Nature

Shah MP (2024b) Development in waste water treatment research and processes: applied technologies for clean up of environmental contaminants. Elsevier

Shah MP (2024c) Aerobic and anaerobic microbial treatment of industrial wastewater. CRC Press

Shah MP (2025) Extraction of precious metals from industrial waste using microbial technology. Springer Nature

Shimoda K, Hamada H (2009) Bioremediation of bisphenol A and benzophenone by glycosylation with immobilized marine microalga Pavlova sp. Environ Health Insights 3:EHI.S2758. https://doi.org/10.4137/EHI.S2758

Staples CA, Dorn PB, Klecka GM, O'Block ST, Harris LR (1998) A review of the environmental fate, effects, and exposures of bisphenol A. Chemosphere 36(10):2149–2173. https://doi.org/10.1016/s0045-6535(97)10133-3

Sun F, Kang L, Xiang X, Li H, Luo X, Luo R, Lu C, Peng X (2016) Recent advances and progress in the detection of bisphenol A. Anal Bioanal Chem 408(25):6913–6927. https://doi.org/10.1007/s00216-016-9791-6

Tarafdar A, Sirohi R, Balakumaran PA, Reshmy R, Madhavan A, Sindhu R, Binod P, Kumar Y, Kumar D, Sim SJ (2022) The hazardous threat of bisphenol A: toxicity, detection and remediation. J Hazard Mater 423:127097. https://doi.org/10.1016/j.jhazmat.2021.127097

Tchen R, Tan Y, Boyd Barr D, Barry Ryan P, Tran V, Li Z, Hu Y-J, Smith AK, Jones DP, Dunlop AL, Liang D (2022) Use of high-resolution metabolomics to assess the biological perturbations associated with maternal exposure to bisphenol A and bisphenol F among pregnant African American women. Environ Int 169:107530. https://doi.org/10.1016/j.envint.2022.107530

Tillett T (2009) Bisphenol A, Chapter 2: New data shed light on exposure, potential bioaccumulation. Environ Health Perspect 117(5):A210–A210. https://doi.org/10.1289/ehp.117-a210b

Trivedi J, Chhaya U (2022) Bioremediation of bisphenol A found in industrial wastewater using Trametes versicolor (TV) laccase nanoemulsion-based bead organogel in packed bed reactor. Water Environ Res 94(10):e10786. https://doi.org/10.1002/wer.10786

Yao J, Wang J, Wu L, Lu H, Wang Z, Yu P, Xiao H, Gao R, Yu J (2020) Perinatal exposure to bisphenol A causes a disturbance of neurotransmitter metabolic pathways in female mouse offspring: a focus on the tryptophan and dopamine pathways. Chemosphere 254:126715. https://doi.org/10.1016/j.chemosphere.2020.126715

Zhang Z, Chen X, Rao W, Chen H, Cai R (2014) Synthesis and properties of magnetic molecularly imprinted polymers based on multiwalled carbon nanotubes for magnetic extraction of bisphenol A from water. J Chromatogr B 965:190–196. https://doi.org/10.1016/j.jchromb.2014.06.031

Zhang Y-F, Ren X-M, Li Y-Y, Yao X-F, Li C-H, Qin Z-F, Guo L-H (2018) Bisphenol A alternatives bisphenol S and bisphenol F interfere with thyroid hormone signaling pathway in vitro and in vivo. Environ Pollut 237:1072–1079. https://doi.org/10.1016/j.envpol.2017.11.027

Zhang Y-F, Shan C, Wang Y, Qian L-L, Jia D-D, Zhang Y-F, Hao X-D, Xu H-M (2020) Cardiovascular toxicity and mechanism of bisphenol A and emerging risk of bisphenol S. Sci Total Environ 723:137952. https://doi.org/10.1016/j.scitotenv.2020.137952

Zhao H, Liu M, Lv Y, Fang M (2022) Dose-response metabolomics and pathway sensitivity to map molecular cartography of bisphenol A exposure. Environ Int 158:106893. https://doi.org/10.1016/j.envint.2021.106893

Zhu M, Li Y, Niu Y, Li J, Qin Z (2020) Effects of bisphenol A and its alternative bisphenol F on Notch signaling and intestinal development: a novel signaling by which bisphenols disrupt vertebrate development. Environ Pollut 263:114443. https://doi.org/10.1016/j.envpol.2020.114443

Zhuang Y, Zhou M, Gu J, Li X (2014) Spectrophotometric and high performance liquid chromatographic methods for sensitive determination of bisphenol A. Spectrochim Acta Part A Mol Biomol Spectrosc 122:153–157. https://doi.org/10.1016/j.saa.2013.11.015

Chapter 11
Exploring the Effectiveness of Biological Remediation Techniques in the Cleanup of Crude Oil Spill Sites

Priyanka Singh, Samriddhi Singh, and Meenu Singh

Abstract Crude oil spills pose a significant environmental threat and require effective cleanup strategies. This research focuses on applying bioremediation techniques to mitigate the ecological impacts of a crude oil spill at a sensitive site. The leak, caused by a ruptured pipe, contaminated soil and water and threatened local ecosystems and water resources. Bioremediation technology, which uses indigenous microorganisms to break down hydrocarbons, was chosen because it is environmentally friendly and cost-effective. Key steps include site assessment, microbial screening, and the development of customized bio-stimulation strategies to increase microbial activity. Monitoring techniques such as gas chromatography and microbial community analysis are used to assess remediation progress. The results showed a significant reduction in hydrocarbon concentrations during remediation, demonstrating the effectiveness of the bioremediation approach. Challenges such as fluctuating environmental conditions and substrate limitations were addressed through adaptive management strategies. The study underscores the importance of integrating bioremediation into spill response plans, highlighting its potential as a sustainable solution for crude oil contamination. Overall, this research provides valuable insights into the practical application of bioremediation techniques for restoring ecosystems impacted by crude oil spills.

Keywords Bioremediation · Crude oil spill · Ecosystem restoration · Biostimulation · Microbial degradation · Bioaugmentation

P. Singh · S. Singh (✉) · M. Singh (✉)
Department of Biotechnology, School of Engineering, IILM University, Greater Noida, U.P., India
e-mail: samriddhi.singh.bt26@iilm.edu

M. Singh
e-mail: meenu.singh@iilm.edu

P. Singh
e-mail: priyanka.singh@iilm.edu

© The Author(s), under exclusive license to Springer Nature Singapore Pte Ltd. 2026
S. Dey and M. P. Shah (eds.), *Microbial Applications in Industrial Wastewater Treatment*, Advances in Wastewater Research,
https://doi.org/10.1007/978-981-96-9128-9_11

11.1 Introduction

11.1.1 Overview of Crude Oil Spills and Their Environmental Impact

Crude oil is a naturally occurring, unrefined petroleum substance composed of hydrocarbon deposits and various other materials. It is extracted from underground reservoirs and redefined into various products such as gasoline, diesel, kerosene, and petrochemicals. This versatile energy source plays a crucial role in the global economy, powering industries, transportation, and heating. Crude oil spills, a catastrophic event that poses a significant threat to the environment, have garnered global attention due to their detrimental impact on ecosystems and human health. It refers to the accidental release of petroleum into the environment, typically occurring during transportation, extraction, or storage processes. These spills can be caused by human error, equipment failure, natural disasters, or deliberate acts of sabotage. The frequency and distribution of crude oil spills vary globally, with regions like the Gulf of Mexico, the Niger Delta, and the Persian Gulf experiencing higher incidences due to their significant oil production and transportation activities. The environmental impact of crude oil spills is profound, with immediate effects on marine life and ecosystems. Oil spills coat marine organisms, such as fish, birds, and mammals, leading to suffocation, poisoning, and reduced reproductive success. Furthermore, the long-term consequences include biodiversity loss, habitat destruction, and disruptions in the food chain. Oil contamination can persist in the environment for years, affecting water quality and soil fertility and posing risks to human health through the consumption of contaminated seafood.

11.1.2 Importance of Effective Remediation Techniques

Effective remediation techniques for oil spills are critical for mitigating the extensive environmental, economic, and social impacts of such disasters. Swift and efficient responses help minimize damage to marine and coastal ecosystems, which can suffer from the toxic effects of crude oil, leading to long-term harm to wildlife, including fish, birds, and marine mammals. Effective remediation also aids in preserving the livelihood of communities reliant on fishing, tourism, and other coastal industries. Moreover, prompt cleanup can prevent oil from spreading to larger areas, reducing the overall environmental footprint of the spill. By employing advanced and effective remediation techniques, we can ensure that oil is removed from the environment as quickly and thoroughly as possible, thus aiding in the restoration of affected habitats and reducing the time needed for natural recovery processes. This not only helps in preserving biodiversity and maintaining ecosystem services but also in upholding public health and safety and safeguarding economic resources.

11.1.3 Introduction to Biological Remediation as a Sustainable Solution

Biological remediation, or bioremediation, offers a sustainable solution for environmental cleanups by leveraging natural processes to degrade, detoxify, or transform pollutants into harmless byproducts. This approach uses microorganisms and plants to break down pollutants in soil, water, and air. This can effectively restore ecosystems without the need for harmful chemicals or extensive physical intervention. By enhancing or accelerating these natural processes, bioremediation not only reduces pollution but also promotes biodiversity and ecosystem resilience. Moreover, it is cost-effective and minimizes environmental disruption, making it an attractive option for long-term environmental management and sustainability. Its application ranges from oil spill response to heavy metal detoxification, demonstrating its versatility and potential to address various forms of pollution while aligning with ecological conservation goals.

11.1.4 Purpose and Scope of the Chapter

The goal of authoring a book chapter on bioremediation strategies for oil spills is to unveil the ecological approach that stands as an antidote for one of the gravest forms of environmental contamination in a detailed and comprehensive way. This chapter intends to enlighten the readers on the mechanisms, efficiency, and field application that come along with using bioremediation while dealing with oil spill management. We will not only go through the scientific rationale underpinning bioremediation but also paint a picture through successful case studies, delving into the choice of biological agent ranging from bacteria, fungi, all the way to plants. The gamut covered involves looking at existing bioremediation technologies, teasing out challenges impeding their implementation (and likely solutions) and keeping an eye out for further innovative and improved prospects. These details are the focus of the chapter, with the aim of providing information on the potential and constraints of bioremediation for use as an ecological restoration and pollution control tool. The intended audience is researchers, environmental professionals, policymakers, and students. The chapter strongly promotes (without any compromise) wider utilization of bioremediation in coordination with the integration into environmental managerial practices as well.

11.2 Background and Literature Review

11.2.1 *Historical Context of Crude Oil Spills*

Humans have utilized petroleum for thousands of years. Initially used in medicine and weapons, the material did not come into use until the Industrial Revolution, when it became valued as an illumination fuel plus lubricant—then later as an energy source that replaced wood, coal, and animal power. Liquid petroleum had some significant advantages over other energy sources of the times: it was concentrated, and it could be transported easily from one place to another. As oil use boomed, its refined products came to be used to fuel automobiles and a worldwide system of wells, ships, storage terminals grew along with pipelines (Shah 2024a; Rafferty 2023). There are occasions where oil is released into the environment due to extracting oil from the ground and moving it to refineries and beyond—caused by old equipment or bad luck. The biggest disaster has resulted in ten million of gallons of oil spilled, clogging fisheries, killing and injury animals, and decreasing tourism income. The following list in Table 11.1 contains the top nine most damaging oil spills.

11.2.2 *Overview of Traditional Remediation Methods*

An important risk to the ecosystem, marine life, and coastal towns is oil spill-related marine contamination. When oil leaks into the oceans, it forms an oil slick that upsets the surrounding environment by spreading quickly across the water surface. Due to the coating that this pollution causes on their bodies, marine species such as fish, birds, and marine mammals may suffer greatly. This impact includes impaired breathing, movement, and body temperature regulation. Planktons, larvae, and other fundamental components of the marine food web can be impacted by the poisonous components of oil, which can also infiltrate the water column. Oil pollution of beaches, mangroves, and marshlands causes long-term ecological harm as well as financial losses for coastal communities that depend on fishing and tourism. Although there have been oil spills for many years, sadly, cleaning methods have not improved much in that period. Techniques for responding to oil spills include chemical treatments like dispersion, thermal methods like in-situ burning, mechanical means like booms, sorbents, and skimmers, and natural approaches like bioremediation (Shah 2024b; Pete et al. 2021) Booms and skimmers are still often utilized because they serve as physical barriers and collectors to stop oil from spreading and clearing the water's surface. To facilitate natural biodegradation, chemical dispersants are also used to break down oil into tiny droplets. Furthermore, biological agents that consume oil, such as bacteria, are being used to hasten the breakdown of oil. Two common biological processes are used to clean up oil spills: bioaugmentation, which includes the addition of certain types of bacteria strains that can break down oil components, and biostimulation, which involves adding nutrients or oxygen to

Table 11.1 Represents the description of oil spill incident

S. No	Oil spill incident in history	Description
1	The *Amoco Cadiz* Oil Spill (1978)	The Amoco Cadiz, a very large crude carrier (VLCC), ran aground off the coast of Brittany, France, on March 16, 1978. This resulted in the spilling of its entire cargo of sixty-nine million gallons of light crude oil into the sea. About 321 km of French coast was polluted and causes extensive damage to marine life, local fishing industries, and costal habitats
2	The *Castillo de Bellver* Oil Spill (1983)	On August 6, 1983, the Spanish tanker Castillo de Bellver caught fire and broke into two off the coast of South Africa. Approximately 110,000 tons of oil were released into the Atlantic Oceans. The spill had significant environmental impact, through much of the oil dispersed in the open sea
3	The Incidents at the Nowruz Oil Field (1983)	During Iraq-Iran war, the Nowruz Oil field in the Persian Gulf, experienced multiple spills. On February 10, 1983, an oil platform was attacked, caused a major spill. Before it was eventually sealed in September 1983, the well spewed over 1,500 barrels (63,000 gallons) of oil into the Persian Gulf per day Later that year, a tanker collision exacerbated the situation, releasing an estimated 733,000 barrels of crude oil into the gulf
4	The Kolva River Spill (1994)	An oil pipeline break in the Russian Arctic due to corrosion was the source of the Kolva River oil spill. A dike kept the oil contained around the burst site for eight months, but eventually broke way, causing some 84 million gallons of oil to leak into the Kolva River. Oil contaminated an area of 186 square kilometers, or around seventy-two square miles, of marshes and tundra. Millions of gallons of oil were spilled into the environment via pipeline breaches in the area, some of which ended up in the Kolva River
5	The Mingbulak (or Fergana Valley) Oil Spill (1992)	The largest inland oil spill occurred on March 2, 1922, in Uzbekistan's Fergana valley. It took two months for the oil to catch fire and burn until the well pressure decreased. Over eighty-eight million gallons of the projected 88 million gallons released were kept out of the fire by berms and dikes and caused long-term environmental damage
6	The *Atlantic Empress* Oil Spill (1979)	The Atlantic Express, another VLCC, collided with Aegean Captain off the coast of Trinidad and Tobago on July 19, 1979. The accident caused the release of about ninety million gallons of oil into the Atlantic Oceans, making it the largest spill in the history. The spill leads to serve marine pollution

(continued)

Table 11.1 (continued)

S. No	Oil spill incident in history	Description
7	The *Ixtoc 1* Oil Spill (1979)	Between June 1979 and March 1980, the Ixtoc 1 catastrophe in Mexico spilled up to 140 million gallons of crude oil into the Bay of Campeche. The spill was caused by an explosion that occurred on the Ixtoc 1 platform, which was drilling exploratory wells in 164 feet (about fifty meters) of water A buildup of gas and oil in the pipe due to the drilling mud not moving properly was what triggered the explosion. The blowout preventer was bypassed, and a slurry of mud, oil, and natural gas surged up the pipe as workers tried to remove the drill so that material could flow back down the line and cover the hole. There was a leak of between 126 and 140 million gallons into the southern portion of Gulf of Mexico for about 9 months
8	BP's *Deepwater Horizon* Oil Spill (2010)	On April 20, 2010, an explosion on the Deepwater Horizon oil rig, operated by BP in the Gulf of Mexico, caused the rig to sink and a subsea blowout to release around 134 million gallons of oil were released in American Gulf Coast, spanning from Texas to Florida, over 2,100 km (1,300 miles), was covered with oil
9	The Persian Gulf War Oil Spill (1991)	During the Gulf War, Iraqi forces intentionally released oil from several tankers and oil terminals into the Persian Gulf. Estimate the spill's size vary, but it is believed to be between 380 and 520 million, making it the largest oil spill in history. The environmental damage was extensive, affecting coastal ecosystem and marine life

**Reference* (Rafferty 2023)

encourage the growth of naturally occurring microorganisms that break down oil into the surrounding.

11.2.3 Introduction to Biological Remediation

One essential tactic for dealing with oil spills is biological remediation, also known as bioremediation, which uses the environment's natural processes to break down contaminants and repair the ecosystem. Physical and chemical procedures are not employed in any situation involving an oil spill because of the potential for environmental pollution, their excessive cost, and their low adsorption capability. Therefore, as a backup strategy, inexpensive biodegradable biomaterials are frequently employed to remove oil spills from the surface (Wolok et al. 2020). Because biological techniques utilize live organisms to carry out the rehabilitation of contaminated sites—particularly the cleaning of crude oil spills in the detoxification of these environments—they are favored over artificial methods that use many local microorganisms in water and soil (Ndimele et al. 2018). These microbes change dangerous pollutants like carbon dioxide and water into less hazardous or non-toxic molecules by breaking down the components of the oil. An environmentally friendly method of cleaning up oil spills from the marine environment is to use bioremediation techniques. This technique, known as bioaugmentation or biostimulation, applies naturally occurring or genetically modified microorganisms to the contaminated location and/or enriches the polluted environment with nutrients (Baniasadi and Mousavi 2018). Although bioremediation takes longer than certain mechanical techniques, it is a vital tool in the control of marine oil spills since it may accomplish complete, long-term cleaning.

11.2.4 Summary of Key Research Studies on Biological Remediation Techniques

Important studies on biological cleanup strategies for oil spills have shown how effective and promising these approaches may be in reducing environmental harm. Research has indicated that the introduction of certain microbial strains that break down oil, known as bioaugmentation, accelerates the rates at which hydrocarbons degrade in polluted settings. The significance of maximizing nutrition and oxygen levels to enhance the natural microbial activity and promote more effective oil breakdown has been brought to light by research on biostimulation. Research on the use of genetically modified microbes has shown encouraging findings, demonstrating a quicker and more thorough breakdown of oil constituents. Furthermore, research on the synergistic benefits of mixing bioremediation with other techniques, such mycoremediation—which uses fungi—and phytoremediation—which uses plants to absorb and break down contaminants—has shown increased cleanup results. All things considered, these investigations highlight the efficiency of biological cleanup methods and their indispensable position in all-encompassing plans for responding to oil spills.

11.3 Types of Biological Remediation Techniques

11.3.1 Bioremediation

Oil spills are often cleaned up using bioremediation, an eco-friendly method that makes use of microorganisms' innate capacity to break down and purify toxins. This process converts the poisonous hydrocarbons present in oil into less hazardous compounds like carbon dioxide and water by using bacteria, fungi, and algae. Natural bioremediation has been utilized for ages by civilizations to remediate wastewater; however, purposeful usage for the reduction of hazardous pollutants is a more recent development (Adams et al. 2015). Bioaugmentation—introducing certain bacteria that break down oil to the polluted area—and biostimulation—optimizing circumstances by supplying oxygen or nutrients to encourage the activity of native microorganisms—are two ways to improve the process. Because bioremediation is affordable, environmentally friendly, and suitable for a variety of settings—including soils, shorelines, and aquatic bodies—it is beneficial. Bioremediation provides a comprehensive, long-term solution for oil spill cleaning, albeit one that often proceeds more slowly than some mechanical techniques. It also makes a substantial contribution to the restoration of impacted ecosystems.

11.3.1.1 Definition and Mechanism

The term "bioremediation" refers to the process of employing living things—microorganisms—as nutrition or energy sources to remove toxins from the environment (Baba 2016). George Robinson is partially responsible for the application of bacteria in contemporary bioremediation (Shah 2024c; Adams et al. 2015). His work has been crucial in expanding the application of microbial processes to reduce pollution in the environment, especially when it comes to cleaning up oil spills. By his research and development work, Robinson contributed to the creation of methods that break down and detoxify oil, the most favored agent of hydrocarbon breakdown for bioremediation may have been bacteria due to their propensity for reproduction and widespread distribution (Macaulay 2014). The principal approach is bioaugmentation, in which the polluted site is inoculated with certain strains of oil-degrading bacteria to expedite the breakdown process.

The goal of this strategy is to increase the pace or degree of degradation of complex pollutants by introducing microorganisms that break down contaminants (Goswami et al. 2018). A further process is biostimulation, which is the addition of nutrients, such as oxygen, nitrogen, and phosphorus, to the natural microbial activity in order to provide the ideal circumstances for microbial growth and activity. These microbes eventually degrade the hydrocarbons by using them as a source of carbon and energy. Furthermore, the cooperative use of phytoremediation, in which plants aid in the uptake and degradation of pollutants in oceans. Five different methods of phytoremediation—phytostabilization, phytoextraction, phytodegradation, rhizofiltration, and phytovolatilization—can be used to remove contaminants (Nemati 2024). Together, these systems enable an extensive and long-lasting cleaning of oil spills, aiding in the restoration of contaminated areas.

11.3.1.2 Microorganisms Involved in Bioremediation

Microorganisms are found all across the biosphere due to their remarkable metabolic capacity and ease of growth in a variety of environmental settings (Abatenh et al.

2017). They may use a variety of organic and inorganic molecules as sources of carbon, energy, and nutrition because of their metabolic flexibility. Using living things to remove or neutralize pollutants from a contaminated region makes advantage of different microorganism's metabolic capacities, such as those of bacteria, fungus, and algae. This approach depends on encouraging the development of certain microflora or microbial consortia that are native to the polluted locations and capable of carrying out required tasks (Singh 2014). It is possible to separate microorganisms from practically any type of environment. Microbes are adaptable and may develop in a variety of environments, including subzero temperatures, intense heat, desert climates, water, an abundance of oxygen, anaerobic circumstances, the presence of dangerous substances, and waste streams (Kensa 2011).

Multiple crucial elements that govern the outcome of pollutant degradation impact the efficacy of microbial bioremediation. Temperature, pH, and moisture content are important environmental factors that affect microbial activity and metabolic rates. Certain parameters have ideal ranges for these factors. In nutrient-deficient settings, supplementation is frequently necessary due to the critical role that nutrients, especially nitrogen and phosphorus, play in microbial development and function. The most beneficial microbes for the bioremediation process will either be aerobic or anaerobic depending on the availability or absence of oxygen. Active members of microbial consortiums are defined as microorganisms that perform biodegradation in a variety of settings (Singh 2014). Table 11.2 lists the microorganisms that are involved in the breakdown of hydrocarbons.

List of microorganisms that important for oil remediation is mentioned in Table 11.3.

Table 11.2 Contains a detailed information about microorganisms and hydrocarbon interaction

Microorganisms	Hydrocarbon	References
Penicillium chrysogenum	Monocyclic aromatic hydrocarbons, benzene, toluene, ethyl benzene and xylene, phenol compounds	Abatenh et al. (2017)
Gleophyllum striatum MTCC-1117	Polycyclic aromatic hydrocarbons pyrene, acenaphthene, anthracene, dibenothiophene and 9-methyl anthracene	Yadav et al. (2011)
Coprinellus radians	PAHs, methylnaphthalenes, and dibenzofurans	Abatenh et al. (2017)
Rohodococcus sp. Strain Q15	Hydrocarbons	Salleh et al. (2003)
Mycobacter calcoaceticus	Naphthalene	Salleh et al. (2003)
Mycobacterium sp.	Pyrene	Salleh et al. (2003)
Phanerochaete chrysosporium	Biphenyl and triphenylmethane	Abatenh et al. (2017)

Table 11.3 Represents microorganisms which degrade oil

Microorganisms	Compound	References
Fusarium sp.	Oil	Abatenh et al. (2017)
Aspergillus niger, Candida glabrata, Candida krusei and *Saccharomyces cerevisiae*	Crude oil	Burghal et al. (2016)
Pseudomonas aeruginosa, P. putida, Arthobacter sp. and *Bacillus* sp.	Diesel oil	Sukumar and Nirmala (2016)
Pseudomonas cepacia, Bacillus cereus, Bacillus coagulans, Citrobacter koseri and *Serratia ficaria*	Diesel oil, Crude oil	Abatenh et al. (2017)
Alcaligenes odorans, Bacillus subtilis, Corynebacterium propinquum, Pseudomonas aeruginosa	Oil	Singh et al. (2013)
B. brevis, P. aeruginosa KH6, *B. licheniformis* and *B. sphaericus*	Crude oil	Abatenh et al. (2017)

11.3.2 Phytoremediation

11.3.2.1 Role of Plant in Oil Degradation

Using living plants to remove or break down pollutants from the environment is known as phytoremediation, and it is an environmentally friendly and sustainable technique (Angulu et al. 2018). This method makes use of plants' innate capacity to take in, retain, and cleanse pollutants via a variety of processes. Utilizing plants to remove pollutants can help cut down on the requirement for costly machinery, labor-intensive procedures, and energy-intensive methods. In addition to metabolizing organic pollutants and promoting their microbial breakdown in the root zone, plants can be employed to sequester inorganic and organic pollutants (Shah 2025; Arthur 2005). Numerous pollutants, such as metals, pesticides, explosives, and oil, can be cleaned up by plants. Additionally, the plants aid in retaining contaminants at the locations by preventing wind, rain, and groundwater from moving them away (Etim 2012). Several approaches are included in phytoremediation, each of which is suited to a particular kind of pollution and its surroundings. By using plants' inherent abilities to absorb, collect, detoxify, or stabilize contaminants, these methods take advantage of such abilities. Table 11.4 gives the brief description of the different phytoremediation techniques. Through a process called phytoremediation, plants significantly contribute to the deterioration of oil in maritime environments. Due to its advantages over chemical techniques, phytoremediation has grown in popularity since the 1980s (Hussein et al. 2022). Reports regarding the in-situ phytoremediation of submerged sediments polluted with PCBs and PAHs are unknown (Huesemann et al. 2009).

Table 11.4 Shows different phytoremediation techniques

Phytoremediation	Description	References
Phytoextraction	This technology refers that plants absorb metals from soil and translocation them to the harvestable shoots where they accumulate. The roots and shoots are subsequently harvested to remove the contaminants from the soil	Moosavi and Seghatoleslami (2013)
Phytostabilization	Through precipitation in the rhizosphere or accumulation by roots, phytostabilization seeks to keep pollutants out of the vadose zone. By doing this, off-site pollution caused by their movement through soil dispersion, leaching, and erosion from wind and water is prevented	Bolan et al. (2011)
Phytodegradation	Using microorganisms and plants to absorb, digest, and break down the organic pollutant is known as phytodegradation. This method involves decontaminating soil that has been polluted with organic substances by using plant roots in conjunction with microorganisms. The term "phytotransformation" also applies	Moosavi and Seghatoleslami (2013)
Rhizofilteration	The process of using plant roots to absorb, concentrate, and precipitate harmful metals from contaminated ground water or polluted effluents is known as rhizofiltration. According to most studies, plants used in phytoremediation should only store metals in their roots. This is because, in all forms of phytoremediation, the plants' root systems interact with the pollutants or contaminated site to remove the pollutants and restore the region to a pollution-free state	Rawat (2012)
Phytovolatilization	Through the process of phytovolatilization, pollutants are drawn up from the soil, changed into a volatile state, and then transpired into the atmosphere by means of plants. As expanding trees and other plants absorb water and the organic and inorganic pollutants, phytovolatilization takes place	Singh and Ghosh (2005)

11.3.2.2 Types of Plant Used and Their Effectiveness

By taking in pollutants (mostly water soluble) through their roots and moving them through different plant tissues where they can be digested, sequestered, or volatilized, plants function as solar-driven pumping and filtration systems (Moosavi and Seghatoleslami 2013). There have been reports of over 400 plant species from at least 45 plant groups hyperaccumulating metals (Lasat 2002). *Brassica juneca* is one of the species of plants that warrants special attention among the others since several tests have demonstrated its importance to the phytoextraction process of heavy metals from soil. Research has shown that *B. juncea* has a significant potential to collect Cd, primarily in the shoots, where a Cd level of 1450 μg Cd/g dry weight was detected. This is three times more than the 555 μg/g dry weight observed in *Brassica napus* (Moosavi and Seghatoleslami 2013). There are other various kinds of plant species which are used for the treatment of oil contaminants detail for the same is provided in Table 11.5.

11.3.3 Mycoremediation

Mycoremediation is a type of bioremediation where pollutants from the environment, such as oil pollution in marine environments, are broken down or sequestered using fungus. The technique known as "mycoremediation" uses fungus to remove pollutants from the environment. These fungi grow long threads called hyphae that mate with rocks, soil, and tree roots to form a filamentous body (Biswas 2023). The synthesis of extracellular enzymes like lignin peroxidases, manganese peroxidases, and laccases by fungi gives them a special capacity to degrade complex hydrocarbons. The difficult-to-break compounds present in oil may be broken down by these enzymes and converted into less dangerous materials. This process involves certain mechanisms for the degradation such as biosorption, bioaccumulation, bioconversion, and biodegradation (El-Ramady et al. 2022). Furthermore, the fungal mycelium has the ability to form a biofilm on oil surfaces, which increases the biodegradation process by giving enzymes a lot of surface area to work with. In addition to being efficient, the use of fungus in mycoremediation is also eco-friendly as it eliminates the need for chemical dispersants and other conventional cleanup techniques that might have a negative influence on the environment.

11.3.3.1 Fungi and Their Role in Breaking Hydrocarbon

Organic molecules that are made entirely of hydrogen and carbon atoms are known as hydrocarbons. They are important in the context of oil pollution since they are the main components of natural gas and petroleum. Land and water pollution caused by hydrocarbons are one of the significant problems in developing countries. Among the microorganisms previously identified from soil, water, and sediment that have

Table 11.5 Represents the collective data of different types of plant species with their effectiveness

Plant species	Effectiveness
Seagrasses (e.g., *Zostera* spp. and *Thalassia* spp.)	Seagrasses are very good in stabilizing sediments and giving microorganisms that break down oil a place to live. Their vast root systems aid in microbial activity and soil oxygenation, which accelerates the decomposition of hydrocarbons. Seagrass meadows can help marine ecosystems recover by drastically lowering the amount of oil present in polluted regions
Salt Marsh Plant (e.g., *Spartina alterniflora* and *Juncus roemerianus*)	Plants that grow in intertidal zones and have a high resistance to saltwater are known characteristics of salt marsh plants. They are essential in capturing oil and stopping its spread. These plants' oxygen- and organic-rich rhizosphere is home to a thriving bacterial population that breaks down oil. Because of this, salt marsh plants—especially those found near coasts—are very useful in bioremediation projects
Mangroves (e.g., *Rhizophora* spp. and *Avicennia* spp.)	Because of their intricate root systems, which offer a large surface area for microbial colonization, mangroves are very successful at degrading oil. Mangroves' pneumatophores, or aerial roots, allow gas exchange, which supports the maintenance of aerobic conditions that support the microbial breakdown of hydrocarbons. Additionally, mangroves serve as physical barriers, keeping oil contained and preventing it from getting to more vulnerable places. In tropical and subtropical areas, where they create thick forests along coasts, their role is particularly important
Microalgae (e.g., *Chlollera* spp. and *Dunaliella* spp.)	Hydrocarbons can be immediately absorbed by microalgae and then changed by metabolic processes. Additionally, during photosynthesis, they create oxygen, which helps bacteria degrade organic matter aerobically. Certain species of microalgae have the ability to break down oil components, which helps to reduce the amount of oil pollutants in the water column overall. Even though they could have a more limited effect than bigger plants, they are nevertheless an important part of the overall phytoremediation plan

been contaminated by hydrocarbons are fungi. Numerous fungi have been identified in areas polluted by petroleum hydrocarbons in recent years (Tasiu Mahmud 2022). Particularly, the crude oil has hundreds of distinct hydrocarbons, ranging from C1 (methane) to more than C30. Saturate, Aromatic, Resin, and Asphaltene (SARA) analysis reveals that, except for gases, the majority of aliphatic hydrocarbons in petroleum are alkanes. These include open-chain linear or branched alkanes and cycloalkanes, which are also referred to as naphthene (Daccò et al. 2020). Since fungi have enzyme systems, they are uniquely able to break down complicated hydrocarbon compounds. Extracellular enzymes that can dissolve the hard, intricate structures of hydrocarbons include lignin peroxidases, manganese peroxidases, and laccases, which are produced by fungi, especially white-rot fungus. White rot fungi (WRF) are excellent candidates for use in bioremediation because they can break down a wide range of organic pollutants that are not easily broken down by other microorganisms (Boyle et al. 1998). Fungi also possess vast mycelial networks that may infiltrate polluted surfaces and soils, expanding the surface area available for contact with contaminants and speeding up the degradation process. Fungi are important agents in environmental rehabilitation efforts because of their large natural attenuation and cleaning of hydrocarbon contaminants through these methods.

11.4 Mechanisms of Action

11.4.1 Microbial Metabolisms of Hydrocarbon

In ecosystems affected by oil spills and other hydrocarbon sources, the biodegradation of organic pollutants depends critically on the microbial metabolism of hydrocarbons. Hydrocarbons may be broken down by a variety of microorganisms, such as bacteria, fungi, and archaea, and used as sources of carbon and energy. Hydrocarbon-utilizing bacteria and fungi are widely distributed in both terrestrial and aquatic ecosystems, but their proportion to the overall heterotrophic community varies greatly. According to reports, their frequencies range from 6 to 82%, 0.13 to 50% for soil bacteria, and 0.003 to 100% for marine bacteria (Colwell 1990). Initial hydrocarbon activation is frequently followed by a sequence of oxidation and reduction processes that transform the hydrocarbon into intermediate molecules. These intermediate compounds are then further metabolized to produce biomass, CO_2, and water. The accelerated disappearance of petroleum that contaminates fields and streams, the degradation of some synthetic and natural rubber products, the spoiling of cooling oils, the deterioration of oiled or asphalt-surfaced highways, and the alteration of petroleum or its products stored in the presence of water can all be attributed to the microbial oxidation of hydrocarbons (Zobell 1946). Especially, when it comes to cleaning up contaminated areas and oil spills, this metabolic capacity is crucial for reducing environmental contamination.

11.4.1.1 Aerobic and Anaerobic Degradation Pathway

Both aerobic and anaerobic mechanisms, each involving unique microbial activities, are involved in the breakdown of hydrocarbons. However, hydrocarbon biodegradation slows down and becomes less extensive in anaerobic environments, and the range of substrates that are broken down usually narrows down (Grishchenkov 2000). The principal electron acceptor in aerobic circumstances is oxygen. A sequence of enzymatic processes is how microorganisms, including bacteria and fungus, decompose hydrocarbons. On the other hand, microbes use other electron acceptors, including nitrate, sulfate, or carbon dioxide, in anaerobic conditions, where oxygen is not present. Aerobic conditions provide the most fast and thorough breakdown of most organic contaminants. The enzymatic activity of oxygenase and peroxidases catalyzes the first intracellular assault of organic contaminants, which involves the oxidative process of activation and incorporation of oxygen.

Organic pollutants are progressively transformed into central intermediary metabolic intermediates by peripheral degradation mechanisms, such as the tricarboxylic acid cycle. A few examples of the central precursor metabolites used in the biosynthesis of cell biomass include acetyl-CoA, succinate, and pyruvate. Gluconeogenesis synthesizes sugars that are needed for a variety of biosynthesis and growth (Essien and Olajire 2014). Likewise, in anaerobic degradation that starts with the hydrocarbon being activated, usually by adding fumarate or other chemicals, and then proceeds through a series of reduction processes. This results in the creation of intermediate molecules, which are then refined into other simpler chemicals such as carbon dioxide and methane. To help with the bioremediation of polluted areas, both routes are essential for the natural attenuation of hydrocarbons in the environment.

11.4.1.2 Enzymatic Breakdown Processes

Enzymatic hydrocarbon degradation occurs in marine settings when certain bacteria use different enzymes to break down hydrocarbons. The hydrocarbons are transformed into alcohols, aldehydes, and carboxylic acids under aerobic circumstances by enzymes like monooxygenases and dioxygenases adding oxygen to the hydrocarbons. After further metabolism, carbon dioxide and water are produced from these intermediates. When there is no oxygen present, other enzymes help activate hydrocarbons by reacting with substitute electron acceptors such as nitrate or sulfate. This produces simpler molecules like carbon dioxide and methane. Anaerobic conditions are created when this process occurs. To bioremediate hydrocarbon contaminants in marine habitats, this enzymatic activity is necessary.

11.4.2 Plant–Microbe Interactions

11.4.2.1 Rhizoremediation and Root-Associated Microbial Activity

Rhizoremediation, or the employment of plants and the microbes they are linked within the root zone, is one of the new techniques in bioremediation. It is particularly good at breaking down organic pollutants and has benefits for both ecology and engineering (Hoang et al. 2021). Plant exudates can increase microbial activity in the rhizosphere because of the substances they produce through their roots. Under this process, plants interact with soil organisms to carry out the breakdown, detoxification, or immobilization of contaminants in polluted soils (Hassand et al. 2024). In this mutualistic interaction, microorganisms have a favorable environment to develop in while plants profit from improved nutrient absorption and protection from harmful chemicals. Plants can enhance the degradative capacity of rhizospheric microbes in several ways, such as through increased microbial population densities generally, specialized increases in microbial communities that break down petroleum hydrocarbons, increased expression of catabolic genes, increased catabolic gene transfer horizontally, and improved hydrophobic hydrocarbon bioavailability (Ibok 2019). Furthermore, some plants can improve microbial community conditions, stabilize the soil, and lessen erosion—all of which help to promote the breakdown of hydrocarbons. Oil pollution in polluted regions may be effectively and sustainably addressed by the synergistic technique known as phytoremediation.

11.4.2.2 Plant Uptake and Translocation of Oil Component

Hydrocarbons are absorbed by plants and moved throughout the system by means of translocation and uptake of oil components. Plants can absorb certain oil components through their roots when there is an oil spill. These compounds could include hydrocarbons that are lighter, more soluble, and able to enter root tissues. Not only do plants remove pollutants from the environment, but they also slow down groundwater hydraulics and stop pollution from spreading (Rafiee et al. 2017). Certain hydrocarbons may be transformed by plants into less harmful substances, which helps in detoxification. The kind of hydrocarbon, the species of plant, and the surrounding surroundings are some of the variables that affect this process' effectiveness. These factors affect the plant's ability to absorb and relocate hydrocarbons. Although this capacity is helpful for phytoremediation, it needs to be managed carefully to prevent phytotoxicity or food chain contamination. Plants aid in the natural attenuation of pollutants in polluted areas by absorbing and relocating oil components. This procedure aids in the breakdown of dangerous materials and lessens the effect they have on ecosystems. Nonetheless, maximizing the efficacy of phytoremediation initiatives requires careful consideration of site-specific factors and the selection of suitable plant species. The bioremediation of oil spills is mostly dependent on plant absorption and translocation, which helps the contaminated regions recover better.

11.4.3 Fungal Degradation

The breakdown of hydrocarbons by different fungi, which release enzymes that decompose complex organic substances, is known as fungal degradation of oil spills. It has been found that fungi can break down oil in both marine and terrestrial environments (Simister et al. 2015). Enzymes like lignin peroxidases and manganese peroxidases, which may break down a variety of hydrocarbons, are produced by fungi, including black and white rot species. Fungal colonization and hydrocarbon use as a carbon source are two ways in which fungi contribute significantly to oil spill situations. They assist with environmental detoxification by dissolving these complex molecules into simpler, less hazardous ones. Improved nutrient cycling brought about by this degrading process also improves soil health and benefits the environment. The utilization of fungus in bioremediation techniques is essential because they can break down and access hydrocarbons that bacteria could find difficult to break down. Fungi are very useful in cleaning up oil-contaminated places because of their capacity to adapt and endure in a variety of challenging environments. Its capacity aids in the breakdown of contaminants and promotes the general rehabilitation of habitats that have been impacted. Utilizing fungal activity increases the efficiency of bioremediation operations, allowing oil pollutants to naturally attenuate and encouraging ecological regeneration in contaminated regions.

11.4.3.1 Mycelia Network and Enzymatic Activity

Fungal mycelium networks and their enzyme functions are primary components of biodegradation systems that take care of organic pollutants, among them hydrocarbons which arise from oil spills. The hyphal network (mycelium) extends through the soil and contaminated substrates, facilitating transport between nutrients along with pollutants due to increased surface contact. It is through this vast network that fungi can probe great soil volumes and reach hydrocarbons that may be hidden or out of their reach. Fungi release an extensive range of external enzymes which include but are not limited to lignin peroxidases, manganese peroxidases, and laccases—which act on the high molecular weight components by breaking them down into low molecular weight products (Hammel 2011). These simpler compounds are also less toxic; the enzyme systems act by generating free radicals which tend to react with organic molecules in proximity (thus involving a chain reaction). This is how the biodetoxification process takes place: these oxidative reactions can occur under very mild conditions and degrade almost any kind of organic molecule. The enzymatic activity within the mycelial network not only aids in breaking down pollutants but also enhances nutrient cycling in the soil, facilitating the uptake of nutrients by plants and other microorganisms. This interaction fosters a healthier soil ecosystem, promoting the recovery of contaminated environments. Furthermore, the ability of fungal mycelia to withstand harsh environmental conditions, such as extreme pH levels and limited nutrient availability, makes them particularly effective in bioremediation. Their resilience and adaptability allow fungi to thrive in diverse environments, contributing to the natural attenuation of oil pollutants.

11.4.3.2 Degradation Pathway Specific to Fungi

Fungi employ a specialized degradation pathway to break down organic pollutants like hydrocarbons in environments such as oil spills. Central to this process are the extracellular enzymes secreted by fungi, including but not limited to lignin peroxidases, manganese peroxidases, and laccases. These enzymes play crucial roles in catalyzing oxidative reactions that initiate the breakdown of complex hydrocarbon molecules. Lignin peroxidases and manganese peroxidases, for instance, can oxidize a wide range of aromatic compounds present in hydrocarbons, breaking down their chemical structures. This initial oxidation step is followed by further enzymatic actions that cleave these compounds into smaller, more manageable fragments. These fragments can then be metabolized by the fungi as carbon and energy sources or assimilated into cellular structures. Additionally, fungi possess metabolic pathways that allow them to utilize various components of hydrocarbons for growth and reproduction. This metabolic versatility enables fungi to thrive in environments where hydrocarbons are abundant, contributing to their role in bioremediation efforts. The resilience of fungal species to diverse environmental conditions further enhances their effectiveness in degrading pollutants in contaminated sites. Overall, the degradation pathway specific to fungi involves the coordinated action of extracellular enzymes to degrade hydrocarbons into simpler compounds that can be utilized or further metabolized. This capability underscores fungi's significance in ecological processes and their potential as valuable tools in environmental biotechnology for mitigating the impacts of oil spills and other forms of pollution.

11.5 Factors Influencing the Effectiveness of Biological Remediation

Biological remediation, also known as bioremediation, is an eco-friendly approach that leverages living organisms such as bacteria, fungi, or plants to remove or neutralize pollutants from contaminated environments. The success of bioremediation processes is largely dependent on a variety of environmental conditions that influence the metabolism and growth of microorganisms involved. Key factors include temperature, pH, oxygen levels, and nutrient availability, all of which play crucial roles in determining the effectiveness of the remediation process.

11.5.1 Temperature

Temperature is a critical factor in bioremediation as it directly affects microbial metabolism. Most bioremediative microorganisms have an optimal temperature range within which they perform most efficiently. For instance, bacteria such as *Pseudomonas* species, often used in bioremediation, thrive in moderate temperature ranges between 20 and 30 °C. At temperatures outside this range, microbial activity may decrease, leading to a slower remediation process. On the other hand, some extremophiles are capable of functioning in extreme temperatures, making temperature-tolerant species a valuable asset in diverse environmental contexts.

11.5.2 pH Levels

The pH of the environment significantly affects the survival and enzymatic activity of microbes involved in bioremediation. Most microorganisms prefer a neutral pH, ranging from 6 to 8, for optimal enzymatic functions. Extremes in pH, either highly acidic or basic conditions, can inhibit microbial growth and activity, rendering the remediation process ineffective. Specific organisms, such as acidophilic bacteria, can survive in low pH environments, making them useful for remediation in acidic conditions, such as mine drainage sites.

11.5.3 Oxygen Levels

The availability of oxygen also plays a crucial role in the efficiency of bioremediation processes. Oxygen is essential for aerobic microorganisms, which are typically involved in the degradation of hydrocarbons and other organic pollutants. However, certain environments may lack oxygen, and in such cases, anaerobic bioremediation becomes necessary. Anaerobic bacteria, such as *Clostridium* species, can degrade pollutants in oxygen-deprived environments, such as subsurface soils or sediments, expanding the applicability of biological remediation to diverse settings.

11.5.4 Nutrient Availability

The availability of nutrients such as nitrogen, phosphorus, and carbon is essential for the growth and metabolic activity of microorganisms used in bioremediation. Nutrient-limited environments can significantly reduce the effectiveness of the process, as microorganisms may not have enough resources to support their growth and pollutant degradation activities. Supplementing contaminated environments with appropriate nutrients, known as biostimulation, can enhance microbial activity and, subsequently, the bioremediation process.

11.6 Advances and Innovations in Biological Remediation

Biological remediation, or bioremediation, has emerged as a promising solution for mitigating environmental contamination caused by industrial, agricultural, and municipal waste. The process involves using microorganisms, plants, or enzymes to break down or neutralize harmful pollutants in soil, water, and air. As pollution becomes a more pervasive issue globally, advancements in bioremediation technologies are crucial for enhancing the efficiency and applicability of this approach. Recent innovations, such as genetic engineering, bioaugmentation, and phytoremediation, have significantly improved the effectiveness of bioremediation. This essay explores these innovations and their impact on the field of environmental restoration.

Genetic Engineering of Microorganisms

One of the most significant advances in biological remediation is the genetic engineering of microorganisms to enhance their ability to degrade pollutants. Traditionally, natural microbial communities were employed in remediation processes, but the metabolic capacity of these organisms was often limited. With the advent of genetic engineering, scientists have been able to modify microbial genomes to optimize their degradation pathways for specific pollutants. For example, recombinant bacteria such as genetically modified *Pseudomonas putida* have been engineered to break down toxic compounds like toluene, a common industrial solvent (Ghosal et al. 2016). These genetically modified organisms (GMOs) can be tailored to address a variety of contaminants, from hydrocarbons to heavy metals.

Moreover, gene editing technologies like CRISPR-Cas9 allow for precise manipulation of microbial DNA, enabling the development of microorganisms with enhanced resistance to environmental stressors such as extreme temperatures, pH, and salinity (Alavi and Nejad 2020). These resilient microbes are particularly useful in remediation efforts in harsh or inhospitable environments, such as polar regions or contaminated mine tailings. Such innovations have the potential to revolutionize bioremediation by expanding its applicability to a wider range of contaminated sites.

Bioaugmentation and Biostimulation

Bioaugmentation involves the introduction of specific microbial strains or consortia to contaminated sites to accelerate the degradation of pollutants. This approach has gained attention due to its ability to enhance the natural microbial population's efficiency in breaking down complex or recalcitrant pollutants (Das and Chandran 2011). Recent advancements in microbial isolation and culturing techniques have allowed scientists to identify and utilize highly specialized bacteria and fungi capable of degrading specific contaminants, such as polychlorinated biphenyls (PCBs) or petroleum hydrocarbons.

In tandem with bioaugmentation, biostimulation involves adding nutrients or electron donors to stimulate the indigenous microbial populations' metabolic activity. Recent studies have demonstrated that the combination of bioaugmentation and biostimulation can yield synergistic effects, particularly in environments with limited nutrient availability (Shukla et al. 2010). By optimizing the growth conditions for both introduced and native microbial communities, the remediation process can be significantly accelerated. This innovation has been particularly successful in treating oil spills and sites contaminated with organic solvents.

Phytoremediation

Phytoremediation is an innovative bioremediation technique that uses plants to absorb, sequester, or break down pollutants from soil and water. Certain plants, known as hyperaccumulators, have the unique ability to take up large quantities of heavy metals such as cadmium, arsenic, and lead (Joutey et al. 2013). One of the most well-known examples of a hyperaccumulator is *Thlaspi caerulescens*, a plant capable of extracting zinc and cadmium from contaminated soils.

Recent advances in biotechnology have allowed scientists to genetically engineer plants for enhanced phytoremediation capabilities. By modifying the genes responsible for pollutant uptake and metabolism, researchers have developed plants that can remove contaminants more efficiently and in greater quantities than their natural counterparts (Shukla et al. 2010). Additionally, advances in agronomic practices, such as soil amendments and crop rotation, have improved the efficiency of phytoremediation, making it a viable option for large-scale environmental restoration projects.

Nanotechnology in Bioremediation

The integration of nanotechnology with bioremediation processes represents another exciting area of innovation. Nanoparticles, due to their small size and large surface area, have unique physicochemical properties that make them highly effective in catalyzing the degradation of pollutants. For example, iron nanoparticles have been used in conjunction with microbial communities to enhance the reduction of chlorinated organic compounds, such as trichloroethylene (TCE), a common groundwater contaminant (Ghosal et al. 2016).

Moreover, nanoparticles can serve as carriers for enzymes or bacteria, protecting them from environmental stressors and improving their delivery to contaminated sites. This "nano-bioremediation" approach has shown promise in the treatment of persistent organic pollutants (POPs) and heavy metal contamination in both soil and water environments (Joutey et al. 2013). As the field of nanotechnology continues to advance, it is likely that its integration with bioremediation will lead to even more effective and efficient pollution mitigation strategies.

Biosensors for Monitoring Remediation Progress

Recent innovations in biosensing technology have significantly improved the monitoring of bioremediation progress. Biosensors are analytical devices that use biological elements, such as enzymes or microorganisms, to detect specific pollutants or byproducts of microbial degradation. These devices provide real-time data on pollutant concentrations and microbial activity, allowing for more precise control of the bioremediation process (Das and Chandran 2011).

The development of wireless and portable biosensors has enabled on-site monitoring, reducing the need for labor-intensive sampling and laboratory analysis (Shukla et al. 2010). These innovations have made it possible to track the success of remediation efforts more efficiently and accurately, leading to better decision-making in environmental management. As a result, biosensors have become an invaluable tool for optimizing bioremediation strategies in both urban and rural environments.

11.7 Challenges and Future Directions

Despite the numerous advancements in bioremediation, several challenges remain. One of the main obstacles is the unpredictability of microbial behavior in natural environments. Factors such as temperature fluctuations, moisture content, and competition with native species can affect the success of bioremediation efforts (Alavi and Nejad 2020). To address these challenges, researchers are exploring the use of omics technologies, such as metagenomics and proteomics, to gain a deeper understanding of microbial communities and their interactions with pollutants. These insights will pave the way for more targeted and effective bioremediation approaches.

Furthermore, the development of multifunctional bioremediation systems that combine several techniques, such as phytoremediation and bioaugmentation, is a promising direction for future research. By integrating various biological, chemical, and physical processes, these systems have the potential to address complex contamination scenarios that would otherwise be difficult to remediate using a single approach.

11.8 Future Directions and Needs for Emerging Trends in Biological Remediation: Innovations in Technology and Methodology for Large-Scale Applications

Biological remediation, also known as bioremediation, has demonstrated considerable potential in addressing environmental contamination by leveraging microorganisms, plants, and biological processes to degrade or neutralize pollutants. While the field has made significant advances in recent years, particularly in the areas of genetic engineering, bioaugmentation, and phytoremediation, there remain many challenges and opportunities for improvement. As environmental threats from industrial pollution, agricultural runoff, and hazardous waste grow more acute, emerging trends in bioremediation technology and methodology will play a crucial role in large-scale applications. This essay explores the future directions of biological remediation, focusing on innovations in technology and methodology, and their potential for broad-scale implementation.

Emerging Trends in Bioremediation Technologies

1. **Genetic Engineering for Enhanced Microbial Degradation**

Genetic engineering continues to be a cornerstone of emerging bioremediation technologies. With advancements in gene-editing tools like CRISPR-Cas9, the future of bioremediation will likely see the development of highly specialized microbial strains engineered for targeted degradation of complex pollutants. Microorganisms such as *Pseudomonas putida* have already been modified to degrade toxic compounds such as toluene, but future directions could include the development of microbial consortia designed to metabolize a wide array of persistent organic pollutants (POPs), including polychlorinated biphenyls (PCBs) and polycyclic aromatic hydrocarbons (PAHs) (Ghosal et al. 2016).

Additionally, the integration of omics technologies—such as metagenomics, proteomics, and transcriptomics—will further enhance our understanding of microbial communities involved in remediation. This will enable scientists to design microorganisms with more robust metabolic pathways for pollutant degradation in diverse environmental conditions, from highly saline coastal areas to extreme temperature zones in the Arctic.

2. **Nanotechnology in Bioremediation**

Nanotechnology is another emerging trend that offers exciting possibilities for enhancing bioremediation processes. Nanoparticles, due to their small size and high surface area, can increase the bioavailability of pollutants, making them easier for microorganisms to degrade. For instance, iron nanoparticles have been used to reduce chlorinated compounds like trichloroethylene (TCE) in groundwater. Moving forward, the development of multifunctional nanoparticles could enhance the catalytic activity of microorganisms or enzymes used in bioremediation (Joutey et al. 2013).

Nano-bioremediation systems, where nanoparticles act as carriers for enzymes or bacteria, can significantly improve the efficiency of pollutant breakdown, especially for persistent pollutants like heavy metals and hydrocarbons. The ability to apply these technologies on a larger scale will be pivotal to the future of remediation strategies in industrial and urban settings.

3. **Biosensors and Real-Time Monitoring**

The future of large-scale bioremediation will require efficient monitoring systems to track the progress of pollutant degradation in real time. Innovations in biosensing technologies have already shown promise in offering rapid, cost-effective solutions for on-site monitoring. Biosensors utilize biological elements such as enzymes, antibodies, or microbial cells to detect pollutants and their degradation byproducts in soil, water, and air. As wireless biosensors become more advanced, they will allow environmental engineers to monitor bioremediation efforts remotely and in real time (Das and Chandran 2011).

In the future, the integration of biosensors with artificial intelligence (AI) and machine learning could further optimize bioremediation processes. AI-powered models could analyze data from biosensors to predict the efficiency of remediation efforts and suggest adjustments in real time, ensuring that remediation projects are not only effective but also efficient and sustainable.

Methodological Innovations

1. **Bioaugmentation and Biostimulation for Enhanced Degradation**

The combination of bioaugmentation and biostimulation has emerged as a powerful approach to accelerate the remediation process. While bioaugmentation involves introducing specialized microbial strains to a contaminated site, biostimulation enhances the growth and activity of indigenous microorganisms by providing them with essential nutrients or electron donors (Shukla et al. 2010). Recent innovations in microbial ecology and genomics have allowed scientists to fine-tune this process, making it more targeted and efficient.

Future research could focus on the development of microbial consortia designed for specific types of pollution. These consortia would consist of microorganisms with complementary metabolic pathways, allowing for the complete breakdown of complex pollutants like petroleum hydrocarbons or heavy metals. Additionally, the use of synthetic biology could lead to the creation of microbial communities that are both highly efficient and resilient to environmental stressors such as fluctuating temperatures, moisture levels, or oxygen availability.

2. **Phytoremediation with Engineered Plants**

Phytoremediation, the use of plants to absorb, sequester, or degrade pollutants from soil and water, has shown promise for large-scale environmental restoration projects. Hyperaccumulator plants, such as *Thlaspi caerulescens*, are capable of extracting heavy metals like cadmium and arsenic from contaminated soil. However, the future of phytoremediation lies in the genetic engineering of plants to enhance their capacity for pollutant uptake and degradation (Joutey et al. 2013).

By introducing genes that improve a plant's ability to metabolize or sequester contaminants, researchers can develop plants specifically suited to different types of contamination. Furthermore, biotechnological advances in root architecture and symbiotic relationships between plants and soil microorganisms could further increase the efficacy of phytoremediation strategies, making them more scalable for larger, more complex contamination sites.

Scaling Bioremediation for Large-Scale Applications

The future of biological remediation lies in its ability to scale up for large-scale applications in diverse environments. Large-scale bioremediation projects face challenges such as inconsistent pollutant distribution, environmental variability, and high implementation costs. To overcome these obstacles, future methodologies must focus on cost-effective solutions, such as the use of locally available microbial strains, modular treatment systems, and adaptive management practices.

Innovations in bioreactor design, for instance, could make large-scale bioremediation more feasible. Bioreactors provide controlled environments for microbial growth and pollutant degradation and could be deployed on contaminated sites to enhance the remediation process. Moreover, the development of mobile bioreactor systems could enable on-site treatment of hazardous waste in remote areas, further broadening the scope of bioremediation.

Furthermore, combining various bioremediation techniques, such as microbial degradation, phytoremediation, and nanotechnology, in a single treatment plan could lead to multifunctional systems capable of addressing complex contamination scenarios. These hybrid approaches will likely be necessary to manage large industrial sites, agricultural lands, and urban environments where multiple pollutants coexist.

Conclusion

As bioremediation continues to evolve, emerging trends in technology and methodology will play a vital role in making it a more efficient and scalable solution for environmental restoration. Innovations such as genetic engineering, nanotechnology, biosensing, and synthetic biology are transforming the field, allowing for more precise, targeted, and large-scale applications. As environmental challenges become increasingly complex, the ability to deploy these innovative technologies on a large scale will be essential for the future of pollution mitigation and sustainable development. Moving forward, interdisciplinary collaboration between environmental engineers, biotechnologists, and data scientists will be critical to overcoming the remaining challenges and maximizing the potential of bioremediation for global environmental health.

References

Abatenh E et al (2017) The role of microorganisms in bioremediation—a review. Open J Environ Biol 2(1):038–046

Adams GO et al (2015) Bioremediation, biostimulation and bioaugmention: a review. Int J Environ Bioremed Biodegrad 3:28–39

Alavi M, Nejad M (2020) Discussed how gene editing technologies like CRISPR-Cas9 enable precise manipulation of microbial DNA, resulting in enhanced resistance to environmental stressors such as temperature, pH, and salinity

Angulu C et al (2018) Application of phytoremediation in the management of oil spillage: a review. Glob J Earth Environ Sci 3(3):16–22

Arthur EL (2005) Phytoremediation—an overview. Crit Rev Plant Sci 24(2):109–122

Baba U (2016) Bioremediation: a management tool. J Bioremed Biodegrad

Baniasadi M, Mousavi S (2018) A comprehensive review on the bioremediation of oil spills. In: Microbial action on hydrocarbons. Springer Nature

Biswas D et al (2023) Mycoremediation is a potential strategy for environmental clean-up of heavy metal: a review. J Surv Fish Sci 10(1S):6316–6327

Bolan NS et al (2011) Chapter four—Phytostabilization: a green approach to contaminant containment. In: Sparks DL (ed) Advances in agronomy, vol 112. Academic Press, pp 145–204

Boyle D et al (1998) Factors affecting the degradation of polyaromatic hydrocarbons in soil by white-rot fungi. Soil Biol Biochem 30(7):873–882

Burghal AA, Abu-Mejdad NMJA, Al-Tamimi WH (2016) Mycodegradation of crude oil by fungal species isolated from petroleum contaminated soil. Int J Innov Res Sci Eng Technol 5:1517–1524

Colwell JG (1990) Microbial degradation of hydrocarbons in the environment. Microbiol Rev 54(3):305–315

Daccò C et al (2020) Key fungal degradation patterns, enzymes and their applications for the removal of aliphatic hydrocarbons in polluted soils: a review. Int Biodeterior Biodegrad 147

Das, Chandran (2011) Noted that bioaugmentation has gained attention for its ability to enhance natural microbial populations' efficiency in degrading complex pollutants like petroleum hydrocarbons and PCBs
El-Ramady H et al (2022) Biotechnology of oyster mushroom (Pleurotus ostreatus L.): a sustainable strategy for myco-remediation and bio-fermentation. Sustainability 14(6)
Essien JP, Olajire AA (2014) Aerobic degradation of petroleum components by microbial consortia. Petrol Environ Biotechnol 5:1–22
Etim EE (2012) Phytoremediation and its mechanisms: a review. Int J Environ Bioenergy 2(3):120–136
Ghosal, Ghosh, Dutta, Ahn (2016) Explained that recombinant bacteria like genetically modified *Pseudomonas putida* are now designed to degrade toxic compounds such as toluene, a commonly found industrial pollutant
Goswami M et al (2018) Bioaugmentation and biostimulation: a potential strategies for environmental remediation. J Microbiol Exp
Grishchenkov VG et al (2000) Degradation of petroleum hydrocarbons by facultative anaerobic bacteria under aerobic and anaerobic conditions. Process Biochem 35(9):889–896
Hammel AH (2011) Fungal biodegradation of lignocelluloses. In: Hofrichter M (ed) Industrial applications. The Mycota, vol 10. Springer, Berlin, pp 319–340
Hassand MH et al (2024) Microbial-plant interactions and their role in mitigating of oil pollution: a review. Eur J Theor Appl Sci 2(2):11–22
Hoang SA, Lamb D et al (2021). Petroleum hydrocarbon rhizoremediation and soil microbial activity improvement via cluster root formation by wild proteaceae plant species. Chemosphere 275
Huesemann MH et al (2009) In situ phytoremediation of PAH- and PCB-contaminated marine sediments with eelgrass (Zostera marina). Ecol Eng 35(10):1395–1404
Hussein ZS et al (2022) Phytoremediation of crude petroleum oil pollution: a review. Egypt J Bot 62(3):611–640
Ibok AE (2019) Role of plants and microbes in bioremediation of petroleum hydrocarbons contaminated soils. Int J Environ Bioremed Biodegrad 7(1):1–19
Joutey, Sayel, Bahafid, El Ghachtouli (2013) Emphasized the role of hyperaccumulator plants such as *Thlaspi caerulescens* in phytoremediation, which absorb large quantities of heavy metals from contaminated soils
Kensa VM (2011) Bioremediation—an overview. J Ind Pollut Control 27(2):161–168
Lasat MM (2002) Phytoextraction of toxic metals: a review of biological mechanisms. J Environ Qual 109–120
Macaulay B (2014) Understanding the behaviour of oil-degrading micro-organism to enhance the microbial remediation of spilled petroleum. 247
Moosavi SG, Seghatoleslami MJ (2013) Phytoremediation: a review. Adv Agric Biol 1(1):5–11
Ndimele PE et al (2018) Chapter 24—Remediation of crude oil spillage. In: Ndimele PE (ed) The political ecology of oil and gas activities in the Nigerian aquatic ecosystem. Academic Press, pp 369–384
Nemati BB (2024) Phytoremediation of pollutants in oil-contaminated soils by Alhagi camelorum: evaluation and modeling. Springer Nature
Pete AJ et al (2021) Nano-enhanced bioremediation for oil spill—review. ACS Publication
Rafferty JP (2023) 9 of the biggest oil spills in history. https://www.britannica.com/story/9-of-the-biggest-oil-spills-in-history
Rafiee M et al (2017) Bioaccumulation and translocation factors of petroleum hydrocarbons in Aeluropus littoralis. Environ Health Eng Manag J 4:131–136
Rawat KM (2012) Rhizofiltration: a green technology for remediation of heavy metals. Int J Innov Bio-Sci 2(4):193–199
Salleh AB et al (2003) Bioremediation of petroleum hydrocarbon pollution. Indian J Biotechnol 2:411–425

Shah MP (2024a) Microbial remediation of hazardous chemicals from water & industrial wastewater treatment plant. Springer Nature

Shah MP (2024b) Development in waste water treatment research and processes: applied technologies for clean up of environmental contaminants. Elsevier

Shah MP (2024c) Aerobic and anaerobic microbial treatment of industrial wastewater. CRC Press

Shah MP (2025) Extraction of precious metals from industrial waste using microbial technology. Springer Nature

Shukla, Singh, Sharma (2010) Explored how biosensors, used for real-time monitoring of microbial degradation, have become vital in optimizing bioremediation strategies, especially in remote environments

Simister RL et al (2015) Degradation of oil by fungi isolated from Gulf of Mexico beaches. Mar Pollut Bull 100:327–333

Singh PSR (2014) Microorganism as a tool of bioremediation technology for cleaning environment: a review. Proc Int Acad Ecol Environ Sci 4(1):1–6

Singh SP, Ghosh M (2005) A review on phytoremediation of heavy metals and utilization of it's by products. Asian J Energy Environ 214–231

Singh A et al (2013) Assessment of bioremediation of oil and phenol contents in refinery waste water via bacterial consortium. J Petrol Environ Biotechnol 4

Sukumar S, Nirmala P (2016) Screening of diesel oil degrading bacteria from petroleum hydrocarbon contaminated soil. Int J Adv Res Biol Sci 3:18–22

Tasiu Mahmud IA (2022) Hydrocarbon degradation potentials of fungi: a review. J Environ Bioremed Toxicol 5(1):50–56

Tiwari M et al (2022) Mycoremediation of petroleum hydrocarbon contaminants: an overview. Adv Microbiol 1:106–119

Wolok E et al (2020) Study of bio-material for removal of the oil spill. Arab J Geosci

Yadav M et al (2011) Oxidation of polyaromatic hydrocarbons in systems containing water miscible organic solvents by the lignin peroxidase of Gleophyllum striatum MTCC-1117. Environ Technol 32:1287–1294

Zobell CE (1946) Action of microorganisms on hydrocarbons. Bacteriol Rev 10(1–2):1–49

Chapter 12
Paradigm Shift of Food Industry Waste into High Valued Resources Through Valorisation as an Integrated Sustainable Approach

Manisha Mohapatra and Sanjana Singh

Abstract Food by-products and waste from various food industries are abundant sources of bioactive compounds, nutraceuticals, and naturally occurring substances beneficial to human health. The generation of by-products and waste by various food businesses highlights the critical importance of effective waste management and by-product utilization within the food sector. Research indicates that bioactive compounds such as phenolics, carotenoids, and proteins can be recovered from food industry by-products using advanced technologies. Consequently, there is a paradigm shift toward the sustainable valorization of food sector waste, aligning with circular economy principles aimed at ensuring environmental protection and economic development while minimizing the environmental impact of food production. Future prospects in this field are promising, focusing on innovative techniques and controlled release mechanisms to extract value-added products from food waste. Advanced bioconversion methods, such as immobilized enzyme technology and nanotechnology, offer potential for creating high-value products like biofuels, biopolymers, and bioactive compounds. These methods are cost-effective and environmentally friendly, enhancing the sustainability of food waste management. Techniques involving immobilized enzymes, which are highly stable and efficient under various conditions, are particularly promising. Despite significant efforts to minimize and valorize food waste, comprehensive evaluations of the economic, environmental, and social impacts are still limited. Successful food waste management requires a trans-disciplinary approach for waste-to-wealth validation along with application of the 4R concepts (Reduce, Reuse, Recycle, and Recover) to combat global hunger and promote sustainability. Addressing food waste at its source and

The original version of the chapter has been revised. A correction to this chapter can be found at https://doi.org/10.1007/978-981-96-9128-9_17

M. Mohapatra (✉) · S. Singh
Department of Biotechnology, Trident Academy of Creative Technology (TACT), Infocity, Chandrasekharpur, Odisha, India
e-mail: manisha.mohapatra@trident.ac.in

© The Author(s), under exclusive license to Springer Nature Singapore Pte Ltd. 2026, corrected publication 2026

S. Dey and M. P. Shah (eds.), *Microbial Applications in Industrial Wastewater Treatment*, Advances in Wastewater Research,
https://doi.org/10.1007/978-981-96-9128-9_12

exploring advanced approaches like Waste-Energy and Waste-Value on a global scale can significantly mitigate the impact of food waste, making valorization a viable strategy for transforming a major problem into valuable resources.

Keywords Food waste valorisation · Sustainable approach · Value added products · Waste-to-wealth

12.1 Introduction

Food waste valorization is increasingly essential in the food industry as it addresses the dual challenge of waste management and resource conservation. The food industry generates substantial waste, particularly from the processing of fruits, foods, meat, and dairy products. These by-products, often rich in valuable bioactive compounds, have traditionally been discarded or used as low-value feed or fertilizers. However, growing awareness of sustainability and resource efficiency has shifted the focus toward valorization, where waste is transformed into high-value products. For instance, components such as fibers, proteins, and phytochemicals from fruit peels can be repurposed as nutraceuticals or functional food ingredients. By adopting waste valorization practices, the food industry not only reduces its environmental footprint but also creates new revenue streams through the development of innovative products (Sha et al. 2023). The need for food waste valorization in the industry is underscored by increasing global concerns over food security, environmental sustainability, and economic viability. With approximately one-third of food produced globally being wasted annually, the implications for resource depletion and environmental degradation are significant. Valorizing food waste helps mitigate these issues by reducing the volume of waste sent to landfills, cutting down on greenhouse gas emissions, and conserving natural resources. Moreover, it aligns circular economy principles, where waste is seen as a resource to be looped back into the production cycle. This not only enhances the sustainability of food production but also responds to consumer demand for eco-friendly practices and products. Hence, food waste valorization has become a critical strategy for the industry to ensure long-term sustainability and competitiveness (Baysal and Ülkü 2022; FAO 2024).

Waste valorization is increasingly seen as a vital strategy to tackle the rising challenges of waste management, environmental degradation, and resource scarcity. As global population growth and urbanization accelerate, the volume of waste produced continues to surge, exerting significant pressure on landfills and natural ecosystems. Traditional disposal methods, such as landfilling and incineration, not only contribute to pollution of air, soil, and water but also lead to deprivation of valuable resources that could be repurposed. Waste valorization, on the other hand, seeks to elevate from waste streams by transforming them into useful products, energy, or materials, thereby alleviating the strain on landfills and reducing environmental impacts. By embracing waste valorization, societies can move toward more sustainable waste management systems that enhance resource efficiency, promote circularity, and protect the environment (Leporini et al. 2020). The current chapter deals with the need of Valorization as sustainable approach for converting waste-to-wealth along with describing different available techniques for food waste valorization. The topic also covers different holistic recommendation for industrial food waste valorization dealing with current challenges and future prospectus.

12.2 Food Waste Issues: Global Scenario

Over the past decade, the food sector has seen tremendous growth, positioning it as one of the most rapidly expanding industries worldwide. However, this growth has also brought forth significant challenges, particularly in the areas of food safety and waste management. According to Baysal and Ülkü (2022), approximately 1/ 3 rd of global food produced is wasted each year, equating to around 1.6 billion tons. The Food and Agriculture Organization (2024) further emphasizes the environmental impact of this waste, noting that it contributes to a carbon footprint equivalent to 3.3 billion tons of CO_2 emissions. Using a general equilibrium framework, the total costs of food wastage are measured as the difference in aggregate net welfare between the current food system, which includes food wastage, and an idealized system with reduced wastage. The optimal level of food wastage is identified where this welfare difference is maximized, balancing the benefits and costs of reducing wastage. Achieving zero food wastage is not economically optimal, so the goal is to find a level of wastage that maximizes overall social welfare (Sha et al. 2023). However, the level of food wastage is achieved when the cost of further reducing waste exceeds the benefits gained from such reductions. Beyond economic considerations, non-economic factors also play a role in food wastage, such as the consumer satisfaction derived from having a broad range of choices (Liu et al. 2023). However, psychological research suggests that too much choice can impair decision-making, indicating that while variety enhances utility; it can also lead to inefficient outcomes and higher levels of waste (Fig. 12.1).

The initial approximation of the full costs of food wastage includes both the direct internal and external costs associated with food production that is ultimately lost or wasted at various stages of the value chain. These costs are directly linked with the amount of food that is discarded or wasted. An inefficient food system, which wastes more food, requires increased production to meet consumption levels, thus placing greater strain on natural resources. Beyond the direct costs, this inefficiency also leads to increased pressure on resource stocks, not just on the quantities used, making these relative costs more complex to evaluate. They depend on assessing the potential scarcity of these resources (Hoehn et al. 2023).

A significant portion of this waste is generated during food processing, where by-products from the conversion of raw materials into useful products become waste. These by-products are generally categorized into animal waste (from meat, seafood,

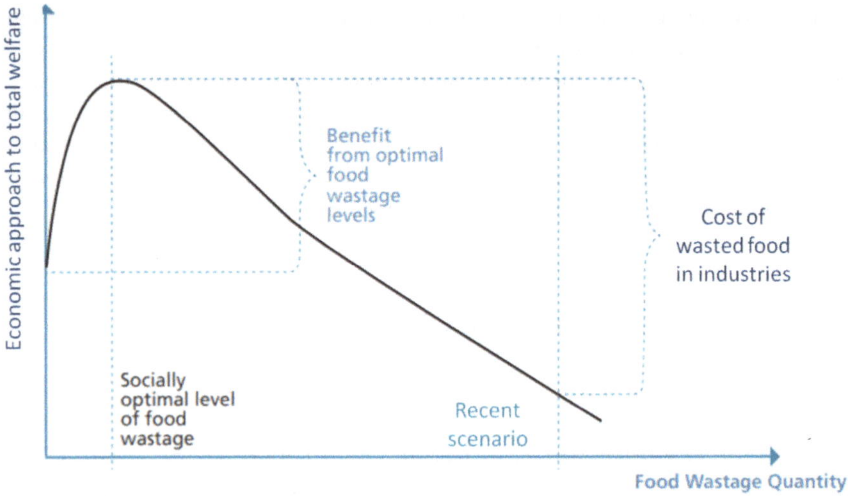

Fig. 12.1 Economic modalities related with food wastage quantities

livestock, and dairy processing) and vegetable waste, depending on the source (Osorio et al. 2021). Food waste occurs at various steps of production, including industrial manufacturing to distribution, and other activities. Notably, household domestic work accounts for around 42% of food waste, while 39% originates from food processing industries, 14% from the food service sector, and 5% during distribution. In India, with a population exceeding 1.4 billion, more than 0.5 kg of organic waste is produced per capita daily (Paulraj et al. 2019). Among food processing industries, the fruit and vegetable sector is particularly wasteful, generating approximately 25–30% waste such as peels, seeds, skins, shells, cores, pulp, and pomace (Rifna et al. 2021). Preventative measures, such as separating valuable components like fibers, aromatic compounds, phytochemicals, proteins, and polysaccharides, can allow these by-products to be utilized as nutraceuticals (Shah 2025; Russo et al. 2021). Despite various initiatives to therapeutically harness food waste, particularly from fruits and vegetables, most agro-industrial waste is still used in large quantities as animal feed or fertilizers (Rudra et al. 2015).

12.3 Food Waste Valorization as a Reservoir for Production of Value Added Compounds

Waste valorization refers to the process of extracting value from waste materials instead of merely discarding them into landfills, incinerating them, or releasing them into the environment. This approach is integral to promoting sustainability by reducing waste generation, conserving resources, and encouraging more efficient and environmentally friendly waste management practices (Campos et al. 2020). A key method of waste valorization is recycling, which involves collecting and processing waste materials such as plastics, glass, paper, and metals to create new products or raw materials. Recycling not only reduces the demand for virgin resources but also lowers energy consumption and greenhouse gas emissions associated with the extraction and manufacturing of these materials. Another approach is upcycling, which transforms waste materials into value-added products with enhanced quality or functionality. Examples include repurposing discarded textiles into new clothing or converting food waste into biofuels. Despite its potential, waste valorization faces challenges such as the perishability of certain materials, logistical complexities, and disposal methods. Nonetheless, fruit and vegetable residues, rich in bioactive components, are seen as having significant untapped potential, especially in response to the growing global demand for high-value products (Sha et al. 2023). Additionally, waste valorization research often focuses on integrating these processes into circular economy frameworks to minimize waste generation, conserve resources, and advance sustainable development. Overall, this field plays a critical role in improving sustainable waste management practices and transitioning toward a more resource-efficient and ecologically aware society (Leporini et al. 2020; Shah 2024a; Muscat et al. 2021; Sharma et al. 2022; Sha et al. 2023).

Industrial food residues, including whole plants and/or different plant parts, fruits, seeds, generate billions of tons of waste throughout various stages of the food processing, such as production, handling, storage, processing, packaging, distribution, marketing, and consumption. This waste has significant negative effects on both the environment and human health, highlighting the need for innovative methods to convert these wastes into high-value-added products (Yadav et al. 2024). Industrial food residues, including peels, pulp, seeds, shells, pomace, husks, straws, barks, bran, and leaves (Prasad et al. 2020), as well as the inedible parts of fruits (peels, pulp, seeds, pomace) and vegetables are rich in organic materials that are important sources of phytonutrients and bioactive compounds (Table 12.1). Bioactive compounds are natural biological substances or phytoconstituents that can trigger physiological responses in living organisms and offer numerous health benefits. Industrial food waste contains a variety of these compounds, such as phenolics, peptides, polyphenols, tannins, alkaloids, carotenoids, terpenes, flavonoids, flavanols, anthocyanins, and essential oils, along with carbohydrates, enzymes, fatty acids, dietary fibers, lipids, biopolymers, nitrogenous compounds, minerals, vitamins, amino acids, chemicals, and other phytoconstituents (Zuin et al. 2018; Campos et al. 2020; Shah 2024b; Haq et al. 2021).

Food by-products such as pressed pulp, spent peel, leaves, seeds, and secondary juice are valuable sources of bioactive compounds, pectins, cellulose, and more (Mahato et al. 2021). In response, various strategies have been proposed to enhance the management of citrus fruit waste. Recently, there has been a transition from a linear economic model to a more sustainable circular model in managing fruit and vegetable by-products (Campos et al. 2020). In the traditional linear economy, resources are extracted, transformed into products, consumed, and eventually disposed of as waste, following a "take-make-dispose" approach (Elisha 2020). In contrast, the circular economy emphasizes recovering valuable materials from waste

Table 12.1 Value-added bioactive compounds obtained by food industrial waste valorization

Sl. No	Name of Bioactive compound	Structure	References
1	Pectin	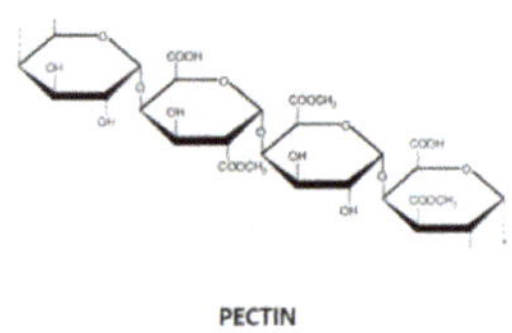	(Leporini et al. 2020)
2	Cellulose	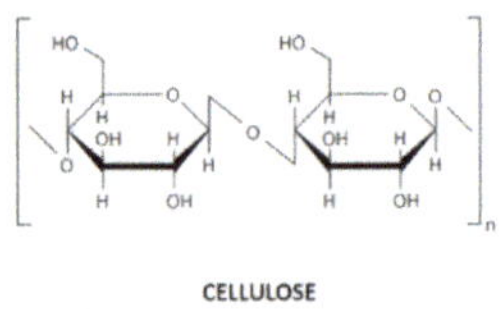	(Maugeri et al. 2019)
3	Monoterpenes	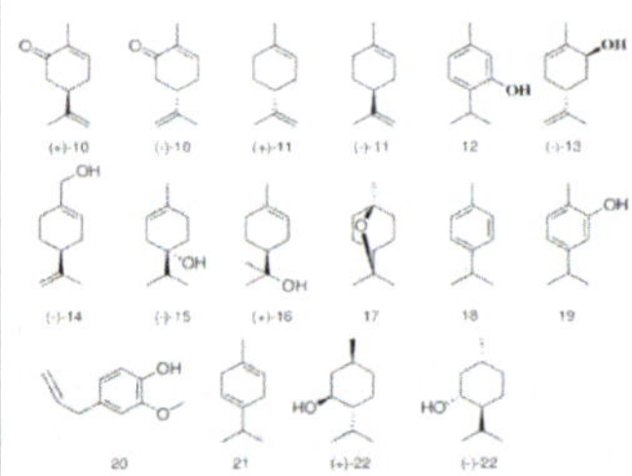	(Mandalari et al. 2017)
4	Xanthophylls	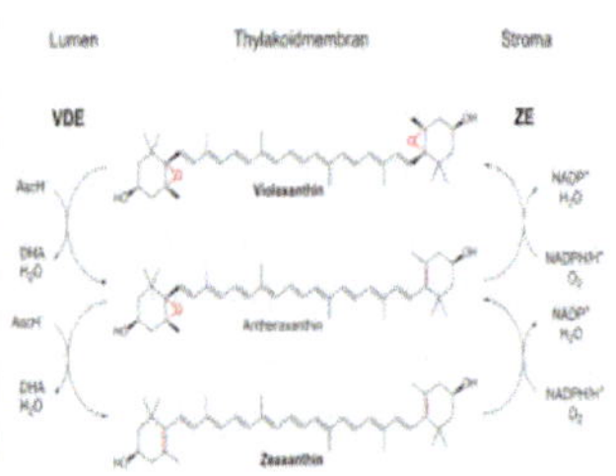	(Ferlazzo et al. 2016)
5	Carotenes	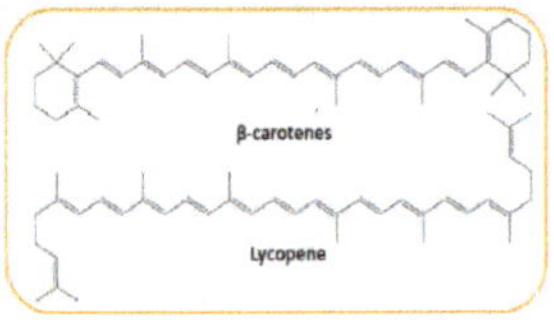	(Leporini et al. 2020)

(continued)

Table 12.1 (continued)

Sl. No	Name of Bioactive compound	Structure	References
6	Naringin	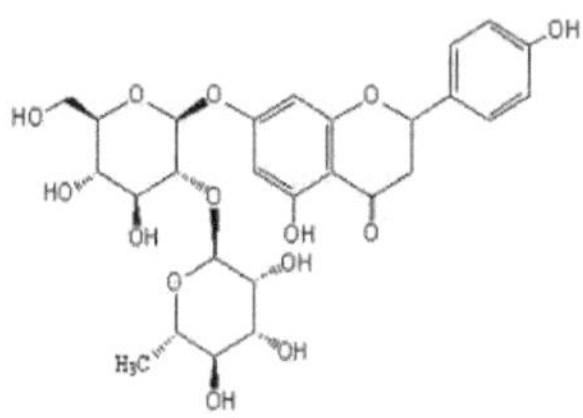	(Maugeri et al. 2019; Leporini et al. 2020)
7	Rutin	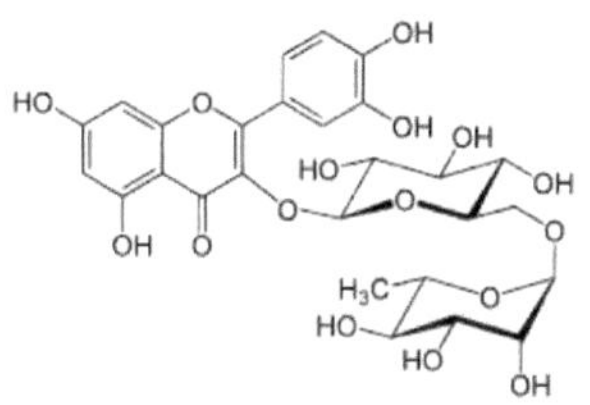	(Leporini et al. 2020)
8	Flavanols	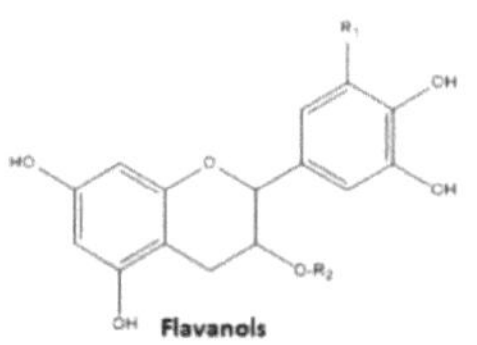	(Mandalari et al. 2017; Maugeri et al. 2019)
9	Isoflavones	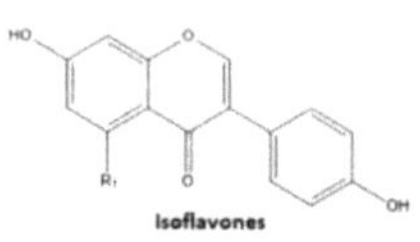	(Leporini et al. 2020)
10	Anthocyanins	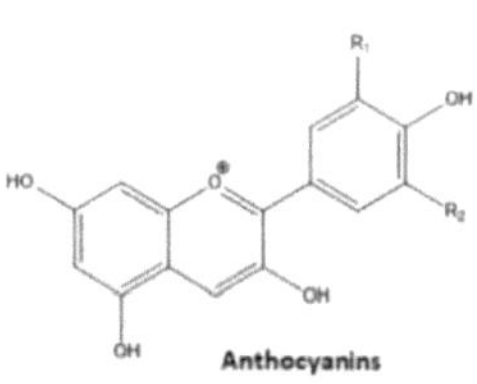	(Leporini et al. 2020)
11	Chlorogenic acid	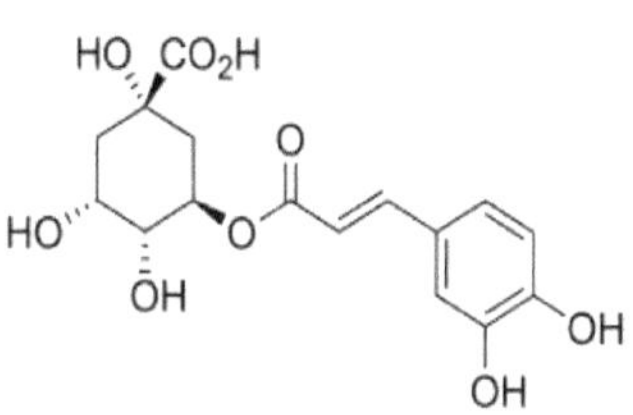	(Maugeri et al. 2019; Leporini et al. 2020)

(continued)

Table 12.1 (continued)

Sl. No	Name of Bioactive compound	Structure	References
12	Gallic acid	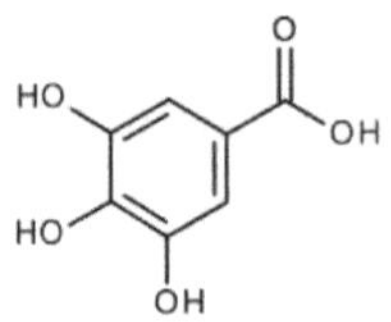	(Leporini et al. 2020)
13	Caffeic acid	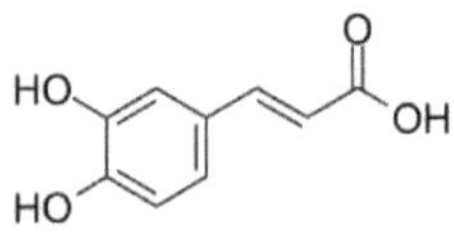	(Ferlazzo et al. 2016)
14	Vitamin C	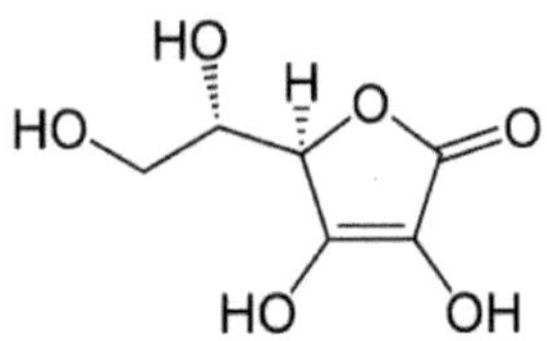	(Sharma et al. 2022)
15	Lipids	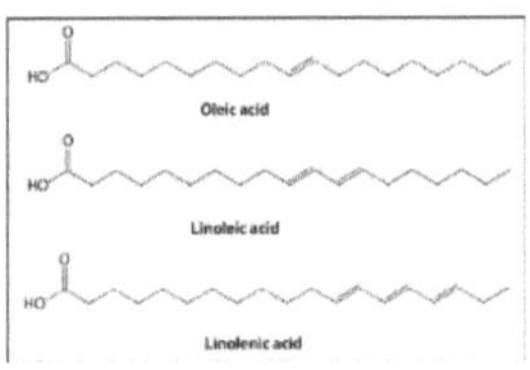	(Chavan et al. 2018)
16	Catechin	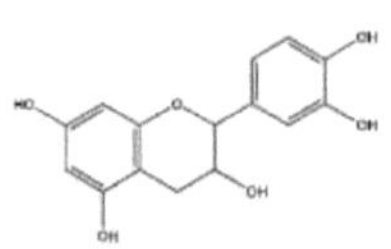	(Muscat et al. 2021)
17	Ferulic acid	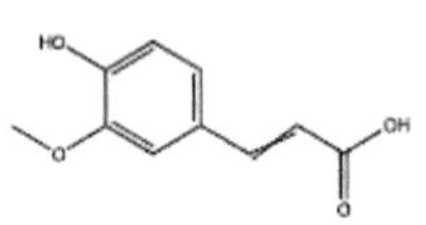	(Maugeri et al. 2019)

(continued)

Table 12.1 (continued)

Sl. No	Name of Bioactive compound	Structure	References
18	Carvacrol	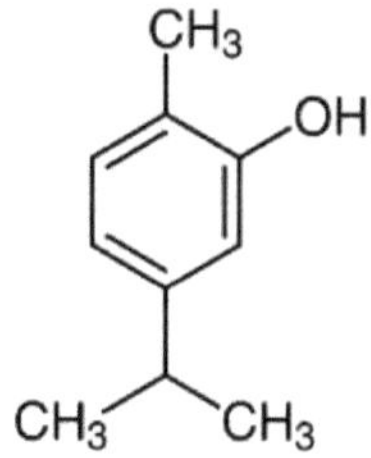	(Martau et al. 2021)
19	Delphinidin	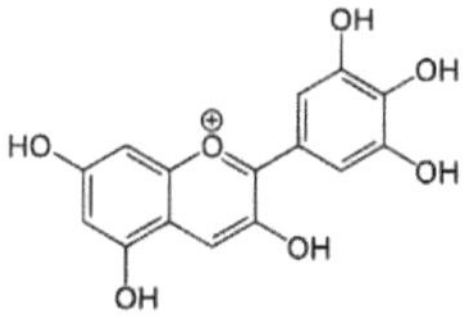	(Iorizzo et al. 2020)
20	Peonidin	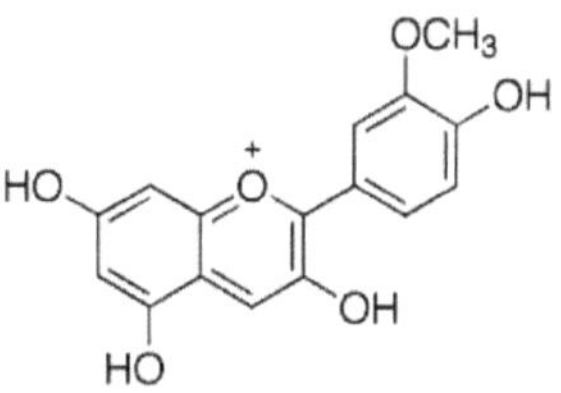	(Iorizzo et al. 2020)
21	Hesperidin	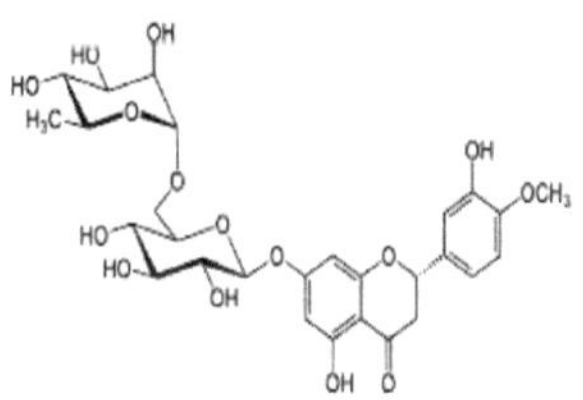	(Victor et al. 2021)
22	Resveratrol	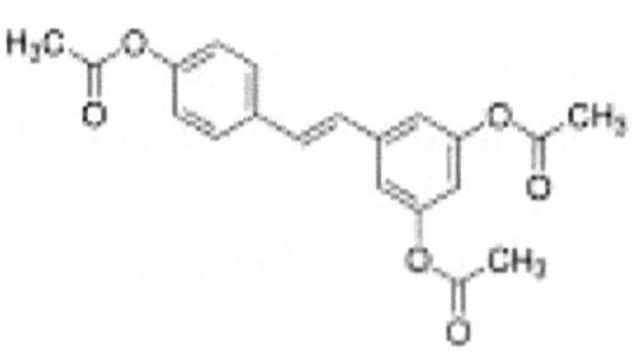	(Pastor et al. 2019)

to start a new production cycle, thereby reducing waste generation and following a "take-make-use" model. Moreover, within this circular economy framework, revalorizing waste can lead to the recovery of compounds with significant biological value (Panwar et al. 2021; Muscat et al. 2021). The fruit industry generates substantial amounts of waste, with apple pomace being a notable example, contributing up to 12 million tons globally each year (Martau et al. 2021). Apple peel is particularly rich in flavonoids, containing about three to six times more than the flesh, including unique compounds like quercetin glycosides that are absent in the flesh (Khalid et al. 2021). Extracts from apple peel exhibit significant antioxidant activity, with key polyphenols such as phloridzin, chlorogenic acid, and quercetin identified in apple pomace, showing antioxidant effects comparable to ascorbic acid (Hobbi et al. 2023). Grape skins and seeds are also rich in resveratrol, a polyphenol known for its antioxidant properties and health benefits. Although resveratrol is found in over 70 plants, grapes remain the primary dietary source, with trans-resveratrol content in grape skins ranging from 1.11 to 12.3 mg/100 g dry weight, and in white grape seeds from 0.18 to 1.42 mg/100 g dry weight (Pastor et al. 2019). During the wine-making process, some resveratrol is transferred to the wine, but a significant amount remains in the grape marc. Cranberry pomace, a by-product of cranberry processing, includes seeds, skins, and stems. Despite being rich in beneficial polyphenols like proanthocyanidins and anthocyanins, it is often discarded due to its unpalatability (Iorizzo et al. 2020).

Fruit waste can also contain significant amounts of dietary fiber, varying depending on the type of food and the part discarded. For example, banana peels are rich in insoluble dietary fiber, which possesses desirable properties such as water-holding, oil-holding, and glucose adsorption abilities (Victor et al. 2021). Additionally, banana peels are a good source of xylooligosaccharides, with residues dissolved in hydrogen peroxide yielding up to 54% after enzymatic digestion. Other fruit by-products, such as guava, pomegranate, and orange pomace, can also be utilized for extracting xylooligosaccharides (Pereira et al. 2022). *Jerusalem artichoke* residues, often discarded as waste, contribute to environmental pollution and resource waste but hold potential for extracting high-purity, high-performance soluble dietary fiber, which has promising applications in the food packaging industry. Similarly, garlic peel, despite having a high dietary fiber content of around 70%, is frequently disposed of as industrial waste. However, its soluble dietary fiber content is only 5.3%, falling short of the 10% minimum required for high-quality dietary fiber. Modifying the composition of dietary fiber derived from garlic peel could enhance its solubility, making it a more viable source of soluble dietary fiber (Zhou et al., 2022).

12.4 Techniques for Food Waste Valorization

Bio-based molecules, natural biopolymers, and phytochemicals can be directly extracted from natural sources rather than being synthesized from petroleum-based chemicals. Once extracted, these bio-based products can serve as high-value items such as food additives, nutraceuticals, therapeutics, and cosmetics. However, the economic feasibility of extracting these high-value components must be carefully considered. To achieve this, appropriate extraction methods should be utilized to fully exploit the waste and recover all valuable components. The choice of extraction techniques varies based on the nature of the food matrix and the specific bioactive ingredient to be extracted. The extraction method and the cellular structure of the by-product significantly influence the recovery rate of the target compound.

Various traditional extraction methods, such as vigorous grinding, maceration, Soxhlet extraction, and supercritical fluid extraction, are generally discouraged for sustainably converting food wastes due to their high energy requirements and costs (Herrero and Ibanez 2018; Shah 2024c; Mozhiarasi 2021). Modern extraction techniques have emerged as greener alternatives, including microwave-assisted, ultrasound-assisted, pressurized solvents, enzymatic extraction, negative pressure extraction, and high-pressure homogenization (Lu et al. 2021). These methods are preferred for their sustainability and lower environmental impact. Moreover, anaerobic digestion is recognized as a robust approach for converting food wastes into valuable products. As such, embracing advanced, sustainable extraction strategies holds great potential for enhancing food waste valorization and reducing toxicity concerns (Zia et al. 2020a, b; Majumder and Annegowda 2021) (Fig. 12.2).

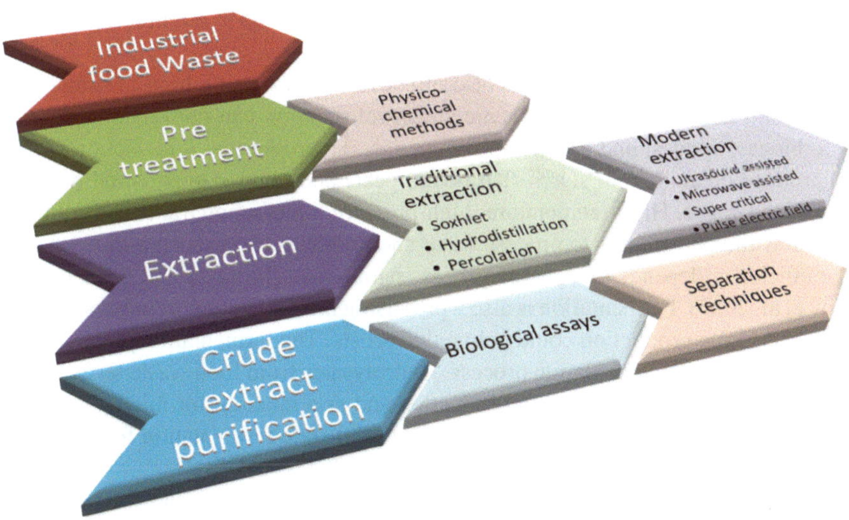

Fig. 12.2 Available approaches for industrial food waste valorization

12.4.1 Conventional Approach

Extraction methods are generally categorized into conventional and non-conventional techniques. Conventional methods, such as solvent extraction, Soxhlet extraction, maceration, and hydrodistillation, typically involve factors like temperature, agitation, and the use of organic solvents. To preserve the integrity of bioactive components, operating factors such as temperature, solid-to-liquid ratio, contact time, particle size, pH, and stirring rate must be carefully controlled. However, conventional methods have several drawbacks, including longer extraction times, the use of large volumes of solvents that generate toxic waste, and the need for additional purification steps to remove solvent residues (Jablonský et al. 2018; Zia et al. 2020a, b).

One of the oldest methods, Pyrolysis is a process that converts food waste into valuable compounds, with or without catalysts. Low-temperature pyrolysis helps in better carbon fixation, while catalytic pyrolysis enhances product yield and quality. The process parameters significantly influence the outcome. Combining industrial food waste with herbal medicine by-products or wood bark during pyrolysis can increase hydrogen yield and reduce harmful chemicals (Kim et al. 2020). Incinerating food waste is energy-intensive due to its high moisture content. Pretreatment methods help reduce moisture before incineration, making it more efficient for power generation. This process can reduce waste volume by 80–85% and minimize landfill use (Zamli et al. 2021).

Hydrothermal carbonization is also effective for converting high-moisture food waste into useful products. Biogenic steam reformation can convert mixed waste into biogas, syngas, and other useful materials. Hydrothermal liquefaction is used to produce ethanol from starch-rich potato peels and recover energy and nutrients from food waste. High-temperature, high-pressure liquefaction transfers significant nitrogen and phosphorus to process water and produces energy-dense biocrude, but the efficiency depends on specific process parameters (Motavaf and Savage 2021; Roy et al. 2023). Fermentation is also a promising method for valorizing food waste. Onsite enzyme production could help reduce costs. The fermentation process's efficiency relies on the inoculation process. Hydrolysates from food waste fermentation have been used for biohydrogen production and as a medium for microalgae cultivation. However, the high cost of scaling up solid-state fermentation for producing flavor compounds remains a barrier (Carmona-Cabello et al. 2021).

The study of waste valorization covers a wide array of research and practical applications aimed at transforming different waste into valuable products or resources. Researchers have explored numerous methods and technologies for valorizing different waste streams, including organic waste, agricultural residues, industrial by-products, and municipal solid waste. These studies typically involve analyzing the composition and properties of the waste, identifying potential value-added products or processes, and assessing the economic, environmental, and social impacts of these valorization approaches. Common techniques include biological methods like composting, anaerobic digestion, and bioconversion, along with physical and chemical processes such as pyrolysis, hydrothermal treatment, and the extraction of valuable compounds.

12.4.2 Modern Approach

In contrast, modern or non-traditional extraction methods are designed to be more efficient, cleaner, and environmentally friendly, with reduced solvent consumption and extraction times, as well as lower energy usage. Advanced extraction methods like microwave-assisted, pulsed electric field-assisted, pressurized liquid, ultrasound-assisted, supercritical fluid, and enzyme-assisted extractions have been extensively studied for their superior extraction efficiency. Given the limitations of conventional solvents, such as high cost, toxicity, and unfavorable melting points, there has been a growing interest in developing greener and more cost-effective alternatives (Jablonský et al. 2018). These solvents have a high capacity to dissolve biomass, making them effective for valorizing food wastes (Zia et al. 2020a, b).

Future developments in extraction technologies will play a crucial role in maximizing the recovery of bioactive compounds from food waste. Innovations in enzymatic extraction, supercritical fluid extraction, and other green technologies will enable more efficient and cost-effective recovery of valuable nutrients such as phenolics, carotenoids, proteins, and fibers. These technologies will not only enhance the yield but also preserve the integrity and bioavailability of the extracted compounds (Cecilia et al. 2019). Encapsulation techniques will continue to evolve, allowing for better protection, stability, and controlled release of bioactive compounds derived from food waste. Nanotechnology and micro-encapsulation enable targeted delivery and improved bioavailability of these compounds. This will open new avenues in the nutraceutical and functional food industries, where encapsulated bioactives can be incorporated into a wide range of products, offering health benefits to consumers (Martins et al. 2023).

Industrial food wastes are a significant source of natural bioactive compounds, which can be recovered through valorization technologies and utilized in various sectors like food, therapeutics, and the environment due to their health benefits and low toxicity (Ali et al. 2021). However, these bioactive compounds face challenges such as instability, low solubility, and sensitivity to environmental factors, which can diminish their effectiveness. Encapsulation technology, including micro-encapsulation and nano-encapsulation, has emerged as a solution to protect these compounds, improve their bioavailability, and ensure controlled release (Sobel et al. 2014). Nanoencapsulation, in particular, offers enhanced solubility, stability, and sustained release, making it a promising method for optimizing the delivery of bioactive compounds in diverse applications (Verónica et al. 2021; Yadav et al. 2024). The use of micro- and nanocapsules enhances the absorption and controlled release of bioactive compounds from food waste, offering benefits such as extended shelf life due to their antimicrobial and antioxidant properties. Most encapsulation techniques start with emulsion formation and vary in their top-down or bottom-up approaches, with ionic gelation being commonly used. This approach supports sustainability and offers economic potential within a circular economy (Martins et al. 2023; Borah et al. 2023).

Food waste valorization aims to convert discarded materials into valuable products like biopolymers, biofuels, and bioactive compounds. Advanced chemical and microbial extraction methods with immobilized enzymes show promising results due to their operational stability, reusability, and resistance to temperature and pH changes. These enzymes enhance the efficiency of bioconversion processes, and various immobilization techniques, such as physical adsorption, entrapment, and covalent bonding, have been explored using support media like alginate, glass beads, and clay minerals. For example, β-galactosidase immobilized on a nanocatalytic surface effectively converts carbohydrates from food waste into bioactive galactosides, while pectin from vegetable waste is synthesized using immobilized pectin-degrading enzymes from Bacillus licheniformis on agar–agar, offering a sustainable and cost-effective alternative to traditional chemical methods (Banjanac et al. 2016; Bilal et al. 2019).

12.4.3 Integrated Biorefinary Approach

Integrated valorization methods combine various waste treatment and conversion processes to maximize the value extracted from waste streams, particularly food waste, enhancing resource efficiency and sustainability (Dahiya et al. 2018; Paini et al. 2021). These methods include biological, chemical, and thermal processes, alongside the recovery of valuable by-products. Recent strategies have focused on using horticultural waste, especially food waste, to produce high-value bioactive compounds, bioenergy, and biofuels. The "waste-to-energy" concept has been widely adopted for converting food waste into bioenergy and biofuels, with approaches like landfilling, composting, anaerobic digestion, and enzymatic conversion proving effective. Among these, anaerobic digestion is particularly efficient for large-scale industrial bioconversion of food waste. Various value-added products, including biohydrogen, biodiesel, bioethanol, and volatile fatty acids, have been derived from food waste (Mozhiarasi 2021; Lu et al. 2021).

Biorefinery processes have shown significant potential in converting waste into value-added products, such as nanoparticles from pineapple waste and bioconversion of carrot peel waste into pectin and carotene (Herrero and Ibanez 2018; Majumder and Annegowda 2021). Additionally, greenhouse food waste can be effectively transformed into valuable products using bioconversion and biorefinery approaches, supporting the bioeconomy concept. In Singapore, recent studies have highlighted the conversion of HW into biochar through gasification strategies (Zia et al. 2020a, b). Biorefinery techniques can also convert waste biomass into bioenergy or biofuels via processes like acidogenesis, fermentation, microbial bioconversion, and electrolysis. Microbial fuel cells further offer the ability to generate bioelectricity by oxidizing organic materials. Moreover, HW rich in lignin and fibers can be fermented or bioconverted into bioplastics through microbial or enzymatic decomposition, producing sustainable alternatives to traditional plastics (Villacís-Chiriboga et al. 2020).

12.5 Valorization as Sustainable Solution

Food waste from households and industries poses a significant environmental challenge due to its underutilization and insufficient sustainable recycling practices. Despite being rich in hidden bioactive compounds and valuable products, the conversion of food waste into wealth through sustainable methods remains underexplored. This is largely due to a lack of research and supportive policies. This study highlighted current approaches to food waste valorization, focusing on innovative and sustainable technologies that align with Sustainable Development Goals aiming to maximize the use of unutilized waste streams (Sarker et al. 2024). Food waste valorization techniques focus on transforming food waste into value-added products that can be reintegrated into the food supply chain. Sustainable approaches for food waste valorization include methods such as composting, anaerobic digestion, bioconversion, and upcycling. Food waste, available in solid, semi-solid, and liquid forms, is rich in proteins, carbohydrates, lignin, lipids, cellulose, and other valuable compounds. For instance, liquid food waste like potato-processing wastewater, apple residue sludge, and dairy industry whey contains these substances in diluted forms (Hartikainen et al. 2018; Chen et al. 2020). Due to this composition, food waste is considered a cost-effective and high-potential second-generation raw material (Springmann et al. 2018). The key to any food waste valorization process lies in understanding its composition (Willett et al. 2019). Food waste composition varies based on region, season, and dietary habits, typically containing over 50% total sugars, more than 40% starch, 45% proteins, and less than 2% cellulose. These components can be utilized in the production of biofuels, bulk chemicals, enzymes, food supplements, nutraceuticals, bio-adsorbents (e.g., activated charcoal), antimicrobial products (e.g., antibiotics), bio-based fertilizers, and bioplastics, among other applications (Chen et al. 2020). The resource recovery from food waste through these sustainable techniques aims to convert waste into products that can be reintroduced into various stages of the food supply chain. Valorization strategies recognized as sustainable in waste management include composting, anaerobic digestion, bioconversion, and upcycling to create value-added derivatives such as fertilizers, animal feed, and food additives (Hartikainen et al. 2018; Ng et al. 2020).

12.6 Available Policies and Holistic Recommendations for Industrial Food Waste Valorization

Valorization aims to utilize unavoidable by-products from the food supply chain that are not directly intended for human consumption. This can be achieved through methods like feeding animals, processing ingredients, or applying other high-value strategies, following the food-use hierarchy. The approach emphasizes creating integrated policies to minimize food waste, promoting sustainable food systems, and improving supply chain collaboration through data transparency and impact measurements. Priority should be given to high-impact food products, such as bread, fruits, vegetables, and animal-based items, focusing on areas with the most significant potential for waste reduction and considering environmental, economic, and social benefits. Additionally, efforts should address the absorption of food system surpluses, encourage year-round availability of produce, and promote comprehensive use of animal products while exploring the effects of dietary changes on waste prevention (Bos-Brouwers et al. 2020). Various global and regional studies have highlighted the nutritional, environmental, and economic challenges posed by food waste, also known as food loss and waste, which is exacerbated by inadequate recycling practices in households and kitchens across many countries (Hartikainen et al. 2018). The COVID pandemic has further strained food waste management policies, making food loss and waste a global concern due to poor handling during lockdowns. Consequently, global entities like UNEP, FAO, and IFPRI have revised regulations to address food waste, aiming to enhance food security, reduce hunger, and promote sustainable waste management through technologies and the 4R (reduce, reuse, recycle, recover) approach (IFPRI 2016; Teigiserova et al. 2020; UNEP 2023).

12.7 Challenges and Future Prospects

One of the significant challenges in the valorization of food industry waste is the lack of scalable and efficient technologies. While innovative extraction techniques like ultrasound-assisted extraction, enzymatic hydrolysis, and microwave-assisted extraction have shown promise in lab settings, their large-scale application is often hindered by high costs, energy consumption, and operational complexities. Additionally, conventional methods still dominate many industries due to their simplicity and established infrastructure, making the transition to advanced technologies slow and cumbersome (Liu et al. 2023). The economic viability of valorization processes is a crucial factor for widespread adoption. The cost of implementing novel technologies, acquiring specialized equipment, and optimizing extraction processes can be prohibitive, especially for small and medium-sized enterprises (SMEs). Additionally, the market value of the extracted compounds may not always justify the investment, particularly in regions where there is limited demand for high-value-added products. This economic barrier is a significant challenge that must be addressed through innovative business models, financial incentives, and subsidies (Khan et al. 2022). The lack of clear regulations and standardized practices for the valorization of food waste poses another significant challenge. Without uniform guidelines and quality standards, industries may struggle to ensure the safety, efficacy, and marketability of their valorized products. This is particularly critical in sectors like food, pharmaceuticals, and cosmetics, where stringent safety regulations govern product development. The absence of regulatory frameworks can also hinder the global trade of valorized products, limiting their market potential (Negi et al. 2021). The integration of food waste valorization into existing supply chains is a complex task. Coordinating the collection, transportation, and processing of food waste requires robust logistical frameworks that are not always in place. The perishability of certain food wastes further complicates the process, necessitating efficient handling and processing to avoid degradation. Additionally, ensuring a consistent and reliable supply of waste materials can be challenging, particularly for industries that rely on seasonal or geographically concentrated production (Mo et al. 2018).

As the global food industry continues to evolve, the need for sustainable practices has never been more pressing. The paradigm shift toward a circular economy is gaining momentum, with the valorization of food waste and by-products at the forefront of this movement. The future of industrial food valorization holds immense potential, driven by technological advancements, increased awareness of environmental concerns, and the demand for healthier, eco-friendly products. Here, we explore the key prospects that are likely to shape the future of this field. The valorization of food waste and by-products will become an integral part of sustainable food systems. Industries will increasingly adopt zero-waste policies, where by-products are repurposed into valuable resources, reducing environmental impact. With growing consumer awareness of sustainability and health, the demand for products made from valorized food waste is expected to rise. Companies that prioritize transparency, sustainability, and innovation will have a competitive edge in the market. Education and communication strategies will be crucial in fostering consumer acceptance of these products, as they transition from niche markets to mainstream consumption. The future of industrial food valorization will rely heavily on collaboration between academia, industry, and government. Knowledge sharing and open innovation platforms will be vital in accelerating progress and overcoming challenges in the field (Yadav et al. 2024).

12.8 Conclusion

The paradigm shift from waste to value through food waste valorization represents a promising and necessary evolution for the food industry. Despite the challenges, the potential benefits are substantial, offering economic, environmental, and societal advantages. By addressing technological, economic, regulatory, and perceptual barriers, and leveraging future prospects such as technological advancements, circular economy integration, and collaborative innovation, the food industry can fully realize the potential of waste valorization. This integrated sustainable approach not only contributes to waste reduction and resource efficiency but also aligns with global efforts to achieve sustainability and climate goals. The future of industrial food valorization is promising, with significant advancements on the horizon. By embracing sustainable practices and leveraging technological innovations, the food industry can transform waste into wealth, contributing to a healthier planet and population. As the world moves toward a more circular economy, the valorization of food waste will become a cornerstone of sustainable food production, driving both economic and environmental benefits for years to come.

References

Ali EA, Nada AA, Al-Moghazy M (2021) Self-stick membrane based on grafted gum Arabic as active food packaging for cheese using cinnamon extract. Int J Biol Macromol 189:114–123. https://doi.org/10.1016/j.ijbiomac.2021.08.071

Banjanac K, Carević M, Ćorović M, Milivojević A, Prlainović N, Marinković A, Bezbradica D (2016) Novel β-galactosidase nanobiocatalyst systems for application in the synthesis of bioactive galactosides. RSC Adv 6:97216–97225. https://doi.org/10.1039/C6RA20409K

Baysal SS, Ülkü MA (2022) Food loss and waste: a sustainable supply chain perspective. In: disruptive technologies and eco-innovation for sustainable development; Ulas A ed. IGI global: Hershey, PA, USA, pp 90–108

Bilal M, Asgher M, Cheng H, Yan Y, Iqbal HMN (2019) Multi-point enzyme immobilization, surface chemistry, and novel platforms: a paradigm shift in biocatalyst design. Crit Rev Biotechnol 39(2):202–219. https://doi.org/10.1080/07388551.2018.1531822

Borah MS, Tiwari A, Sridhar K, Narsaiah K, Nayak PK, Stephen Inbaraj B (2023) Recent trends in valorisation of food industry waste and by-products: encapsulation and in vitro release of bioactive compounds. Foods (Basel, Switzerland) 12(20):3823. https://doi.org/10.3390/foods12203823

Bos-Brouwers H, Burgos S, Colin F, Graf V (2020) Policy recommendations to improve food waste prevention and valorisation in the EU. https://eu-refresh.org/sites/default/files/D3.5%20Policy%20recommendations_v.2.pdf

Campos DA, Gomez-Garcia R, Vilas-Boas AA, Madureira AR, Pintado MM (2020) Management of fruit industrial by-products-a case study on circular economy approach. Molecules 25:320

Carmona-Cabello M, García IL, Papadaki A, Tsouko E, Koutinas A, Dorado MP (2021) Biodiesel production using microbial lipids derived from food waste discarded by catering services. Bioresour Technol 323:124597. https://doi.org/10.1016/j.biortech.2020.124597

Cecilia JA, García-Sancho C, Maireles-Torres PJ, Luque R (2019) Industrial food waste valorisation: a general overview. In: Bastidas-Oyanedel JR, Schmidt J (eds) Biorefinery. Springer, Cham. https://doi.org/10.1007/978-3-030-10961-5_11

Chavan P, Singh AK, Kaur G (2018) Recent progress in the utilization of industrial waste and by-products of citrus fruits: a review. J Food Process Eng 41:e12895

Chen C, Chaudhary A, Mathys A (2020) Nutritional and environmental losses embedded in global food waste, Resour, Conserv Recycl 160:104912. https://doi.org/10.1016/j.resconrec.2020.104912.

Dahiya S, Kumar AN, Shanthi JS, Chatterjee S, Sarkar O, Mohan SV (2018) Food waste biorefinery: sustainable strategy for circular bioeconomy. Bioresour Technol 248:2–12. https://doi.org/10.1016/j.biortech.2017.07.176

Elisha OD (2020) Moving beyond take-make-dispose to take-make-use for sustainable economy. Int J Sci Res Educ 13:497–516

FAO (2024) Food wastage foot print impacts on natural resources. Accessed 17 Feb 2024. https://www.fao.org/news/story/en/item/196402/icode/

Ferlazzo N, Cirmi S, Calapai G, Ventura-Spagnolo E, Gangemi S, Navarra M (2016) Anti-inflammatory activity of Citrus bergamia derivatives: Where do we stand? Molecules 21:1273

Haq IU, Qaisar K, Ali N, Fatima A, Hamid M, Yong X, Muhammad WM, Umer R, Wan A, Wan AKG (2021) Advances in valorisation of lignocellulosic biomass towards energy generation. Catalysts 11:309

Hartikainen H, Mogensen L, Svanes E, Franke U (2018) Food waste quantification in primary production–the Nordic countries as a case study. Waste Manage 71:502–511. https://doi.org/10.1016/j.wasman.2017.10.026

Herrero M, Ibanez E (2018) Green extraction processes, biorefineries and sustainability: recovery of high added-value products from natural sources. J Supercrit Fluids 134:252–259. https://doi.org/10.1016/j.supflu.2017.12.002

Hobbi P, Okoro OV, Hajiabbas M, Hamidi M, Nie L, Megalizzi V, Musonge P, Dodi G, Shavandi A (2023) Chemical Composition, antioxidant activity and cytocompatibility of polyphenolic compounds extracted from food industry apple waste: potential in biomedical application. Molecules 28:675

Hoehn D, Azquez-Rowe IV, Kahhat R, Margallo M, Laso J, Fernandez-Ríos A, Ruiz-Salmon I, Aldaco R (2023) A critical review on food loss and waste quantification approaches: Is there a need to develop alternatives beyond the currently widespread pathways?, resources. Conserv Recycl 188:106671

IFPRI (2016) Global Food Policy Report. DC, Search PubMed, Washington

Iorizzo M, Curaba J, Pottorff M, Ferruzzi MG, Simon P, Cavagnaro PF (2020) Carrot anthocyanins genetics and genomics: Status and perspectives to improve its application for the food colorant industry. Genes 11:906

Jablonský M, Škulcová A, Malvis A, Šima J (2018) Extraction of value-added components from food industry based and agro-forest biowastes by deep eutectic solvents. J Biotechnol 282:46–66. https://doi.org/10.1016/j.jbiotec.2018.06.349

Khalid MU, Shabbir MA, Mustafa S, Hina S, Quddoos MY, Mahmood S, Maryam Y, Faisal F, Rafique A (2021) Effect of apple peel as an antioxidant on the quality characteristics and oxidative stability of mayonnaise. Appl Food Res 1:100023

Khan MKI, Asif M, Razzaq ZU, Nazir A, Maan AA (2022) Sustainable food industrial waste management through single cell protein production and characterization of protein enriched bread. Food Biosci 46:101406

Kim S, Lee Y, Lin KY, Hong A, Kwon EE, Lee J (2020) The valorisation of food waste via pyrolysis. J Clean Prod 259:120816. https://doi.org/10.1016/j.jclepro.2020.120816.

Leporini M, Loizzo MR, Sicari V, Pellicanò TM, Reitano A, Dugay A, Deguin B, Tundis R (2020) Citrus clementina hort. juice enriched with its by-products (Peels and Leaves): chemical composition, in vitro bioactivity, and impact of processing. Antioxidants 9:298

Liu Z, de Souza TSP, Holland B, Dunshea F, Barrow C, Suleria HAR (2023) Valorisation of food waste to produce value-added products based on its bioactive compounds. Processes 11:840. https://doi.org/10.3390/pr11030840

Lu H, Yadav V, Bilal M, Iqbal HMN (2021) Bioprospecting microbial hosts to valorize lignocellulose biomass–Environmental perspectives and value-added bioproducts. Chemosphere 288:132574. https://doi.org/10.1016/j.chemosphere.2021.132574

Mahato N, Sharma K, Sinha M, Dhyani A, Pathak B, Jang H, Park S, Pashikanti S, Cho S (2021) Biotransformation of citrus waste-i: Production of biofuel and valuable compounds by fermentation. Processes. https://doi.org/10.3390/pr9020220

Majumder P, Annegowda HV (2021) Fruit and vegetable by-products: novel ingredients for a sustainable society. In valorisation of agri-food wastes and by-products. Elsevier, pp 133–156. https://doi.org/10.1016/B978-0-12-824044-1.00006-4

Mandalari G, Bisignano C, Cirmi S, Navarra M (2017) Effectiveness of citrus fruits on helicobacter pylori. Evid.-Based Complementary Altern. Med. eCAM2017, 8379262

Martau GA, Teleky BE, Ranga F, Pop ID, Vodnar DC (2021) Apple pomace as a sustainable substrate in sourdough fermentation. Front Microbiol 12:3850

Martins VFR, Pintado ME, Morais RMSC, Morais AMMB, (2023) Valorisation of Micro/ nanoencapsulated bioactive compounds from plant sources for food applications towards sustainability. Foods 12:32. https://doi.org/10.3390/foods12010032

Maugeri A, Cirmi S, Minciullo PL, Gangemi S, Calapai G, Mollace V, Navarra M (2019) Citrus fruits and inflammaging: a systematic review. Phytochem Rev 18:1025–1049

Mo WY, Man YB, Wong MH (2018) Use of food waste, fish waste and food processing waste for China's aquaculture industry: Needs and challenge. Sci Total Environ 613:635–643

Motavaf B, Savage P (2021) Effect of process variables on food waste valorisation via hydrothermal liquefaction. ACS ES&T Eng 1(3):363–374. https://doi.org/10.1021/acsestengg.0c00115

Mozhiarasi V (2021) Overview of pretreatment technologies on vegetable, fruit and flower market wastes disintegration and bioenergy potential: Indian scenario. Chemosphere 288:132604. https://doi.org/10.1016/j.chemosphere.2021.132604

Muscat A, de Olde EM, Ripoll-Bosch R, Van Zanten HHE, Metze TAP, Termeer CJAM, van Ittersum MK, de Boer IJM (2021) Principles, drivers and opportunities of a circular bioeconomy. Nat. Food. 2:1–6

Negi T, Kumar Y, Sirohi R, Singh S, Tarafdar A, Pareek S, Awasthi MK, Sagar NA (2021) Advances in bioconversion of spent tea leaves to value-added products. Bioresour Technol 346:126409

Ng HS, Kee PE, Yim HS, Chen PT, Wei YH, Chi-Wei Lan J (2020) Recent advances on the sustainable approaches for conversion and reutilization of food wastes to valuable bioproducts. Bioresour Technol 302. https://doi.org/10.1016/j.biortech.2020.122889

Osorio LLDR, FlorezLopez E, Grande-Tovar CD (2021) The potential of selected agri-food loss and waste to contribute to a circular economy: applications in the food. Cosmetic and Pharmaceutical Industries. Molecules 26:515

Paini J, Benedetti V, Ail SS, Castaldi MJ, Baratieri M, Patuzzi F (2021) Valorisation of wastes from the food production industry: a review towards an integrated agri-food processing biorefinery. Waste Biomass Valorisation 13:31–50. https://doi.org/10.1007/s12649-021-01467-1

Panwar D, Saini A, Panesar PS, Chopra HK (2021) Unraveling the scientific perspectives of citrus by-products utilization: progress towards circular economy. Trends Food Sci Technol 111:549–562

Paritosh K, Kushwaha SK, Yadav M, Pareek N, Chawade A, Vivekanand V (2017) Food waste to energy: an overview of sustainable approaches for food waste management and nutrient recycling. BioMed Res Int 2017:2370927

Pastor RF, Restani P, Di Lorenzo C, Orgiu F, Teissedre PL, Stockley C, Ruf JC, Quini CI, Garcìa Tejedor N, Gargantini R (2019) Resveratrol, human health and winemaking perspectives. Crit Rev Food Sci Nutr 59:1237–1255

Paulraj CRKJ, Bernard MA, Raju J, Abdulmajid M (2019) Sustainable waste management through waste to energy technologies in India-opportunities and environmental impacts. Int. J. Renew. Energy Res 9:309–342. https://doi.org/10.20508/ijrer.v9i1.9017.g7586

Pereira BS, de Freitas C, Contiero J, Brienzo M (2022) Enzymatic production of Xylooligosaccharides from Xylan solubilized from food and agro-industrial waste. BioEnergy Res. 15:1195–1203

Prasad M, Ranjan R, Ali A, Goyal D, Yadav A, Singh TB, Shrivastav P, Dantu PK (2024) Efficient transformation of agricultural waste in India. In: contaminants in agriculture: sources, impacts and management; Springer: Berlin/Heidelberg, Germany, pp 271–287

Rifna EJ, Misra NN, Dwivedi M (2021) Recent advances in extraction technologies for recovery of bioactive compounds derived from fruit and vegetable waste peels: a review. Crit Rev Food Sci Nutr 1–34

Roy P, Mohanty AK, Dick P, Misra M (2023) A review on the challenges and choices for food waste valorisation: environmental and economic impacts. ACS Environmental Au 3(2):58–75. https://doi.org/10.1021/acsenvironau.2c00050

Rudra S, Nishad J, Jakhar N, Kaur C (2015) Food industry waste: mine of nutraceuticals. Int. J. Sci. Environ. Technol. 4:205–229

Russo C, Maugeri A, Lombardo GE, Musumeci L, Barreca D, Rapisarda A, Cirmi S, Navarra M (2021) The second life of citrus fruit waste: a valuable source of bioactive compounds. Molecules 26:5991. https://doi.org/10.3390/molecules26195991

Sarker A, Ahmmed R, Ahsan SM, Rana J, Ghosh MK, Nand R (2024) A comprehensive review of food waste valorisation for the sustainable management of global food waste. Sustainable Food Technol. 2:48–69. https://doi.org/10.1039/D3FB00156C

Sha SP, Modak D, Sarkar S, Roy SK, Sah SP, Ghatani K, Bhattacharjee S (2023) Fruit waste: a current perspective for the sustainable production of pharmacological, nutraceutical, and bioactive resources. Front. Microbiol 14. https://doi.org/10.3389/fmicb.2023.1260071

Shah MP (2024a) Microbial remediation of hazardous chemicals from water & industrial wastewater treatment plant. Springer Nature

Shah MP (2024b) Development in waste water treatment research and processes: applied technologies for clean up of environmental contaminants. Elsevier

Shah MP (2024c) Aerobic and anaerobic microbial treatment of industrial wastewater. CRC Press

Shah MP (2025) Extraction of precious metals from industrial waste using microbial technology. Springer Nature

Sharma P, Vishvakarma R, Gautam K, Vimal A, Gaur VK, Farooqui A, Varjani S, Younis K (2022) Valorisation of citrus peel waste for the sustainable production of value-added products. Biores Technol 351:127064

Sobel R, Versic R, Gaonkar AG (2014) Introduction to microencapsulation and controlled delivery in foods. In: Gaonkar AG, Vasisht N, Khare AR, Sobe RL, editors. Microencapsulation in the food industry. Elsevier; Amsterdam, The Netherlands, pp 3–12

Springmann M, Wiebe K, Mason-D'Croz D, Sulser TB, Rayner M, Scarborough P (2018) Health and nutritional aspects of sustainable diet strategies and their association with environmental impacts: a global modelling analysis with country-level detail, Lancet Planet. Health 2:e451–e461. https://doi.org/10.1016/S2542-5196(18)30206-7

Teigiserov DA, Hamelin L, Thomsen M (2020) Towards transparent valorisation of food surplus, waste and loss: clarifying definitions, food waste hierarchy, and role in the circular economy. Sci Total Environ 706. https://doi.org/10.1016/j.scitotenv.2019.136033

United Nations Environment Programme (2023). Accessed 10 Aug 2023. https://wedocs.unep.org/20.500.11822/25194

Verónica MP, Tupuna-Yerovi DS, González Z, Ruales J (2021) Encapsulation of bioactive compounds from fruit and vegetable by-products for food application–a review. Trends Food Sci Technol 116:11–23

Victor MM, David JM, Cortez MV, Leite JL, da Silva GS (2021) A high-yield process for extraction of hesperidin from orange (Citrus sinensis L. osbeck) peels waste, and its transformation to diosmetin, a valuable and bioactive flavonoid. Waste Biomass Valoriz. 12:313–320

Villacís-Chiriboga J, Elst K, Van Camp J, Vera E, Ruales J (2020) Valorisation of byproducts from tropical fruits: extraction methodologies, applications, environmental, and economic assessment: a review (Part 1: general overview of the byproducts, traditional biorefinery practices, and possible applications). Compr. Rev. Food Sci. Food Saf. 19(2):405–447. https://doi.org/10.1111/1541-4337.12542

Willett W, Rockström J, Loken B, Springmann M, Lang T, Vermeulen S, Murray CJ (2019) Food in the Anthropocene: the EAT–Lancet commission on healthy diets from sustainable food systems. Lancet 393:447–492. https://doi.org/10.1016/S0140-6736(18)31788-4

Yadav S, Malik K, Moore JM, Kamboj BR, Malik S, Malik VK, Arya S, Singh K, Mahanta S, Bishnoi DK (2024) Valorisation of Agri-food waste for bioactive compounds: recent trends and future sustainable challenges. Molecules 29:2055. https://doi.org/10.3390/molecules29092055

Zamli AF, Mahmood WMFW, Ghopa WAW, Lim MT (2021) The research on food waste pre-treatment technology for incineration in Malaysia. J Environ Treat Technol 9:139–147

Zhou X, Liu L, Li J, Wang L, Song X (2020) Extraction and characterization of cellulose from jerusalem artichoke residue and its application in blueberry preservation. Foods 11:1065

Zia M, Ahmed S, Kumar A (2020a) Anaerobic digestion (AD) of fruit and vegetable market waste (FVMW): potential of FVMW, bioreactor performance, co-substrates, and pre-treatment techniques, Biomass Convers. Biorefin 12:3573–3592. https://doi.org/10.1007/s13399-020-00979-5

Zia S, Khan MR, Shabbir MA, Aslam MA, Khan MKI, Nadeem M, Khalil AA, Din A, Aadil RM (2020) An inclusive overview of advanced thermal and nonthermal extraction techniques for bioactive compounds in food and food-related matrices. Food Rev Int 1–31. https://doi.org/10.1080/87559129.2020.1772283

Zuin VG, Ramin LZ (2018) Green and sustainable separation of natural products from agro-industrial waste: challenges, potentialities, and perspectives on emerging approaches. In: chemistry and chemical technologies in waste valorisation. Springer: Cham, Switzerland, vol 376, pp 229–282

Chapter 13
Extraction and Valorization of Nutrients from Wastewater

B. B. Nayak, S. Mounika, Apurba Baruah, and Chunchu Suchith Kumar

Abstract The rising global demand for food and the depletion of natural resources requires innovative strategies for sustainable agriculture. Nutrient recovery from wastewater presents an effective solution by converting waste streams into valuable resources. Each year, over 3 million tonnes of phosphorus and significant nitrogen quantities are lost in sewage, much of which can be recovered using advanced technologies. Wastewater recovery methods like struvite precipitation, membrane filtration, and ion exchange enable efficient nutrient capture, with phosphorus recovery potentially replacing up to 30% of conventional fertilizers. Extremophilic microbes enhance recovery through processes like Enhanced Biological Phosphorus Removal (EBPR) and nitrogen cycling. Global regions, including the European Union and Southeast Asia, have demonstrated significant cost reductions and decreased nutrient runoff through these approaches. Nutrient valorization supports the circular economy and addresses global sustainability goals, including SDGs 6 (Clean Water), 9 (Industry, Innovation), and 12 (Sustainable Consumption). This chapter highlights the environmental, economic, and technological benefits of nutrient recovery, emphasizing its potential to transform agricultural practices and reduce ecological impact.

Keywords Nutrient recovery · Phosphorus extraction · Extremophilic microbes · Wastewater valorization · Circular economy

---

B. B. Nayak (✉)
School of Sciences, Woxsen University, Telangana, India
e-mail: nayak.96agri@gmail.com

S. Mounika
Department of Agronomy, Professor Jayashankar, Telangana Agricultural University, Hyderabad, Telangana, India

A. Baruah
ICAR-ATARI Zone VI, Guwahati, Kahikuchi, Assam, India

C. S. Kumar
Department of Agronomy, School of Agriculture, SR University, Warangal, Telangana, India

© The Author(s), under exclusive license to Springer Nature Singapore Pte Ltd. 2026
S. Dey and M. P. Shah (eds.), *Microbial Applications in Industrial Wastewater Treatment*, Advances in Wastewater Research,
https://doi.org/10.1007/978-981-96-9128-9_13

13.1 Introduction

The increasing global population, projected to reach 9.7 billion by 2050 (United Nations, 2019), places immense pressure on food production systems. Sustainable agricultural intensification requires significant inputs of essential nutrients such as nitrogen (N), phosphorus (P), and potassium (K). However, conventional sources of these nutrients, particularly phosphate rock, are finite. Phosphorus, a critical component of fertilizers, is estimated to be depleted within 50–100 years at current consumption rates (Cordell et al. 2009). The global fertilizer industry relies heavily on non-renewable resources, with over 190 million tonnes of chemical fertilizers applied annually (FAO, 2021). Meanwhile, global food demand is expected to increase by 50% by 2050, necessitating a sustainable approach to nutrient management (FAO, 2018). Nutrient recovery from wastewater offers a promising solution to mitigate resource depletion, close nutrient loops, and promote circular economy principles.

13.2 Global Statistics on Nutrient Loss and Recovery Potential

13.2.1 Phosphorus

Phosphorus is a finite resource, with over 80% of mined phosphate rock used for agricultural fertilizers (van Vuuren et al. 2010). However, an estimated 3 million tonnes of phosphorus are lost in sewage annually, primarily in urban wastewater systems (Schroeder et al., 2020). Studies suggest that efficient phosphorus recovery could meet up to 30% of global fertilizer demand (Desmidt et al. 2015). The European Union has implemented phosphorus recycling policies, targeting 75% phosphorus recovery from sewage sludge by 2030 (Egle et al., 2016).

13.2.2 Nitrogen

Globally, around 110 million tonnes of nitrogen are applied as fertilizers each year, yet 50–70% of this is lost due to leaching and volatilization, contributing to groundwater contamination and atmospheric pollution (Galloway et al., 2008). Urban wastewater carries approximately 25–30 million tonnes of nitrogen annually, of which only a small fraction is recovered (Jenkins et al., 2015). Advanced nitrogen recovery technologies such as ammonia stripping and ion exchange have demonstrated potential in capturing up to 85% of wastewater nitrogen for reuse (Maurer et al., 2003).

13.2.3 Potassium

Potassium losses from wastewater are less studied compared to nitrogen and phosphorus. However, global estimates suggest that wastewater contains around 16–20 million tonnes of potassium annually (Schmidt et al., 2017). Unlike phosphorus, potassium does not cause eutrophication, but its recovery can significantly contribute to fertilizer sustainability, especially in potassium-deficient soils.

13.3 Regional Insights

- **Europe**: Countries like Germany and the Netherlands have pioneered phosphorus recovery from wastewater, with large-scale projects achieving over 50% phosphorus reclamation rates (Egle et al., 2016).
- **Asia**: China and India, being the largest consumers of fertilizers, have explored nutrient recovery from wastewater to address growing agricultural demands. India generates over 38 billion litres of domestic sewage daily, containing an estimated 0.2 million tonnes of phosphorus and 1 million tonnes of nitrogen annually (CPCB, 2021).
- **Africa**: Many African countries face soil nutrient depletion due to limited fertilizer access. Wastewater nutrient recovery presents a viable solution to enhance soil fertility and agricultural productivity in these regions (Nziguheba*et al.*, 2010).
- **North America**: The United States and Canada have implemented nutrient management strategies to reduce runoff pollution and promote wastewater nutrient recovery. The United States EPA has set phosphorus reduction goals under the Clean Water Act to minimize eutrophication in major water bodies (EPA, 2018).

13.4 Environmental Impacts of Nutrient Loss from Wastewater

Nutrient losses from wastewater contribute significantly to environmental degradation. Each year, approximately 3.2 million tonnes of phosphorus and 18 million tonnes of nitrogen are discharged into water bodies globally (Withers et al. 2014). Excess nutrient runoff leads to eutrophication, causing harmful algal blooms, hypoxia, and biodiversity loss in aquatic ecosystems (Schindler et al., 2016). The cost of eutrophication-related damages, including fisheries collapse and water treatment expenses, exceeds \$2.2 billion annually in the United States alone (Dodds et al., 2009). In developing nations, untreated wastewater is a major pollution source. In regions like sub-Saharan Africa and South Asia, over 80% of wastewater is discharged without adequate treatment (WWAP 2017). Sustainable wastewater management, including nutrient recovery, can significantly reduce water pollution, mitigate greenhouse gas emissions from fertilizer production, and improve water quality.

13.5 Wastewater as a Source of Recoverable Nutrients

Wastewater is an underutilized resource rich in essential plant nutrients. Domestic, industrial, and agricultural wastewater streams contain substantial amounts of nitrogen and phosphorus, making them viable sources for fertilizer production. Municipal wastewater alone contains an estimated 15–20 mg/L of nitrogen and 2–5 mg/L of phosphorus, which can be recovered using advanced treatment technologies (Shah 2025; Liu et al. 2019). The European Union has recognized wastewater as a strategic nutrient source, with policies promoting phosphorus recycling to reduce dependency on imported phosphate rock (European Commission, 2020). In the United States, wastewater treatment plants collectively produce approximately 7 million tonnes of biosolids annually, with over 50% being used as agricultural fertilizer (USEPA 2018).

13.6 Importance and Quantification of Nutrients in Wastewater

Wastewater is a rich but underutilized source of essential plant nutrients, primarily nitrogen (N), phosphorus (P), and potassium (K). These nutrients enter wastewater systems through various anthropogenic and natural pathways. Key sources include:

1. **Domestic Wastewater**: Household sewage, food waste, and detergents contribute significant quantities of nitrogen and phosphorus. Human urine alone contains approximately 3.5–7.0 g/L of nitrogen and 0.5–2.0 g/L of phosphorus, making it a potent source for recovery (Verstraete et al. 2009).
2. **Industrial Effluents**: Agro-based industries such as dairy processing, slaughterhouses, and fertilizer production release substantial amounts of phosphorus and nitrogen into wastewater streams. For instance, the global food industry discharges over 1.3 million tonnes of phosphorus annually (Cordell et al. 2009).
3. **Agricultural Runoff**: Excess fertilizers used in farming leach into nearby water bodies, contributing to nutrient-rich wastewater. It is estimated that up to 30–50% of applied nitrogen and phosphorus fertilizers are lost to runoff, leading to nutrient accumulation in wastewater (Schoumans*et al.* 2014).
4. **Livestock Waste**: Manure from poultry, cattle, and swine is a major contributor to nutrient-rich wastewater. Globally, livestock waste accounts for nearly 65% of anthropogenic phosphorus discharge into freshwater systems (Bouwman et al., 2013).

5. **Municipal and Stormwater Runoff**: Urban wastewater contains significant nutrient loads from organic waste, decomposing plant matter, and fertilizers applied to lawns and gardens.

Understanding these sources is crucial for implementing targeted nutrient recovery strategies and mitigating environmental risks associated with nutrient loss.

13.7 Agricultural Importance of Recovered Nutrients

Reclaiming nutrients from wastewater presents an opportunity to enhance soil fertility, reduce dependency on synthetic fertilizers, and promote sustainable agriculture. The key benefits include:

1. **Enhanced Soil Productivity:** Nutrients recovered from wastewater, such as struvite (magnesium ammonium phosphate), provide slow-release phosphorus and nitrogen, improving crop yields and soil health (Le Corre et al. 2009).
2. **Cost Reduction for Farmers:** Synthetic fertilizers constitute a significant portion of agricultural expenses. Recovered nutrients can lower input costs by replacing conventional fertilizers. Studies indicate that phosphorus recovery from wastewater can reduce fertilizer costs by up to 20% in some regions (Desmidt et al. 2015).
3. **Reduction in Environmental Pollution:** By recovering nutrients, wastewater treatment facilities prevent nutrient overload in water bodies, reducing the risk of eutrophication, algal blooms, and hypoxic zones (Shah 2024a; Withers et al. 2014).
4. **Improved Crop Nutrient Use Efficiency:** Compared to synthetic fertilizers, recovered nutrients often demonstrate higher efficiency due to their gradual release properties. For example, struvite-based fertilizers exhibit up to 90% phosphorus uptake efficiency in certain crops (Bhuiyan et al. 2008).
5. **Circular Economy Contribution:** Nutrient recovery aligns with the principles of the circular economy, reducing reliance on finite mineral reserves and fostering resource sustainability (Cordell et al. 2011).

13.8 Technologies for Nutrient Recovery

Several technologies have been developed to extract nutrients from wastewater, including:

1. **Chemical Precipitation**

Chemical precipitation is one of the most widely used techniques for nutrient recovery, particularly for phosphorus extraction from wastewater. Among the various chemical precipitation methods, struvite (magnesium ammonium phosphate, MAP) formation has gained significant attention due to its dual benefits of nutrient recovery and waste reduction.

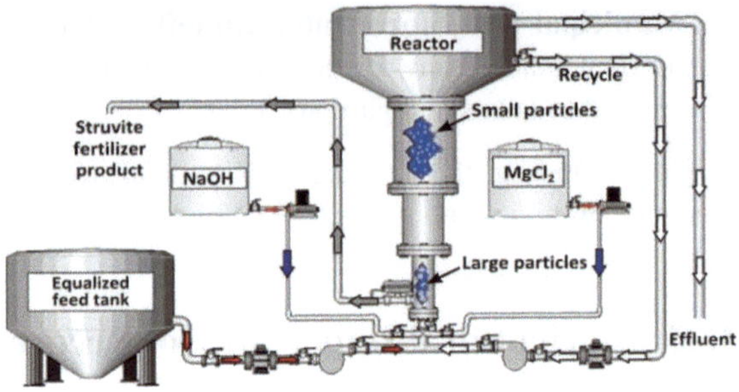

Fig. 13.1 Struvite precipitation process

Struvite Precipitation: This process recovers phosphorus as magnesium ammonium phosphate (struvite), a slow-release fertilizer. Full-scale applications in the Netherlands and Canada have successfully extracted over 90% of phosphorus from wastewater (Le Corre et al. 2009) (Fig. 13.1).

Struvite Formation Process

Struvite precipitation occurs when magnesium, ammonium, and phosphate ions react in a molar ratio of 1:1:1 under controlled pH conditions (typically between 7.5 and 9). The reaction is represented as:

$$Mg^{2+} + NH_4^+ + PO_4^{3-} + 6H_2O \rightarrow MgNH_4PO_4 \cdot 6H_2O \downarrow$$

The process efficiency is influenced by pH levels, temperature, and the presence of competing ions in the wastewater. Various sources such as municipal wastewater, livestock manure, and food industry effluents have been utilized for struvite recovery.

13.9 Agricultural Applications of Struvite

Struvite has been recognized as an effective slow-release fertilizer that gradually supplies essential nutrients to plants. Studies have demonstrated that struvite-derived fertilizers can reduce phosphorus leaching by up to 50% compared to conventional phosphate fertilizers, making it an environmentally friendly alternative (Johnston & Richards, 2022).

Advantages and Challenges

Advantages

- Reduces phosphorus loss from wastewater, mitigating eutrophication risks.
- Enhances nutrient use efficiency in agricultural systems.
- Reduces dependency on non-renewable phosphate rock.

Challenges

- Requires precise pH and ion balance control.
- Competing ions (e.g. calcium) can hinder precipitation efficiency.
- Initial setup and operational costs can be high.

2. Ion Exchange Techniques

Ion exchange is another highly efficient technology used for recovering nitrogen and phosphorus from wastewater. This method relies on synthetic resins or natural zeolites to selectively capture nutrient ions, allowing for their recovery and reuse in agriculture.

13.10 Mechanism of Ion Exchange

Ion exchange resins function by replacing unwanted ions in wastewater with desirable ions attached to the resin. The general reaction for phosphorus and nitrogen removal is:

- Phosphate removal: $\mathbf{R{-}OH + PO_4{}^{3-} \rightleftharpoons R{-}PO_4{}^{3-} + OH^-}$
- Ammonium removal: $\mathbf{R{-}H + NH_4{}^+ \rightleftharpoons R{-}NH_4 + H^+}$.

Natural zeolites, particularly clinoptilolite, have shown high affinity for ammonium ions, making them suitable for nitrogen recovery from wastewater treatment plants (Zhang et al., 2021).

13.11 Agricultural Use of Recovered Nutrients

The captured ammonium and phosphate ions can be released through regeneration processes and transformed into liquid fertilizers. Research suggests that ion-exchange-based fertilizers increase crop nitrogen uptake efficiency by 15–20% compared to synthetic fertilizers (Kumar et al., 2023).

Advantages and Challenges

Advantages

- High selectivity and efficiency in nutrient removal.
- Recovered nutrients can be directly converted into fertilizers.
- Reduces nitrogen volatilization losses.

Challenges

- Resin regeneration requires additional chemical inputs (e.g. acids and bases).
- Limited efficiency in highly contaminated wastewater.
- High operational costs for large-scale applications.
- Membrane Filtration

Membrane filtration technologies, including reverse osmosis (RO) and nanofiltration (NF), provide highly effective means of concentrating and extracting nutrients from wastewater streams.

13.12 Reverse Osmosis (RO) in Nutrient Recovery

RO operates by applying pressure to wastewater, forcing water molecules through a semi-permeable membrane while retaining dissolved salts, nutrients, and contaminants. This process is widely used in wastewater treatment facilities to recover phosphorus and nitrogen.

- **Efficiency:** RO membranes can recover up to 90% of phosphorus and nitrogen from municipal wastewater (Gude et al., 2020).
- **Applications:** The concentrated nutrient stream can be used to formulate liquid fertilizers or reintegrated into agricultural irrigation systems.

13.13 Nanofiltration (NF) for Selective Nutrient Separation

Nanofiltration membranes have larger pore sizes than RO, allowing for selective separation of monovalent and divalent ions. This technology is particularly useful for phosphorus recovery while allowing beneficial nutrients like potassium to pass through, reducing fertilizer formulation costs (Fig. 13.2).

- **Efficiency:** NF achieves up to 85% phosphorus removal and 60–70% nitrogen removal in wastewater treatment plants (Liu et al., 2022).

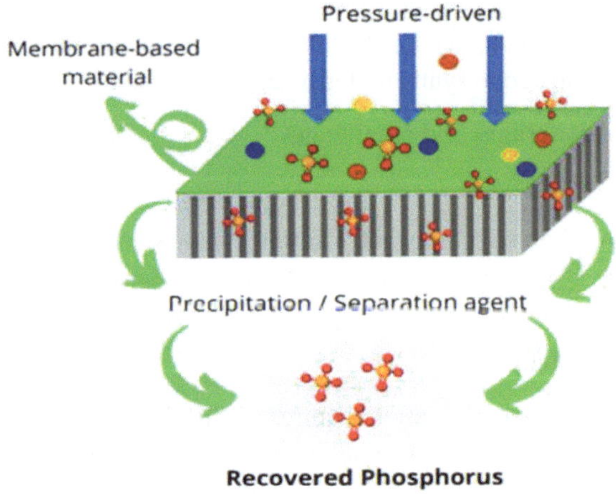

Fig. 13.2 Illustration of membrane-based separation by nanofiltration (NF)

- **Sustainability:** NF membranes require lower energy inputs than RO, making them a cost-effective alternative for large-scale nutrient recovery.

Advantages and Challenges

Advantages

- Produces high-purity nutrient concentrates.
- Reduces pathogen load in wastewater, ensuring safe agricultural reuse.
- Scalable for industrial and municipal wastewater treatment.

Challenges

- High capital and operational costs.
- Membrane fouling can reduce efficiency over time.
- Requires advanced infrastructure and maintenance.
- Biological Processes

Biological processes are fundamental to wastewater treatment, particularly in the removal of phosphorus and nitrogen. These processes rely on microbial communities that facilitate nutrient transformations, making wastewater treatment more efficient and environmentally sustainable. Extremophilic microbes, which thrive in extreme environmental conditions such as high salinity, temperature, or pH, have garnered increasing attention for their role in **Enhanced Biological Phosphorus Removal (EBPR)** and **nitrogen cycling** (Cydzik-Kwiatkowska & Wojnowska-Baryła, 2019).

13.14 Enhanced Biological Phosphorus Removal (EBPR)

EBPR is a widely used biological method for phosphorus removal in wastewater treatment plants. It involves **polyphosphate-accumulating organisms (PAOs)** that take up and store excess phosphorus as intracellular polyphosphate granules under alternating anaerobic and aerobic conditions. The process occurs in two key phases:

- **Anaerobic Phase**: Under oxygen-deficient conditions, PAOs release stored phosphate into the wastewater while simultaneously up taking volatile fatty acids (VFAs) and converting them into intracellular carbon storage compounds such as polyhydroxyalkanoates (PHAs).
- **Aerobic Phase**: When exposed to oxygen or nitrate, PAOs consume the stored carbon and uptake large amounts of phosphate from the wastewater, storing it as intracellular polyphosphate. This results in an overall net removal of phosphorus when excess biomass is periodically wasted from the system.

Extremophilic PAOs, including those adapted to extreme temperature or pH conditions, enhance EBPR efficiency, especially in industrial wastewater treatment where conventional microbial communities struggle to survive.

13.15 Nitrogen Cycling in Wastewater Treatment

Nitrogen removal in biological wastewater treatment involves a combination of nitrification and denitrification processes:

1. **Nitrification**
   - **Ammonia-oxidizing bacteria** (**AOB**) and **ammonia-oxidizing archaea** (**AOA**) convert **ammonium** (**NH_4+**) into **nitrite** (**NO_2^-**).
   - **Nitrite-oxidizing bacteria** (**NOB**) then convert **nitrite** (**NO_2^-**) into **nitrate** (**NO_3^-**).
   - Extremophilic AOB and NOB have been identified in harsh wastewater conditions, increasing treatment efficiency under extreme pH, temperature, and salinity.
2. **Denitrification**
   - Facultative **denitrifying bacteria** reduce **nitrate** (**NO_3^-**), and subsequently to **nitric oxide (NO)**, **nitrous oxide** (**N_2O**), and ultimately **dinitrogen gas** (**N_2**), which is released into the atmosphere.
   - Extremophilic denitrifiers contribute to robust nitrogen removal in challenging wastewater environments, such as saline and high-temperature wastewater streams.
3. **Anammox (Anaerobic Ammonium Oxidation) Process**
   - Anammox bacteria convert **ammonium** (**NH_4^+**) and **nitrite** (**NO_2^-**) directly into **dinitrogen gas** (**N_2**), significantly reducing oxygen and organic carbon requirements compared to conventional nitrification–denitrification pathways.
   - Extremophilic anammox bacteria exhibit high adaptability to fluctuating conditions, making them valuable in wastewater treatment systems.

Comparative Analysis of Recovery Efficiencies Across Technologies

Each of the nutrient recovery technologies discussed has its own strengths and limitations, making them suitable for different wastewater compositions and treatment goals. A comparative assessment of these technologies is summarized in Table 13.1.

Role of Extremophilic Microbes in Wastewater Treatment

Extremophilic microbes are a diverse group of microorganisms that thrive in extreme environmental conditions such as high salinity, temperature, acidity, or alkalinity. Their unique metabolic pathways allow them to play a crucial role in wastewater treatment and nutrient recovery. The use of extremophiles in wastewater treatment has gained significant attention due to their ability to enhance biological nutrient removal (BNR), improve system efficiency, and reduce operational costs (Madigan et al., 2020). One of the primary advantages of using extremophiles is their resilience in harsh wastewater conditions where conventional microbes may struggle to survive. These microbes facilitate the breakdown of organic matter, improve phosphorus and

Table 13.1 Comparative assessment of different nutrient recovery methods

Technology	Nutrient recovered	Recovery efficiency (%)	Primary application	Challenges
Struvite precipitation	Phosphorus, Nitrogen	60–80	Slow-release fertilizers	pH-sensitive, competing ions
Ion exchange	Phosphorus, Nitrogen	70–90	Liquid fertilizers	Resin regeneration costs
Reverse osmosis	Phosphorus, Nitrogen	85–95	Nutrient concentrates	High energy demand, membrane fouling
Nanofiltration	Phosphorus	70–85	Selective nutrient separation	Moderate efficiency for nitrogen recovery
Enhanced biological phosphorus removal (EBPR)	Phosphorus	60–90	Phosphorus removal in wastewater treatment	Requires alternating anaerobic–aerobic conditions, competition with glycogen-accumulating organisms (GAOs)
Anammox process	Nitrogen	80–95	Nitrogen removal in wastewater treatment	Slow bacterial growth, sensitive to operational changes
Nitrification–Denitrification	Nitrogen	70–90	Biological nitrogen removal	High oxygen and carbon source requirement
Microalgal nutrient recovery	Phosphorus, Nitrogen	50–80	Algal biomass for biofertilizers, biofuels	Seasonal efficiency, light dependency, land requirement

nitrogen removal, and contribute to the stabilization of treatment processes (Zhang et al., 2022). Their capacity to withstand toxic elements and recover valuable nutrients makes them indispensable in modern wastewater treatment technologies.

13.16 Enhanced Biological Phosphorus Removal (EBPR) Using Extremophiles

Phosphorus removal from wastewater is a critical step in preventing eutrophication, a phenomenon that leads to excessive algae growth and depletion of oxygen in water bodies. Enhanced Biological Phosphorus Removal (EBPR) is a widely used microbial-driven process that efficiently extracts phosphorus from wastewater and converts it into biomass or recoverable forms such as struvite.

13.16.1 Mechanism of EBPR

EBPR relies on the activity of phosphorus-accumulating organisms (PAOs), which assimilate and store phosphorus as intracellular polyphosphate granules. Under alternating anaerobic and aerobic conditions, PAOs take up volatile fatty acids (VFAs), storing energy as polyphosphate and subsequently releasing it into the wastewater. When exposed to aerobic conditions, these organisms absorb excess phosphorus from the wastewater to replenish their internal reserves, leading to a net reduction in phosphorus concentrations (Oehmen et al. 2019). Extremophilic PAOs, such as Halomonas and Acidithiobacillus species, offer distinct advantages over conventional PAOs due to their ability to function in saline, acidic, or high-temperature environments. Studies indicate that halophilic PAOs can sustain phosphorus uptake even in hypersaline wastewater, which is often a challenge for traditional EBPR systems (Martins et al. 2021).

13.16.2 Extremophiles in EBPR: Case Studies

Recent studies have demonstrated the effectiveness of extremophilic microbes in phosphorus removal across diverse wastewater environments:

- **Haloferax and Halobacterium spp.** have shown high phosphorus uptake efficiency in saline wastewater, making them ideal for treating industrial effluents (González-Martínez et al. 2020).
- **Thermophilic PAOs**, such as Thermus spp., have exhibited phosphorus removal efficiency of up to 80% in wastewater treatment plants operating at temperatures above 50 °C (Li et al. 2022).

- **Acidophilic microbes**, including Acidiphilium and Ferroplasma spp., have been utilized in mining wastewater treatment, efficiently precipitating phosphorus while withstanding low pH conditions (Johnson & Hallberg 2020).

13.17 Microbial Nitrogen Cycling

Nitrogen removal is another essential aspect of wastewater treatment, as excessive nitrogen discharge can lead to water pollution and greenhouse gas emissions. Microbial nitrogen cycling involves several biochemical pathways, including nitrification, denitrification, and anaerobic ammonium oxidation (anammox). Extremophilic microbes have significantly contributed to advancing these processes, enabling more efficient nitrogen removal under varying environmental conditions.

13.17.1 The Anammox Process

Anaerobic Ammonium Oxidation (Anammox) is a highly efficient nitrogen removal process that involves the direct conversion of ammonium (NH4 +) and nitrite (NO2-) into dinitrogen gas (N2), bypassing the need for organic carbon. Anammox bacteria belong to the Planctomycetes phylum and thrive in oxygen-limited environments, making them ideal for wastewater treatment systems with low-energy requirements (Strous et al., 2021) (Fig. 13.3).

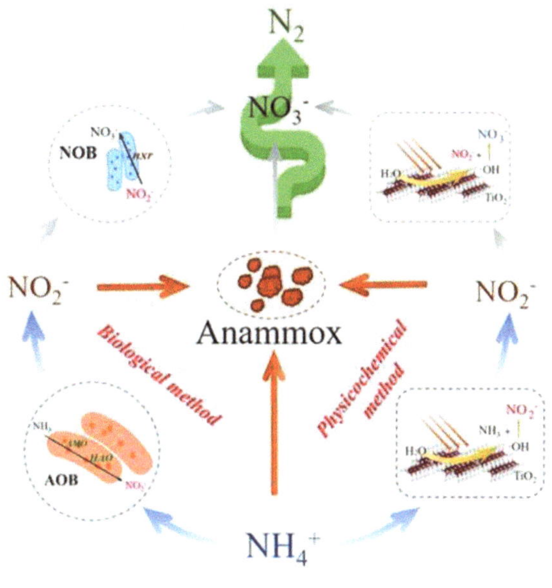

Fig. 13.3 Nitrogen removal pathway of nitrite production and anammox process

Advantages of Anammox Over Traditional Nitrification–Denitrification

- **Energy Savings**: The process does not require aeration, reducing energy consumption by 50–60% compared to conventional nitrification–denitrification systems.
- **Minimal Sludge Production**: Anammox bacteria grow slowly, leading to lower sludge generation and reduced waste handling costs.
- **Higher Nitrogen Removal Efficiency**: Reported nitrogen removal rates exceed 90%, making it a promising alternative for wastewater treatment (Kartal et al. 2020).

13.17.2 Role of Extremophiles in Anammox-Based Nitrogen Removal

Certain extremophilic anammox bacteria have demonstrated remarkable adaptability to varying environmental conditions, enabling nitrogen removal under conditions where conventional microbes struggle:

- **Thermophilic Anammox Bacteria**: Studies have identified *Candidatus Brocadiafulgida* as capable of performing nitrogen removal at temperatures above 45 °C, making it suitable for high-temperature wastewater treatment applications (Shah 2024b; Oshiki*et al.* 2019).
- **Halophilic Anammox Bacteria**: *Scalindua* species thrive in saline environments, efficiently removing nitrogen from industrial and marine wastewater effluents (Ali et al. 2021).
- **Acidophilic Nitrogen-Removing Microbes**: *Nitrospira spp.* have demonstrated high nitrite oxidation rates in acidic conditions, proving effective in mining and tannery wastewater treatment (Shah 2024c; Wang et al. 2022).

13.17.3 Other Microbial Nitrogen Cycling Processes

Besides anammox, other extremophilic microbial processes contribute to nitrogen removal in wastewater treatment:

- **Comammox (Complete Ammonia Oxidation)**: Recently discovered Nitrospira species can convert ammonia directly into nitrate in a single step, offering an efficient alternative to conventional two-step nitrification (Daims *et al.* 2018).
- **Denitrification by Thermophilic Bacteria**: Certain thermophilic species, such as *Geobacillus* and *Bacillus thermodenitrificans*, have shown the ability to perform high-efficiency denitrification under elevated temperatures (Shen et al. 2021).

13.18 Environmental and Economic Benefits

1. **Reduction in Fertilizer Costs and Nutrient Runoff**

Nutrient recovery from wastewater presents a viable and sustainable alternative to conventional fertilizer production. The global demand for synthetic fertilizers, particularly nitrogen (N) and phosphorus (P), has surged due to increasing food production needs. However, the manufacturing of synthetic fertilizers is energy-intensive and contributes to environmental degradation. Nutrient valorization offers a cost-effective solution by reducing dependence on mined phosphorus and industrial nitrogen fixation, both of which have high economic and ecological costs.

2. **Cost Savings for Farmers and the Agriculture Industry**

The economic burden of fertilizers on farmers is significant, with global fertilizer prices fluctuating due to supply chain disruptions, geopolitical tensions, and resource depletion. According to the International Fertilizer Association (IFA), the price of phosphate rock, a key ingredient in phosphorus fertilizers, increased by over 60% between 2020 and 2022 due to resource scarcity and export restrictions. Recovered nutrients from wastewater can serve as an alternative, stabilizing fertilizer costs and making nutrient inputs more affordable, especially for small-scale farmers in developing countries. For instance, studies indicate that struvite-based fertilizers—formed through phosphorus recovery from wastewater—can replace 20–30% of traditional phosphorus fertilizers at a lower cost (Dockhorn, 2009). Moreover, bio-based nitrogen fertilizers derived from ammonium recovery via ion exchange or stripping technologies can reduce nitrogen fertilizer expenses by up to 40% (Batstone et al. 2015). These savings benefit individual farmers and enhance the overall profitability of agricultural production systems.

3. **Mitigating Nutrient Runoff and Eutrophication**

Nutrient runoff from excessive fertilizer use contributes to eutrophication, leading to algal blooms, oxygen depletion, and loss of aquatic biodiversity. The Food and Agriculture Organization (FAO) estimates that over 78% of nitrogen applied in agriculture is lost to runoff, leaching into water bodies and causing environmental damage (FAO, 2021). Nutrient valorization not only reduces this wastage but also ensures that the recovered nutrients are efficiently utilized in a controlled manner, minimizing their escape into the environment. By integrating recovered nutrients into agricultural practices, countries can significantly decrease the incidence of water pollution. A case study from the European Union (EU) revealed that phosphorus recovery projects led to a 40% reduction in nutrient runoff into freshwater ecosystems, improving water quality and reducing eutrophication incidents (Cornel & Schaum 2009). Similarly, research in Southeast Asia indicates that wastewater-derived fertilizers improve soil nutrient balance while reducing nitrogen leaching by up to 35% (Tan et al. 2017).

4. **Contributions to the Circular Economy and Sustainable Development Goals**

Nutrient valorization aligns with the principles of the circular economy, transforming waste into valuable resources and reducing environmental footprints. This transition supports global sustainability efforts, particularly the United Nations Sustainable Development Goals (SDGs), which emphasize resource efficiency, climate resilience, and responsible consumption.

5. **Closing Nutrient Loops and Reducing Waste**

The circular economy promotes the reuse and recycling of materials to create closed-loop systems. Wastewater-derived nutrient recovery exemplifies this approach by recapturing essential nutrients from municipal, industrial, and agricultural effluents. The recovered nutrients are then reintroduced into food production systems, reducing dependency on finite resources like phosphate rock. Countries that have adopted circular nutrient management strategies have witnessed tangible benefits. For instance, the Netherlands implemented the Circular Agriculture Initiative, focusing on phosphorus recycling from wastewater. The programme has successfully reintroduced over 50,000 tonnes of phosphorus per year into agricultural use, reducing reliance on imported fertilizers (van der Hoek et al., 2018). Similar efforts in Japan have demonstrated that nutrient recycling from wastewater can replace up to 25% of synthetic fertilizers while significantly cutting greenhouse gas emissions (Matsubae et al., 2011).

6. **Supporting Sustainable Development Goals**

Nutrient recovery directly contributes to multiple SDGs, reinforcing sustainable agriculture, clean water management, and responsible industrial practices:

- SDG 6 (Clean Water and Sanitation): Nutrient recovery reduces water pollution by limiting the discharge of nitrogen and phosphorus into rivers, lakes, and oceans.
- SDG 9 (Industry, Innovation, and Infrastructure): The development of innovative recovery technologies, such as ion exchange and struvite precipitation, fosters industrial advancements in resource-efficient wastewater management.
- SDG 12 (Responsible Consumption and Production): Nutrient valorization promotes sustainable resource use by reducing waste and maximizing the utility of agricultural inputs.
- SDG 13 (Climate Action): By replacing synthetic fertilizers that require fossil fuels for production, nutrient recovery reduces carbon emissions associated with fertilizer manufacturing and application.

A study by the European Environment Agency (EEA) found that nutrient recycling initiatives reduced greenhouse gas emissions in the EU by 8 million tonnes of CO2 equivalent annually, further contributing to global climate mitigation efforts (EEA, 2020).

7. **Economic and Environmental Synergies**

Nutrient valorization presents a dual advantage: enhancing economic resilience in agriculture while preserving natural ecosystems. The integration of wastewater-derived fertilizers into mainstream agricultural systems provides economic incentives, improves soil fertility, and mitigates adverse environmental impacts.

8. **Enhancing Soil Health and Crop Productivity**

Recovered nutrients, particularly phosphorus and nitrogen, improve soil health by maintaining balanced nutrient levels. Unlike conventional fertilizers, which often lead to soil acidification and microbial imbalances, bio-based fertilizers promote soil biodiversity and enhance long-term productivity (Chowdhury et al. 2014). A comparative field study in Germany demonstrated that crops fertilized with recovered phosphorus from wastewater exhibited similar or higher yields compared to those receiving synthetic fertilizers, indicating the agronomic viability of nutrient valorization (Römer, 2006). Additionally, nutrient recycling can help mitigate soil degradation, a major challenge affecting over 33% of global arable land (UNCCD, 2020). By incorporating organic-rich, wastewater-derived fertilizers, soil structure, and microbial diversity are improved, leading to better water retention and nutrient availability.

9. **Energy and Resource Efficiency**

Traditional fertilizer production is highly energy-intensive. The Haber–Bosch process for nitrogen fixation, for example, consumes approximately 1–2% of global energy supply (Erisman et al. 2008). Recovering nitrogen from wastewater through processes like ammonia stripping or electrochemical conversion significantly reduces energy consumption. Research suggests that wastewater-derived nitrogen recovery could save up to 60% of the energy required for conventional ammonia production (Maurer et al. 2003). Similarly, phosphorus mining depletes finite reserves, with estimates indicating that high-grade phosphate rock may be exhausted within the next 50–100 years (Cordell et al. 2009). By capturing phosphorus from wastewater, countries can extend the lifespan of natural phosphorus reserves while securing long-term agricultural productivity (Fig. 13.4).

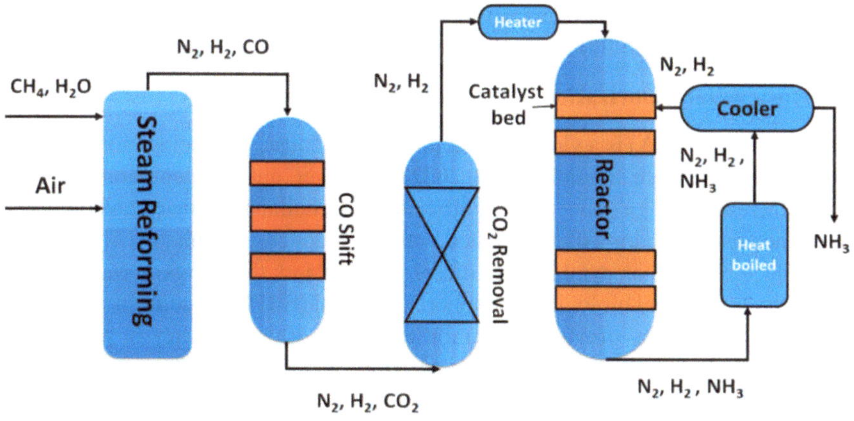

Fig. 13.4 Haber–Bosch process

13.19 Case Studies

1. **Phosphorus Recovery Initiatives of the European Union**

The European Union (EU) has been at the forefront of phosphorus (P) recovery due to its heavy reliance on imported phosphate rock and concerns about environmental degradation caused by P runoff. The EU Circular Economy Action Plan has emphasized the need for sustainable phosphorus management, leading to policies such as the EU Fertilizing Products Regulation (2019/1009) and the Sustainable Phosphorus Platform (ESPP). These initiatives promote the recovery and reuse of P from wastewater, reducing dependency on non-renewable phosphate resources (European Commission, 2020). Several large-scale phosphorus recovery projects have been implemented across the EU. One of the most successful is the Stuttgart Process used in Germany, where phosphorus is recovered from sewage sludge ash using acid extraction techniques. Similarly, the Amsterdam Water Supply (Waternet) has developed a struvite precipitation system, where phosphorus is extracted as magnesium ammonium phosphate (struvite), a slow-release fertilizer with significant agricultural value (van der Velde et al., 2021).

A key EU-funded project, Phosphorus Recycling in Europe (P-REX), tested multiple P recovery techniques across different wastewater treatment plants, showing that up to 50% of the phosphorus in municipal wastewater can be recovered using advanced methods like crystallization and ion exchange (Cornel & Schaum, 2021). The recovered phosphorus has reduced the need for synthetic fertilizers and minimized eutrophication risks in European water bodies. Phosphorus recovery in the EU has significantly decreased fertilizer imports, particularly from phosphate rock-mining countries like Morocco and China. The economic savings from domestic phosphorus recovery are estimated at €200–300 million per year (Smol et al., 2020). Additionally, the EU has seen an 18% reduction in phosphorus runoff into water bodies, contributing to improved water quality and compliance with the Water Framework Directive (2000/60/EC) (European Environment Agency, 2022).

2. **Wastewater Valorization in Agriculture in Southeast Asia**

Southeast Asia (SEA) is a global agricultural powerhouse, producing more than 120 million tonnes of rice annually (FAO, 2022). However, nutrient depletion and water pollution remain major concerns. Countries such as Thailand, Vietnam, and Indonesia have increasingly adopted wastewater nutrient recovery to enhance soil fertility and reduce reliance on synthetic fertilizers. In Thailand, the Bangkok Metropolitan Authority has piloted biological phosphorus removal (EBPR) systems in wastewater treatment plants. The project has demonstrated 40–50% efficiency in phosphorus recovery, producing a biofertilizer that has improved rice and maize yields by 15–20% compared to conventional fertilizers (Chaiwong et al., 2021).

Vietnam's Hanoi Wastewater Management Project integrates anaerobic digestion with struvite precipitation, capturing both nitrogen and phosphorus for agricultural use. The project has led to a 15% cost reduction in fertilizer expenses for local farmers and prevented excess nutrients from polluting the Red River ecosystem (Nguyen et al., 2022). The use of nutrient-enriched wastewater in Indonesia's rice paddies has shown promising results. Research conducted by the Indonesian Agricultural Research Center (IARC) found that controlled irrigation using nutrient-rich treated wastewater led to a 10–12% increase in rice yields while maintaining soil microbial diversity (Sulaiman et al. 2021). Moreover, integrated wastewater valorization programmes have led to a 30% reduction in synthetic fertilizer imports, promoting self-sufficiency in agricultural inputs.

3. **Emerging Projects in Nutrient Recovery for Sustainable Agriculture in Africa**

Africa faces significant challenges related to soil nutrient depletion, with more than 40% of agricultural soils being phosphorus-deficient (IFPRI, 2021). Wastewater recovery has emerged as a critical solution, particularly in countries like South Africa, Kenya, and Ghana, where nutrient extraction from municipal wastewater has gained traction. In South Africa, the Durban Water Recycling (DWR) project has successfully implemented a biological nutrient removal (BNR) system that recovers nitrogen and phosphorus for agricultural use. Farmers using the recovered nutrients have reported a 25% improvement in maize and soybean yields while reducing fertilizer costs by 30% (Munnik et al. 2022). Kenya's Nairobi Sludge Fertilizer Project, a collaboration between the Kenyan government and international NGOs, transforms sewage sludge into nutrient-rich compost. This initiative supplies over 15,000 smallholder farmers with affordable fertilizers, improving soil fertility and crop productivity (Mugo et al. 2023).

The economic impact of nutrient recovery in Africa is substantial. In Ghana, a study by the International Water Management Institute (IWMI) found that utilizing waste-derived fertilizers can increase farmer incomes by 20–30% due to improved crop yields and reduced dependence on costly imported fertilizers (Drechsel et al. 2022). Environmentally, phosphorus recovery has contributed to a 35% reduction in eutrophication risks in key water bodies such as Lake Victoria and the Niger River Basin.

13.20 Comparative Analysis of Global Applications

A comparative assessment of global nutrient recovery projects reveals varying efficiencies based on region and technology. While struvite precipitation is the dominant method in the EU, biological phosphorus removal (EBPR) and anaerobic digestion have been more prevalent in Southeast Asia and Africa due to lower operational costs and adaptability to existing wastewater infrastructure (Cornel & Schaum 2021) (Table 13.2).

Table 13.2 A comparative assessment of global nutrient recovery projects

Region	Primary technology used	Recovery efficiency	Major impact
EU	Struvite Precipitation, Ion Exchange	50–60% P recovery	Reduced P imports, lower runoff
Southeast Asia	EBPR, Anaerobic Digestion	40–50% N & P recovery	Improved crop yields, lower fertilizer costs
Africa	BNR, Composting	30–40% N & P recovery	Enhanced soil fertility, farmer income growth

13.21 Challenges and Opportunities

Despite the success of these initiatives, challenges persist. High initial investment costs, regulatory hurdles, and public perception issues remain barriers to widespread adoption. However, increasing international collaborations, technological advancements, and policy incentives are expected to drive further expansion of wastewater nutrient recovery worldwide (Smolet al. 2020).

While the application of extremophilic microbes in wastewater treatment holds immense promise, several challenges remain in scaling up these technologies:

- **Optimization of Growth Conditions**: Many extremophilic microbes require specific environmental conditions that must be carefully controlled in large-scale operations.
- **Process Integration with Existing Treatment Systems**: Combining extremophile-driven processes with conventional wastewater treatment infrastructure requires innovative engineering solutions.
- **Economic Viability and Cost Reduction**: High initial investment costs for developing extremophilic microbial systems may pose a barrier to widespread adoption.
- **Regulatory and Policy Considerations**: The implementation of microbial-assisted wastewater treatment must align with local and global environmental regulations to ensure safe nutrient recovery and reuse.

13.22 Future Perspectives and Scaling up Nutrient Valorisation

Despite its numerous benefits, scaling up nutrient recovery faces challenges related to technological costs, regulatory frameworks, and market acceptance. Addressing these challenges requires policy support, research investments, and public–private partnerships.

- Policy Interventions: Governments must implement incentives such as subsidies for wastewater-derived fertilizers and stricter regulations on nutrient discharge to promote nutrient recovery.
- Technological Advancements: Innovations in microbial-assisted recovery, nanotechnology-based filtration, and decentralized nutrient recovery units can enhance efficiency and affordability.
- Public Awareness and Farmer Adoption: Educational initiatives and demonstration projects can encourage farmers to adopt wastewater-derived fertilizers and integrate them into existing agricultural practices.

13.23 Challenges and Future Directions

1. **Technical and Economic Barriers to Implementation**

Despite significant advancements in nutrient recovery technologies, several technical and economic challenges hinder large-scale adoption. One of the primary obstacles is the high initial capital investment required for setting up advanced recovery facilities. Technologies like struvite precipitation, ion exchange, and membrane filtration involve costly infrastructure and specialized equipment, limiting their accessibility, especially in developing countries (Batstone et al. 2019). Another technical challenge is the variability in wastewater composition. The concentration of nutrients in wastewater fluctuates based on source, season, and treatment processes, making it difficult to standardize recovery methods. For instance, municipal wastewater has different nutrient profiles compared to agricultural or industrial effluents, requiring tailored recovery strategies (Shu et al. 2021).

Energy consumption is also a major concern. Many nutrient recovery processes, particularly membrane filtration and electrochemical techniques, are energy-intensive, raising operational costs. Research is ongoing to optimize these technologies for lower energy use and higher efficiency, but widespread adoption remains limited due to economic constraints (Egle et al., 2016). Additionally, recovered nutrients like struvite (magnesium ammonium phosphate) and ammonium sulphate must meet agricultural quality standards to be viable alternatives to conventional fertilizers. Contaminants such as heavy metals and pharmaceuticals present in wastewater can impact the safety of recovered products, requiring additional purification steps that further increase costs (Cieślik & Konieczka, 2017). From an economic standpoint, market acceptance and pricing pose challenges. Farmers and fertilizer manufacturers may be hesitant to adopt recovered nutrients due to a lack of awareness or concerns about efficacy compared to synthetic fertilizers. Establishing financial incentives, subsidies, and demonstration projects can help bridge this gap and promote large-scale adoption.

2. **Policy Frameworks, International Collaborations, and Research Gaps**

Policy support plays a crucial role in facilitating the integration of nutrient recovery into mainstream wastewater treatment systems. However, many countries lack comprehensive regulations that mandate or incentivize nutrient recovery. While the European Union's Circular Economy Action Plan promotes phosphorus recovery from wastewater, similar policies are not widely implemented in other regions (European Commission 2020). The lack of harmonized standards for recovered nutrients also creates regulatory uncertainty. Inconsistent guidelines on nutrient content, contaminant limits, and application methods hinder commercialization. Establishing global certification frameworks can enhance confidence in these products and ensure safe agricultural use. International collaborations are essential to accelerate research and knowledge sharing. Organizations like the Global Phosphorus Research Initiative (GPRI) and partnerships under the United Nations Environment Programme (UNEP) have been instrumental in promoting sustainable nutrient management. However, stronger multi-sectoral partnerships involving academia, industries, and policymakers are needed to drive large-scale implementation (Withers et al. 2015).

Research gaps remain in optimizing recovery technologies for diverse wastewater sources, improving cost-effectiveness, and integrating nutrient recovery into circular bioeconomy models. For example, combining wastewater treatment with bioenergy production through anaerobic digestion could enhance sustainability by simultaneously recovering nutrients and generating renewable energy (Rossi et al. 2021).

3. Innovations in Microbial-Assisted Nutrient Recovery

Microbial technologies offer promising solutions to address technical challenges in nutrient recovery. Extremophilic microbes, known for their resilience in harsh environments, have shown potential in biological phosphorus removal and nitrogen cycling. For instance, the anammox (anaerobic ammonium oxidation) process has demonstrated high nitrogen removal efficiency with lower energy requirements compared to conventional nitrification–denitrification methods (Kartal et al. 2010). Emerging synthetic biology approaches are also revolutionizing microbial nutrient recovery. Engineered microbial consortia with enhanced phosphorus and nitrogen uptake capabilities are being developed to improve efficiency. For example, genetically modified polyphosphate-accumulating organisms (PAOs) can increase phosphorus recovery rates, making the process more economically viable (Seviour et al. 2019).

Another innovation is the integration of algae-based systems for nutrient valorization. Algae can effectively absorb nitrogen and phosphorus from wastewater while producing biomass for biofuels and animal feed. Research on optimizing algal strains for wastewater treatment is gaining momentum, with promising results in pilot-scale applications (Christenson & Sims 2011). Moreover, advances in microbial electrochemical systems (MES) are being explored for nutrient recovery. These systems use bacteria to facilitate the electrochemical extraction of nutrients, reducing energy consumption while producing value-added products like hydrogen gas (Kuntke et al. 2018).

13.24 Conclusion

Nutrient recovery from wastewater is a transformative solution for sustainable agriculture, reducing reliance on conventional fertilizers, minimizing environmental pollution, and promoting circular economy principles. Technologies like struvite precipitation, ion exchange, and membrane filtration effectively recover nitrogen and phosphorus, enhancing soil fertility while mitigating nutrient loss. Scaling these technologies globally requires overcoming technical, economic, and policy barriers through collaborative research, innovation, and supportive regulations. Future efforts should focus on optimizing recovery efficiencies, lowering operational costs, and advancing microbial-assisted nutrient extraction. Strengthening policy frameworks, incentivizing private-sector investments, and integrating AI-driven monitoring systems can further enhance process efficiency and scalability. A multi-stakeholder approach—involving researchers, policymakers, and agricultural practitioners—is essential for long-term success. Advancing nutrient recovery aligns with global sustainability goals, particularly SDGs 6 (Clean Water), 9 (Industry & Innovation), and 12 (Sustainable Consumption & Production), ensuring a more resilient and resource-efficient agricultural future.

References

Álvarez JA, Armstrong E, Méndez ML (2018) Advances in biological phosphorus removal processes for wastewater treatment. Water Res 142:113–124

Batista JR, Rounds SA, Pratt J (2019) Phosphorus recovery through struvite precipitation: applications in agricultural fertilizers. Environ Engin Sci 36(5):537–550. Cieślik B, Konieczka P, Namieśnik J (2018) Phosphorus recovery from wastewater: the possibilities and challenges. J Cleaner Product 198:1228–1240

Batstone DJ, Hülsen T, Mehta CM, Keller J (2020) Transformation of wastewater treatment to resource recovery: Economic feasibility and impacts on greenhouse gas emissions. Environ Sci Technol 54(4):2626–2634

Cordell D, Drangert JO, White S (2009) The story of phosphorus: global food security and food for thought. Glob Environ Chang 19(2):292–305

De-Bashan LE, Bashan Y (2004) Recent advances in removing phosphorus from wastewater and its future use as fertilizer. Water Res 38(19):4222–4246

European Commission (2021) Circular economy action plan: For a cleaner and more competitive Europe. Publications Office of the European Union

Garfí M, Flores L, Ferrer I, García J (2017) Agricultural use of nutrient-rich sludge from wastewater treatment: Impacts on soil properties and crop productivity. Agric Water Manag 191:67–77

Guedes A, Amado AM, Carvalho AP (2019) The role of extremophilic microorganisms in biotechnological applications and nutrient recycling. Appl Microbiol Biotechnol 103(22):9057–9071

Jiang Y, Ye L, Yuan Z (2020) Enhancing biological phosphorus removal from wastewater through extremophilic bacteria: Advances and future prospects. Biores Technol 303:122976

Kumar M, Rawat R, Verma P (2021) Wastewater valorization for nutrient recovery: case studies from Southeast Asia. Environ Technol Innov 24:101946

Liu Y, Chadwick D, Withers PJA, Hu Y (2019) Nutrient recovery from wastewater and its agricultural applications. Agric Syst 176:102761

Mayer BK, Baker LA, Boyer TH, Drechsel P, Gifford M, Hanjra MA, Parameswaran P, Stoltzfus J, Westerhoff P, Rittmann BE (2016) Total value of phosphorus recovery. Environ Sci Technol 50(13):6606–6620
Molinos-Senante M, Hernández-Sancho F, Sala-Garrido R (2011) Economic feasibility study for phosphorus recovery processes. Resour Conserv Recycl 55(4):450–457
Puyol D, Batstone DJ, Hülsen T, Astals S, Peces M, Krömer JO (2017) Resource recovery from wastewater by biological technologies: Opportunities, challenges, and prospects. Front Microbiol 8:1566
Reddy KR, Kadlec RH, Flaig E (1999) Phosphorus retention in treatment wetlands. Water Sci Technol 40(3):67–73. https://doi.org/10.1016/S0273-1223(99)00441-3
Schoumans OF, Bouraoui F, Kabbe C, Oenema O, van Dijk KC (2015) Phosphorus management in Europe in a changing world. Ambio 44(2):180–192
Smol M, Adam C, Preisner M (2020) Circular economy model framework in the European water and wastewater sector. J Mater Cycles Waste Manage 22(1):200–213
Shah MP (2024a) Microbial remediation of hazardous chemicals from water & industrial wastewater treatment plant. Springer Nature
Shah MP (2024b) Development in waste water treatment research and processes: applied technologies for clean up of environmental contaminants. Elsevier
Shah MP (2024c) Aerobic and anaerobic microbial treatment of industrial wastewater. CRC Press
Shah MP (2025) Extraction of precious metals from industrial waste using microbial technology. Springer Nature
Tao M, He Z, Chen X (2022) Microbial-assisted nutrient recovery from wastewater: advances in anammox and phosphorus removal processes. J Environ Manage 305:114255
United Nations (2020) Sustainable development goals report 2020. United Nations Publications. https://unstats.un.org/sdgs/report/2020/
Withers PJA, Sylvester-Bradley R, Jones DL, Healey JR, Talboys PJ, Chadwick DR (2014) Sustainable phosphorus use in agriculture. Philosophical Transact Royal Society B: Biological Sci 369(1639):20130371
Yuan Z, Pratt S, Batstone DJ (2012) Phosphorus recovery from wastewater through microbial processes. Biores Technol 122:419–428

Chapter 14
Nanotechnology-Enhanced Microbial Electrochemical Systems for Heavy Metal Recovery and Removal from Industrial Effluents

Priyanka Singh, Mahi Singh, and Meenu Singh

Abstract Heavy metals, extensively utilized in industries like real estate, mining, construction, and transport, pose significant environmental risks due to their carcinogenic properties. Untreated heavy metal wastewater can accumulate in aquatic organisms, leading to health issues such as cancer, skin irritation, allergies, and dermatitis upon human consumption. Existing waste treatment technologies like adsorption are inefficient and conventional methods are costly and eco-unfriendly. Microbial fuel cells (MFCs) offer a promising alternative, utilizing electroactive bacteria for heavy metal removal and electricity generation. Metals are bio-electrochemically reduced at the cathode through bacterial action. This chapter explores the principles of MFCs for heavy metal removal, the microbes, and the pathways involved, and compares MFCs with other methods. It includes case studies on groundwater remediation, factors affecting metal recovery, and challenges in MFC operation, providing a comprehensive understanding of microbial electrochemical heavy metal recovery.

Keywords Microbial fuel cells (MFCs) · Heavy metal removal · Bio-electrochemical reduction · Industrial effluents · Environmental remediation

P. Singh · M. Singh (✉) · M. Singh (✉)
Department of Biotechnology, School of Engineering, IILM University, Greater Noida, UP, India
e-mail: mahi.singh.bt26@iilm.edu

M. Singh
e-mail: meenu.singh@iilm.edu

P. Singh
e-mail: priyanka.singh@iilm.edu

© The Author(s), under exclusive license to Springer Nature Singapore Pte Ltd. 2026

S. Dey and M. P. Shah (eds.), *Microbial Applications in Industrial Wastewater Treatment*, Advances in Wastewater Research,
https://doi.org/10.1007/978-981-96-9128-9_14

14.1 Introduction

Removing heavy metals from industrial wastewater is a serious environmental problem that needs to be solved sustainably and efficiently. An effective method for recovering and removing heavy metals from microbial electrochemical systems is the integration of nanotechnology. The elimination of ions containing heavy metals from the environment and the use of zero-valent iron at the nanoscale and their composites for this purpose were covered by Zou et al. (2016). The study demonstrated how well heavy metal ions may be extracted from water that is contaminated using zero-valent iron at the nanoscale and its composites. Using activated carbon made from leftover coconut buttons, Anirudhan and Sreekumari (2011) studied the absorbent elimination of metallic ions of metal from industrial effluents. The study shed light on the possibility of using waste coconut buttons to create activated carbon, which can be a powerful adsorbent for removing heavy metals from industrial effluents.

Biosorption was highlighted by Beni and Esmaeili (2020) as an effective technique for eliminating toxic metals from wastewater from industries. The study demonstrated the promise of biosorption as a low cost and ecologically benign technique for removing heavy metals from industrial effluents. An extensive review of the immobility of microbial cells for wastewater biotreatment was presented by Bouabidi et al. (2018). The research findings highlighted the importance of immobilized microbial cells in wastewater biotreatment, and hefty metal removal, was highlighted by the research findings. Bio-derived materials act as an environmentally friendly method of recovering and reusing precious and crucial metals. The research brought attention to the possibility of using bio-derived materials as environmentally friendly substitutes for recovering and reusing precious and important metals, supporting the circular economy concept.

Additionally, Diep et al. (2018) talked about the bioaccumulation of heavy metals by employing genetically modified bacteria for heavy metal removal. The results of the study highlighted the ability of genetically modified microbes to bioaccumulate heavy metals, offering a long-term solution for heavy metal elimination. The study has yielded valuable insights into various technologies and approaches for the recovery and removal of heavy metals from industrial effluents. These include activated carbon, graphene-based microbots, nanoscale zero-valent iron, the process of biosorption embedded microbe cells, bio-derived materials, and genetically modified microorganisms. Future studies should examine the possibility of synergistic effects of integrating these technologies for improved heavy metal recovery and removal, as well as the sustainable development and scalability of microbial electrochemical systems enhanced by nanotechnology for heavy metal removal.

Large amounts of industrial effluent including heavy metals have been produced as a result of growing industrialization, and these pollutants seriously endanger both human health and the environment. The potential of microbial electrochemical systems boosted by nanotechnology towards the restoration and elimination of toxic metals from industrial effluents has been investigated in several studies. The establishment of metabolism for bioextractive applications which focused on the recovery and removal of heavy metal ions for use in downstream purification and refinement as opposed to disposal. This implies an important change in the use of bioaccumulation to remove heavy metals, creating new avenues for the effective and long-term treatment of industrial waste water. In order to recover metals, nutrients, sulfur, hydrogen, and heat from industrial effluents, Rajoria et al. (2022) gave a thorough analysis of electrochemical methods, including microbial electrochemical technologies. This analysis highlights the need for additional research to refine and grow up these technologies for real-world applications, while also underscoring the intriguing possibilities of microbiological electrochemical systems in overcoming the intricate problems related to eliminating heavy metals and recovery.

Moreover, research on contemporary nano-enhanced methods for metal and dye removal by Batool et al. (2014) shows how nanotechnology might improve the effectiveness of heavy metal removal procedures. The results of this work offer important new perspectives on how to use microbial electrochemical systems with nanotechnology to create cutting-edge therapeutic methods. There are still unanswered questions that need to be investigated further within the context of nanotechnology-enhanced microbiological electrochemical technologies for heavy-metal extraction and reduction from industrial effluents. First and foremost, a better comprehension of the interplay of heavy metal ions, microbes, and nanomaterials in the setting of microbiological electrochemical systems is required. This knowledge is essential for maximizing the functionality and design of metal recovery systems.

Furthermore, more research is required to determine the viability and affordability of microbial electrochemical systems boosted by nanotechnology for use in industrial settings. This involves the creation of affordable techniques for the synthesis of nanomaterials as well as the evaluation of these systems' long-term stability and performance under practical operating circumstances. In conclusion, there is a lot of potential for developing effective and sustainable methods for the recovery and removal of heavy metals from industrial effluents by combining nanomaterials with microbial electrochemical systems. To progress the field and facilitate the application of these novel treatment options in practice, future research ought to concentrate on filling in the identified knowledge gaps.

14.2 Principles of Microbial Fuel Cells (MFCs)

Microbial fuel cells, or MFCs, are a new technology with a lot of potential for use in wastewater and wastewater treatment, bioremediation, energy production, and the creation of value goods (source). The promise of microbial electrochemistry for environmental remediation and the production of energy that is environmentally friendly has spurred tremendous interest in these applications. We will incorporate and synthesize important research findings in this overview of the literature to provide readers a thorough understanding of the fundamentals of bacterial fuel cells and to suggest possible avenues for future investigation.

The ability of MFCs to produce useful products is another important feature. The potential of microalgae and the optimization of them for economically viable and renewable enterprises producing feed, biofuels, and other goods were explored by Khan et al. (2018). This result implies that MFCs as can be used to produce valuable products by growing microorganisms, providing opportunities for further studies into how to optimize MFCs for particular product synthesis. In his examination of the scaling up of bioelectrochemical systems, including microbial fuel cells.

The literature study offers significant insights into the diverse applications of metal-free batteries (MFCs), such as energy generation, wastewater and waste treatment, bioremediation, and valuable product manufacturing. The results also show that more research is required to address the scaling issues of MFCs, investigate the possibility of product synthesis, and optimize the electrical processes. Future research can take advantage of these information gaps to further establish the fundamentals of microbial fuel cells and realize their full potential in the production of sustainable energy and environmental remediation.

Microbial fuel cells, or MFCs, are a potential technology for producing energy sustainably since they can produce electricity using microorganisms' metabolic processes. The goal of this review of the literature is to present a summary of the state of knowledge on MFCs by combining and condensing the results of various important studies.

Within the framework of microbial electrosynthesis, Rabaey and Rozendal (2010) reexamine the electrical pathway for microbial production. The authors stress the significance of comprehending MFC electrochemical processes and how to use them to support microbial development. This demonstrates the possibilities of MFCs for processes including bioelectrosynthesis in addition to energy generation. Their analysis clarifies the microbial communities that are a part of MFCs and how they can move electron to solid-state electrodes. The writers talk about how electroactive microbes propel electron transfer activities, which are essential to MFC operation.

The significance of the production of biofilm in MFCs and the function of electrically charged microorganisms in the creation of biofilm are emphasized by the authors. This emphasizes how important it is to comprehend the microbial ecology of MFCs. The research sheds light on the basic mechanisms regulating the passage of electrons between bacteria and MFC electrodes. Optimizing the performance and efficiency of MFCs requires a thorough understanding of these electron transfer pathways.

14.3 Heavy Metals: Types, Structures, and Effects

Because of their toxicity, environmental durability, and capacity for bioaccumulation, metallic substances are well-known environmental contaminants. Their anthropogenic origins include mining and other industrial and agricultural operations, while their sources from nature include volcanic eruptions and the weathering of rocks that contain metals. Heavy metals are widely distributed in the environment as a result of their numerous industrial, household, agricultural, medical, and technical uses. This has led to worries about the possible consequences of heavy metals on both individuals and the ecosystem. Due to ongoing sewage discharge and untreated industrial effluents, harmful heavy metal contamination of terrestrial and aquatic ecosystems has significantly grown in recent decades (Shah 2025; Jan et al. 2015).

The levels of heavy metals in drinking water are still over the guidelines established by international regulatory bodies, even though technology is advancing at a rapid pace (Rehman et al. 2018). The sustainability of the environment and public health are seriously threatened by this. Because of growing concerns about the safety of agricultural products, pollution from heavy metals in the soil has become a global environmental problem (Mitra et al. 2022). In farming soil and plants, metals such as lead are also known to have ecological concerns and implications for human health (Alengebawy et al. 2021). The main cause of soil contamination, which includes toxic substances and other chemicals, is waste from the mining, chemical, and metal processing sectors, as well as other related businesses (Li et al. 2019).

Heavy metal detoxification in the surroundings mostly involves stabilizing the metals in place or removing them from the matrix, such as clay. This shows that to cure heavy metal-contaminated soils and reduce the threats that heavy metal pollution poses to the environment and human health, efficient bioremediation approaches are required. Although many heavy metals have biotoxic effects on human biochemistry, other heavy metals are nevertheless important as trace components. A comprehensive comprehension of the mechanisms underlying heavy metal contamination is necessary, encompassing information on levels, states of oxidation, sources, and remediation strategies (Shah 2024a; Li et al. 2019).

After a thorough analysis of the literature, it can be concluded that heavy metals are pervasive environmental pollutants that have a big impact on both ecological balance and human health. The results emphasize how urgently more study is needed to investigate the mechanisms underlying heavy metal toxicity, create counter-defense systems, and design efficient techniques for remediating heavy metal contamination. The body of research that has already been done to address the problems caused by heavy metal contamination in various environmental matrices, such as water, groundwater, and agricultural products, is further highlighted in the literature. The discovery of creative and long-lasting ways to lessen the contamination of heavy metals and its detrimental impacts on the health of humans and the environment should be the main focus of future research endeavors.

A class of metallic elements with a high density that become dangerous or poisonous in low quantities is known as heavy metals. It has been discovered that these heavy metals, which include aluminum, cadmium and lead, copper, nickel, copper, and zinc, have detrimental impacts on both human health and the environment. Numerous public health initiatives have been implemented to manage, stop, and mitigate metal toxicity that manifests at different intensities. Drinking water containing lead, arsenic, nickel, chromium, and mercury is especially dangerous since these elements can impair the body's primary metabolic process, oxidatively damage proteins and DNA, and cause a host of disorders.

Because of the possibility of accumulation through the food chain, the accumulation of heavy metals like cadmium, mercury, lead, arsenic, and chromium within soil poses a serious threat to the well-being and health of both humans and other living things. In addition to direct exposure, heavy metal pollution also has indirect consequences that are passed down through the supply chain. The human body may be negatively impacted by eating fish that has hazardous levels of heavy metals like cadmium, lead, arsenic, and mercury. Thus, it is crucial to keep an eye on the levels of metals in fish flesh to make sure that food safety laws are being followed. This emphasizes how crucial it is to uphold laws and carry out routine inspections to lessen the risk of heavy metal exposure through food intake.

It's crucial to remember that not all metals are dangerous. The body needs certain heavy metals for important functions including copper, nickel, iron, magnesium, magnesium, chromium, selenium, manganese, and zinc. People can experience both acute and long-term toxicities from heavy metals such as cadmium, arsenic, mercury, lead, chromium, and silver. This contrast highlights the necessity for a comprehensive knowledge of the various ways that heavy metals affect human health. In summary, the examined literature offers a thorough understanding of the cytotoxicity and health consequences of heavy metals on both human and environmental health. Even though our understanding of the causes and consequences of toxicity from heavy metals has advanced significantly, there are still areas of unknown information that need to be resolved.

Prospective avenues for future research may encompass examining the enduring consequences of extended exposure to low concentrations of heavy metals, devising enhanced remediation tactics, and investigating possible interplays among distinct heavy metals that may lead to detrimental health consequences. This literature review, taken as a whole, emphasizes the complex nature of the toxicity of heavy metals and the necessity of ongoing study as well as public health initiatives to lessen the harmful impacts on the environment and human health.

14.4 Microbes and Pathways for Heavy Metal Reduction

14.4.1 Biosorption

It has been determined that the dispersal and discharge of heavy metals from industrial wastewater can have detrimental impacts on the environment by contaminating groundwater and surface water supplies. Both ecology and human health are at risk because of this. Significant benefits from biosorption include low cost, accessibility, profitability, simplicity of use, and high efficiency—especially at low concentrations. Furthermore, it has been demonstrated that biosorption is an effective and environmentally friendly method of removing contaminants from effluent (Elgarahy et al. 2021).

Similarly, Javanbakht et al. (2014) underlined the possibility of adsorption for environmental remediation by discussing the processes of heavy metal removal employing microorganisms as biosorbents. The study advanced our knowledge of the effectiveness and mechanisms behind biosorption by shedding light on the substances, their outcomes, science, and modeling simulation techniques used in biosorption processes. Just 3% of Earth's water resources are fresh water and industrial and human activity generates and release waste products that contain metals into water bodies, rendering them unavailable and endangering ecosystem health and human health (Shah 2024b; Bashir et al. 2018).

This emphasizes even more how urgent it is to create sustainable and efficient processes for removing heavy metals from wastewater. There are still unanswered questions despite the body of research demonstrating the effectiveness and benefits of biosorption for the removal of heavy metals. For example, more investigation is required into the precise mechanisms and interactions involved in the biosorption processes, as well as the creation of new biosorbents with improved heavy metal removal capacities. Further advances in this sector would come from comparative research assessing the environmental impact and cost-effectiveness of biosorption technologies in large-scale applications.

In summary, there is a lot of research in the literature that points to biosorption's promise as an effective and environmentally friendly method of removing heavy metals from effluent. Future studies should concentrate on filling in any remaining gaps in understanding and creating creative fixes to improve the sustainability and efficacy of adsorption techniques for environmental cleanup.

14.4.2 Bioaccumulation

Heavy metals can pose a health risk to consumers as they build up in the surroundings and contaminate food systems. As a result of this buildup, numerous bodily tissues and organs experience a range of harmful consequences (Balali-Mood et al. 2021). Moreover, heavy metal pollution of soils impacts plant development and, in severe cases, might result in plant mortality (Yan et al. 2020). Plants have defenses against the toxicity of heavy metals, including the ability to chelate metals and increase their capacity for metal detoxification. Food chains and webs are contaminated by heavy metals that enter living things via the abiotic environment and then accumulate in the biota at various trophic levels (Shah 2024c; Kumar et al. 2020).

There is a serious risk to human and environmental health from the biological accumulation of heavy metals. Entire mechanics of heavy metal biological accumulation and its impacts on many ecosystems still need to be fully understood, though. Future studies should concentrate on investigating the effects of non-essential dangerous metalloids and heavy metals on wildlife and human health, as well as their trophic transfer, bioaccumulation, which and biomagnification in food chains and webs. In addition, more research is required to determine the effects of heavy metal on the bioaccumulation on certain ecosystems and the possible health hazards connected to eating contaminated food items.

14.4.3 Bioleaching

Traditional chemical and physical repair techniques are expensive and produce a lot of chemical waste. As a result, there is increasing interest in investigating more affordable, environmentally benign, and alternative heavy metal remediation techniques. One such technique is bioleaching, which recovers and extracts metals from polluted soils and solutions using plants and microorganisms. With an emphasis on the function of rhizobia, plant-growth-promoting microorganisms (PGPM), and microbiological methods for heavy metal bioleaching, this review of the literature attempts to summarize the most recent discoveries on the biological and molecular mechanisms underlying plant–microbe–metal interactions.

Microbes and plants cohabit or compete for existence, and their harmonious interactions are essential for the adaptation of plants to environments that are metalliferous. Plant–microbe interactions are a possible option for increasing microbe-assisted phytoremediation because of the complex communication systems created between soil microorganisms and plant root exudates. Through a variety of processes, such as the breakdown of mineral nutrients, the synthesis of substances that promote plant growth, and the secretion of particular enzymes, PGPM, such as helpful fungi and bacteria, have been identified to mitigate metal harmful effects and induce plant growth (Hou et al. 2020). Additionally, through processes like acidity, the process of precipitation the chelation process, complexation, and redox reactions, PGPM can alter the bioavailability of metals in soil.

Because rhizobia can fix nitrogen in legumes, legumes are essential to the nitrogen cycle in both agriculture and the environment. It has been determined that a key element supporting the phytoremediation of soil contaminated with metals is the legume-rhizobia symbiosis. It has been observed that in metal-contaminated environments, a variety of legume species predominate. Furthermore, legume development may be aided while metal toxicity is reduced by the rhizobia's ability to fix nitrogen, solubilize phosphorus, synthesize phytohormones, and produce siderophores. But it's unknown how sensitive the legume-rhizobia symbiosis is to metals and how metal toxicity affects their relationship, pointing to future research.

Microorganisms are promising options for heavy metal biological leaching because they possess a range of metal sequestration mechanisms with enhanced metal biosorption capabilities. The use of biosensor microbes acts as an economical and environmentally benign method of extracting and recovering metals from solutions containing living or dead biomass and its constituent parts. This demonstrates the potential of microbiological instruments for the bioleaching of heavy metals and points to a future line of inquiry for studying the biosorption abilities of microbes in metal-contaminated settings.

The fields of environmental science and biological remediation have studied the biological transformation and breakdown of pollutants by plants and microorganisms in great detail. In addition to highlighting knowledge gaps and outlining possible future research objectives in this field, the goal of this literature review is to combine and synthesize the available research discoveries on the tolerance, and degradation, and health impacts of heavy metals. The capacity of plants and microbes to tolerate and degrade heavy metals sheds light on the processes by which plants and microbes can withstand and break down environmental heavy metals. The authors talk about these biological agents' ability to lessen the damaging impacts of heavy metals and how to use them in bioremediation techniques.

In a similar vein, Dixit et al. (2015) concentrated on the biological remediation of soil and aquatic habitats for heavy metals. Their synopsis of the guiding concepts and standards of basic procedures clarifies the possibility of using bioremediation methods to remove heavy metals from polluted places. To create efficient bioremediation techniques for heavy metal contamination, the authors stress the significance of comprehending the biological processes of biotransformation and degradation. The processes by which silicon mitigates the toxicity of heavy metals in plants were investigated by Adrees et al. (2015). Their review sheds light on how silicon helps to lessen the harmful impacts of toxic metals on the physiological functioning of plants. The authors stress the possibility of using silicon-based treatments to increase plants' resistance to stress caused by heavy metals, which would aid in the phytoremediation of contaminated sites.

Engwa et al. (2019) examined the underlying processes and health consequences of heavy metal intoxication in humans in an analogous setting. Their analysis clarifies the processes via which the body becomes intoxicated with heavy metals and the resulting health consequences. Developing measures to reduce the health concerns associated with heavy metal exposure requires an understanding of the mechanisms leading to the toxicity of heavy metals in humans. In general, the body of research on heavy metal tolerance, and degradation, and health impacts in biological systems is quite informative. Still, some areas of knowledge need to be explored further. For example, further research is required to clarify the precise processes by which plants and microbes detoxify and change heavy metals. Furthermore, the environment's long-term consequences of biotransformation processes need further study.

The subject of heavy metal cleanup and biotransformation should progress by solving these knowledge gaps through future research goals. The existing research on biosynthesis and heavy metals concludes by highlighting the processes and health impacts of heavy metal exposure in humans, as well as the capacity of microbes and plants to tolerate and degrade heavy metals. A thorough understanding of biological transformation processes is essential to creating long-term solutions to remove heavy metal pollution and safeguard the environment and public health. In general, the body of research on heavy metal acceptance, degradation, and health impacts in biological systems is quite informative. Still, some areas of knowledge need to be explored further.

For example, further research is required to clarify the precise processes by which plants and microbes detoxify and change heavy metals. Furthermore, further research

Table 14.1 Microbial pathways for heavy metal removal

Pathway	Description	Examples
Absorption	The uptake of heavy metal ions onto microbial cell surfaces or internal structures	**Bacteria**: *Pseudomonas aeruginosa, Bacillus cereus*
Biosorption	Passive process where microbial cells, dead or alive, adsorb heavy metals from the environment	**Fungi**: *Aspergillus niger, Penicillium chrysogenum*
Bioaccumulation	Active uptake and internal accumulation of heavy metals by living microbial cells	**Microalgae**: *Chlorella vulgaris, Spirulina platensis*
Biotransformation	Microbial conversion of heavy metals into less toxic forms through biochemical processes	**Bacteria**: *Desulfovibrio desulfuricans* (reduces Cr(VI) to Cr(III))
Bioreduction	Reduction of heavy metals from higher oxidation states to lower, often less toxic, states	**Bacteria**: *Geobacter sulfurreducens* (reduces Fe(III) to Fe(II))
Biosolubilization	Microbial processes that increase the solubility of insoluble metal compounds, facilitating their removal	**Fungi**: *Fusarium oxysporum* (solubilizes metal phosphates)
Bioprecipitation	Formation of insoluble metal salts or sulfides by microbial metabolic activities, leading to metal removal	**Bacteria**: *Sulphate-reducing bacteria* (precipitate metals as sulfides)
Metalloid detoxification	Conversion of metalloids into less toxic forms by microbial action	**Bacteria**: *Arthrobacter* sp. (detoxifies arsenic compounds)

is needed to determine the long-term impacts of biological transformation processes on both the environment and public health. The subject of heavy metal cleanup and biotransformation should progress by solving these knowledge gaps through future research goals. The research on biological transformation or heavy metals concludes by highlighting the processes and health impacts of heavy metal exposure in humans, as well as the capacity of microbes and plants to tolerate and degrade heavy metals. An in-depth comprehension of biotransformation pathways is essential to creating long-term solutions to remove pollution from heavy metals and safeguard the environment and public health (Table 14.1).

14.5 Factors Affecting Metal Recovery in Microbial Fuel Cells (MFCs)

An apparatus called a fuel cell made up of microbial cells is used to produce power and remove heavy metals from industrial effluents. It involves taking out the electrons, contaminants, and organic stuff. Metallic elements like sulfide and vanadium are eliminated from the wastewater as a result of the products produced during the MFC operation being deposited. According to Fu and Wang (2011), these techniques include membrane filtration, precipitation of chemicals, coagulation-flocculation, adsorption, and electrochemical treatment technologies. However, there has been a lot of focus on creating innovative, clean methods for recovering of metals to make the processing of these elements both affordable and environmentally benign. Several factors influence the efficacy of MFCs in the elimination of heavy metals during this operation, both positively and negatively. The first and most crucial element that plays a significant role in raising the MFCs' efficiency is.

14.5.1 Environmental and Operational Factors

The MFC system's design is thought to be the most crucial component in its operation. Other environmental issues and operational factors that may be present include (i) substrate; (ii) microorganisms and their metabolism; (iii) the mechanism of electron transfer in an anodic chamber; (iv) the type and material of electrodes; (v) environmental conditions like temperature, salinity, and pH; and (vi) electrode material and shape.

14.5.1.1 Substrate

The simple and complicated substrates that are given to the microorganisms in the MFCs are important for controlling the MFC parameters and impacting the MFCs' effectiveness. The composition of the substrates, or the organic and inorganic substrates, affects MFC. Protons and electrons are discharged into the aqueous solution throughout the substrate's oxidation process (Mathuriya and Yakhmi 2014). Carbon supplies for microbiological growth and metabolism are provided by organic substrates, such as glucose and acetate. Nitrate is an example of an inorganic substrate that is unlikely to promote microbial growth as well, which would lower the effectiveness of metal recovery.

Increased electron transfer efficiency results from bacteria' easier metabolism of simple substrates like acetate. Microbial growth and metal recovery may be impacted by complex substrates (such as wastewater), which may contain compounds that are hard to breakdown or inhibitors. Optimizing the substrate utilized in metal-free capacitors (MFCs) is essential to increasing the effectiveness of metal recovery procedures. To get the greatest results in particular applications, each component must be properly considered.

14.5.1.2 Microorganisms Including Metabolism

The primary factor influencing a microorganism's cell potential is its metabolic pathway. Exoelectrogens, also known as electroactive microorganisms, are prokaryotes that may interact with charged conductive surfaces of electrodes by employing them as electron suppliers or acceptors. Both Gram-positive and gram-negative electricigens have been described. To extract energy from biological material present in wastewater, anaerobic microorganisms that form an electrically charged biofilm are employed as catalysts in MFCs (Nancharaiah et al. 2015).

14.5.1.3 Electron Transfer Mechanism

Shuttles or electron mediators are needed to move the electrons generated in the anodic device to the anode. Electrons generated during the breakdown of organic materials move to the anodes, eventually oxidizing them where they are eventually oxidized by electron takers at the cathode end. Fast motion, high redox capability, environmental availability, long-term viability, and low cost are a few qualities of an excellent electron acceptor (Lu and Li 2012).

14.5.1.4 Electrodes Material and Shape of Electrodes

Anode

The electrodes where electrons exit the cell and undergo oxidation are known as the anode while the electrodes where electrons input the cell and undergo reduction are known as the cathode. A wide range of anode materials, including those based on carbon and metal, have been assessed for improving MFC performance. Carbon is the most widely utilized electrode material because of its crucial characteristics, which include being reasonably priced, manageable, and having a definite surface area. Because of their strong biocompatibility, low cost, and high electrical conductivity, carbon-based materials like graphite rod, graphite felt, graphite plate, and carbon-based paper (Zhang et al. 2012), as well as carbon fabric are currently the most commonly utilized anode materials.

Cathode

The most important and difficult component of successfully implementing MFC technology, especially for wastewater treatment, is the manufacture of cathode components that mix high power generation, columbic efficiency, and lower expenses (Rahimnejad et al. 2015). The oxygen is consumed at the cathode through the oxygen reducing reaction (ORR), which produces hydrogen peroxide or water. Graphite, carbon paper, and carbon cloth that are sold commercially are the most often utilized materials for a cathode.

Abiotic Cathodes

As a terminal electron acceptor, oxygen can receive electron transfers from abiotic cathodes. Since oxygen has a high potential for oxidation, is readily available, inexpensive, sustainable, and doesn't produce any chemical waste, it is the best acceptor of electrons for a metal-free catalyst (MFC). There are two oxidation states of chromium (Cr) that are present in the environment: hexavalent Cr(VI) and trivalent Cr(III). It has been established that Cr(VI) is extremely carcinogenic, mutagenic, and poisonous. Remediation of Cr(VI) is a serious concern due to its broad spectrum of applications, including tanneries and metal plating. It is recognized as one of the most concerning environmental pollutants.

Biotic Cathodes

Abiotic cathodes and bioanodes make up conventional MFCs. For abiotic cathodes to reduce oxygen, a mediator or catalyst—such as platinum—must typically be present. In a comparative investigation, Jang et al. (2013) found that after eight weeks of operation, the current from biotic cathode MFCs grew and stabilized, in contrast to the present from abiotic cathode MFCs, which remained constant. Furthermore, compared to MFCs with abiotic cathodes (257 W/m^3 and 15.6%), biotic cathode MFCs had power density and coulombic efficiencies that were significantly higher (430 W/m^3 and 59.6%).

14.5.1.5 Environmental Factors

Several variables have been shown to affect MFC performance, including the substrate, the anodic chamber's electron transfer mechanism, pH, temperature, terminal electron acceptors arrangement, PEM type, and the material of the electrode.

Impact of Metal Concentration and Types of Recovery Efficiency

Because it depends upon the kind of recovery techniques employed, the metal concentration in both the elimination and the recovery of the various metals in the metal recovery system is crucial. When it comes to metals, recovery refers to the portion of all the metal in the effluent that is recovered and concentrated. The effects of certain heavy metal ions, such as copper, nickel, cadmium, manganese, zinc, chromium, and cobalt, on the energy output of MFCs were examined by Xu et al. (2016a, b).

14.5.2 Optimizing Strategies for Improving MFC Performance

Because many parameters that affect an MFC's effectiveness are still poorly understood, using one in practice is still difficult. Among other factors, the performance of MFCs is determined by the properties of the electrode material, substrate, inoculum, operating conditions, and design constraints. These operating conditions must be designed (Tiquia-Arashiro and Pant 2020; Jadhav et al. 2019).

Several bioelectrochemical constraints, financial difficulties, engineering design limits, and material science limitations hinder the field use of such systems (Roy and Pandey 2016; Pasupuleti and Arumugam 2016). Moreover, the robust and complicated nature of the repeatability and replicability of the results from MFC research necessitates the minimization of error and sensitivity evaluation of the outcomes from such restrictions. By creating complicated prototypes through repeated iterations as well as modeling the system to make it the ideal answer for field applications, these difficult limits can be reduced to some extent (Heidrich and Edwards 2014).

14.6 Challenges in MFC Operation for Metal Recovery

The incapacity of MFCs to treat every metal ion must be their biggest problem. Heavy metal ions such as mercury, vanadium, chromium, copper, and cobalt have all been reduced using the MFC method. It should be mentioned, nevertheless, that metals can act as straight electron acceptors without the need for an external power source only in cases when the decrease is thermodynamically favorable—that is, when the target metal possesses a positive standard potential.

14.6.1 Technical and Operational Challenges

There are several operational and technical difficulties while implementing an MFC system. Sustaining a stable microbial population capable of producing energy is a major operational problem. Variations in environmental factors like pH, temperature, and nutrition availability can all have an impact. Over time, biofouling and electrode deterioration can also reduce system performance and require regular maintenance. Technically speaking, improving the way electrons are transferred between the microorganisms and the electrodes is necessary to achieve high power density. This involves the creation of appropriate electrode materials and membranes that can improve conductivity. There are difficulties in scaling up MFC equipment from experimental to industrial scale, especially when it comes to maintaining consistent performance, being economical, and cutting efficiency while handling the intricacies of large-scale operations.

An external power source is needed to drive electrons to travel through the cathode so as to be lowered in metals whose redox potentials are equal to or lower than those of the anode. These metals are incapable of accepting electrons from the electrodes on their own. Among these metals include zinc, nickel, lead, and cadmium (Wang and Ren 2014). The processing of waste water that included the nickel ions in both the MEC (microorganism electrolysis cell) as well as an MFC in order to contrast the efficacy of the two. They found that nickel cannot be reduced in an MFC due to its electromagnetic the redox capacity of −0.25 V (vs. NHE).

The removal efficiency using the MEC was threefold better than the MFC at an amount of 1,000 mg/L of Ni^{2+}. The outcome of this experiment showed that the bioelectrochemical process in the cathode compartment could not reduce the nickel, and it also indicated that the MEC method has a great deal of promise for effectively removing Ni^{2+} from wastewater. The declining efficiency and removal capacity of the metals in question present another difficulty for the system's operation. The effects of certain heavy metal ions, such as copper, nickel, cadmium and manganese, which zinc, chromium, and cobalt, on the energy output of MFCs were examined by Xu et al. (2016a, b). Every single metal was shown to be harmful to *S. oneidensis* MR-1, which and to exhibit voltage inhibition at certain concentrations.

14.6.2 Scaling Up from Laboratory to Field Applications

Although tubular MFCs were utilized in certain applications, air–cathode or dual-chambered MFCs continue to be the most popular designs for heavy metal removal. PEM is a necessary component of DCMFCs because it physically divides the anode and cathode compartments while allowing protons to diffuse through the anode to the cathode maintains the balance of ions and electrons, decreases the diffusion of oxygen to the anode, and improves CE principles (Lu et al. 2015; Yang et al. 2016).

Yang et al. (2016) created an innovative hybrid MFC stack that utilized singles and dual chamber MFCs for self-sufficient pH management to address the issue of proton buildup in the anode chamber, which negatively impacts both reliability of operation and energy output in DCMFCs. When compared to stacks made up only of SCMFCs or DCMFCs, the hybrid stack produced the highest power output. A unique hybrid MFC stack that utilizes single and double-chambered MFCs as for self-sufficient pH regulation was created by Yang et al. (2016) to address the issue of protons buildup in the anode chamber, which has a detrimental impact on both operational stability and power output in DCMFCs. The hybrid stack produced the most power output when compared to stacks made up only of SCMFCs or DCMFCs. Here are some other methods and approaches that can support this process.

Materials for the Electrode

- Choose materials with a high surface area, excellent conductivity, and durability. Carbon-based materials, such as graphite and carbon fabric, are examples of common materials.
- Use coatings or surface treatments to improve the process of electron transfer and biofilm growth.

Enhancing Reactor Design

- Create modular units that are easily expandable for a system.
- Conduct chamber testing to determine the optimal layout for the intended use.

- Make sure that substrate is distributed evenly and that byproducts are removed effectively to avoid clogging and guarantee reliable operation.

Increasing Microbial Activity

- To increase stability, use an assortment of microbial communities.
- Preserve ideal parameters for microbial growth, such as pH, temperature, and nutrient availability.

14.6.3 Addressing Biofouling and Maintaining Microbial Activity

Membrane fouling is the process in which solute macromolecules, colloidal particles, or particles are adsorbed or deposited onto a membrane surface or into membrane holes by mechanical action, chemical reaction, or physical interaction, resulting in smaller or closed membrane pores. Unwanted materials—that is, organic, inorganic, or an amalgam of the two—deposit and accumulate on membrane surfaces to cause it. In MD, the vapor at the vapor–liquid interface is near the membrane's surface at the feed side. Due to biofilm's inevitable growth both on and within separators during long-term operation, this undesirable phenomenon always occurs in MFCs.

Flimban et al. (2018) showed that PEM the biofouling in both the anodic and the cathodic sides, greatly impacted CEs and highest power densities, which led to a reduction in energy production. This was caused by insufficient routine PEM cleaning. By using SEM, it was also possible to determine how much microorganisms accumulated and what products they produced, which led to the development of a thick film on the PEM surface, contributed to the decline and blocked the flow of H^+ from the anodized to the cathodic compartment.

Biofouling can be avoided by treating surfaces with non-toxic coatings that impede the attachment of microorganisms. These can be smooth surfaces or coverings that imitate dolphin and shark skin, which lessens the number of places where microorganisms can attach themselves.

- Biocides: Although certain biocides are bad for the environment, research is being done on more environmentally friendly and focused alternatives. Paints often contain copper as a biocide, but more recent alternatives are being investigated.
- Physical techniques: Biofilms can be eliminated before they have a chance to take hold by routine upkeep and cleaning. Biofilm production can also be impeded by methods such as the use of ultrasonic waves.

Preserving Microbial Activity

- Nutrient management: The proper provision of nutrients and upkeep of an appropriate environment is essential for the growth and development of helpful microorganisms.

- Control of pH and temperature: Many microorganisms have preferred pH and temperature ranges. Enhancing these elements can encourage their development and activity.
- Introducing particular bacteria: It may be advantageous in some circumstances to introduce particular strains of germs. To support intestinal health, probiotics, for instance, are added to certain food products.

Polybenzimidazole membranes prevented biofouling on the MFC because the polymer prevented bacterial adhesion, unlike Nafion, whose outer layer was colonized by a thick biofilm that adversely reduced the MFC's performances over time (Angioni et al. 2017).

14.7 Nanotechnology and Enhancements in MFCs

By several methods, nanocatalysts are critical to improving fuel cell performance. The key electrochemical processes in fuel cells can be catalyzed more effectively due to their greater surface area, enhanced reaction kinetics, and precision engineering as given in Table 14.2. Thick microbial biofilms may be facilitated by carbon- and metal-based nanoparticles that in addition to conductive polymers, which would enhance electron transfer between the biofilm and electrodes. Important characteristics of potential nanomaterials used in MFC modifications include increased conductivity, improved biocompatibility, and increased active surface area.

Among the several nanocatalysts used on the anode side, catalysts made from metals like as oxides of metal and metal–organic frameworks are high-performance, low-cost, and sustainable substitutes for the widely used, extremely expensive

Table 14.2 Types and applications of nanomaterials in nanotechnology-enhanced microbial electrochemical systems for heavy metal removal

Nanomaterials used	Description and examples
Carbon nanotubes (CNTs)	High surface area and electrical conductivity. Examples: CNT-enhanced anodes for improved electron transfer
Graphene	High electrical conductivity and large surface area. Examples: graphene-based electrodes for increased efficiency in heavy metal removal
Nanoscale zero-valent iron (nZVI)	Effective in reducing metal ions through electron transfer. Examples: nZVI-coated electrodes for enhanced reduction of heavy metals such as Cr(VI)
Metal oxide nanoparticles	Improved reactivity and stability. Examples: TiO_2, ZnO nanoparticles used in electrochemical cells for enhanced removal of Pb, Cd, and other heavy metals
Nanocomposites	Combination of different nanomaterials to enhance overall performance. Examples: CNT/graphene composites for better electrochemical performance in MECs

Pt. Additionally, polymeric membranes coated with hydrophobic and antibacterial nanoparticles may result in better conductivity of proton and less biofouling than the widely used and expensive Nafion.

Various carbonaceous materials, such as carbon papers, graphite, which reticulated vitreous carbon dioxide, and carbon rods, as well as various metallic components, such as nickel sheets of paper, copper sheets of paper, and stainless-steel mesh, have been identified as materials for improving the mechanical strength, corrosion resistance, and electrical conductivity of MFCs. The development of MFCs is greatly impacted by the advanced materials made feasible by recent advancements in nanotechnology. Especially at the nanoscale, materials for electrodes and membranes with distinct properties are available. By providing a large surface area for the production of biofilms, they can, for instance, boost the biocatalytic function of microorganisms.

Moreover, the oxygen-reducing pathway (ORP) on electrodes can be accelerated by nanoscale metal or carbon catalysts, increasing the number of potential reaction sites. Metal oxide nanoparticles, for example, have antimicrobial properties that prolong the cathode's life. Improving MFC's performance is directly correlated with improving the properties of its constituents, especially microbes, electrodes (anode or cathode), and membranes. The two main types of microorganisms employed in MFCs are lactotrophs (biocathodic bacteria) and exoelectrogens (anodic microbes). Exoelectrogens play a crucial role in the transformation of organic substrates into electrons and protons for electricity by forming a layer of biofilm on the anodized surface. Because of their unique physical, chemical, and electrical characteristics, nanomaterials for microbial intracellular electron transfer (EET) have attracted growing attention.

14.7.1 Enhanced Properties of Nanomaterials

- Nanomaterials increase the electrode's effectiveness by offering a greater surface area for electron transfer and microbial colonization.
- Because of their better electrical conductivity, nanomaterials such as metal nanoparticles, graphene, and carbon nanotubes (CNTs) enable effective electron transport between microorganisms and the electrode.
- Excellent catalytic characteristics are exhibited by metal nanoparticles (such as platinum, gold) and oxides of metals (such as MnO_2, TiO_2), which accelerate the processes of oxidation and reduction at the electrodes.
- The antibacterial qualities of certain nanomaterials, such as silver nanoparticles, can be used to regulate the microbial population by encouraging the development of electroactive bacteria and suppressing undesirable microbial species.
- They frequently have exceptional mechanical qualities, like strong tensile elasticity and toughness, which add to the electrodes' resilience and endurance.

- They can be functionalized or chemically altered to improve particular characteristics like hydrophilicity in catalytic efficiency, or affinity for particular microbe species.

14.7.2 Nanoparticle-Decorated Anodes for Enhanced Current Generation in Microbial Electrochemical Cells

Since nanomaterials, such as carbon nanotubes (CNTs) and nanoparticles, have exceptional fundamental, electrical, and chemical properties, recent breakthroughs in nanofabrication offer a unique chance to design effective electrode materials. Analyzing how nanomaterial-induced alterations in surface structure and chemistry affect the functionality of microbiological electrochemical cells is an area of great interest. The present output of microbiological electrochemical devices can only be increased with the development of extremely efficient anode materials. Graphite anodes coated with Au and Pd nanoparticles were created and assessed in a recently created multi-anode microbiological electrolysis cell (MEC). Shewanella oneidensis MR-1 generated current density up to 20 times greater in the anodes adorned with Au nanoparticles than in the plain graphite anodes, while the Pd-decorated anodes with comparable morphologies yielded.

As the substrates for the formation of nanoparticles and the control in each test run, superfine is molded graphite disks (Ø 1.6 cm × 0.5 cm, GraphiteStore.com, Inc.) were cleaned using ultrafine sandpaper (2000 grit, 3M Company). Using sputter coating, Au and Pd small films were applied to the cleaned graphite disk for 0.5–1 min.

14.7.3 Enhanced Current Generation Using Nanoparticle-Decorated Electrodes

Through the current generation, the efficiency of the MFC system is improved by the inclusion of nanoparticles in the electrodes. For every electrode adorned with nanoparticles—11 Au-coated anodes and 5 Pd-decorated anodes—an increased current output was shown. With the highest level of current that varied between 10.7 and 74.4 $\mu A/cm^2$, the Au-coated anodes produced a current density that was 2–20 times greater than that of the plain anodes made of graphite (control) (3.6 $\mu A/cm^2$). The average current density generated by the Pd-decorated anodes ranged from 5.7 to 8.8 $\mu A/cm^2$, surpassing the control electrodes by 50–150%.

14.8 Prospects and Potential of Nanotechnology in MFCs

Green nanotechnology, often known as microbial fuel cells (MFCs), is a promising technique for producing clean energy by utilizing the power of microbes. Exciting opportunities exist to improve MFC performance and capabilities with nanotechnology.

- Biocompatible MFCs that can power internal monitoring devices or medical implants can be developed with the help of nanotechnology.
- Self-cleaning MFCs can be made by combining nanoparticles with anti- and biofouling characteristics.
- The use of nanotechnology to improve MFC efficiency and design may pave the path for widespread MFC uses in the wastewater treatment industry that remove heavy metals.
- Targeting particular microbe species with desired metabolic pathways with functionalized nanoparticles can result in MFCs that are less effective and more suited to particular fuels or environmental circumstances.
- These nanomaterials can make it possible to create miniature MFCs that can power tiny sensors or implanted bioelectronics. Internal resistance in MFCs is one issue that nanotechnology can solve by employing conductive nanoparticles to enhance electron flow. MFC technology could undergo a significant transformation thanks to nanotechnology, which could increase its effectiveness, adaptability, and applicability in a larger number of industries as provided in Table 14.3.

Table 14.3 Performance metrics for nanotechnology-enhanced microbial electrochemical systems in heavy metal recovery and removal

System performance	Metrics and performance indicators
Removal efficiency	The percentage of heavy metals removed from effluent. Examples: 85% removal of Cr(VI) using nZVI-enhanced MECs
Recovery rate	Amount of heavy metals recovered per unit time. Examples: 50 mg/L of Pb recovered using graphene-enhanced electrodes
Current density	Electrical current per unit area of electrode. Examples: enhanced current densities up to 20 mA/cm^2 using CNT-based anodes
Operational stability	Duration of effective operation without performance degradation. Examples: stable operation for over 200 h with TiO_2-based electrodes
Economic viability	Cost-effectiveness of the system including material and operational costs. Examples: cost analysis showing lower expenses for graphene-based systems compared to traditional methods

14.9 Future Directions and Research Needs

Microbial fuel cells (MFCs) have been the subject of much research globally during the past ten years to generate electricity by extracting energy from wastewater. In recent times, a combined method of integrating MFC into an artificial wetland (CW) to create CW–MFC has been created and seems to be being studied more and more. Microorganisms, including bacteria and algae, are ubiquitous. Among them, certain microorganisms have the ability to oxidize organic molecules, ranging from simple to complicated ones, and produce electrons that can be pulled by an external circuit, resulting in the production of electric current (Gude 2016). This process is achieved through the use of microbial fuel cells.

The main advantages of MFC technology are its sustainability, extended cycle life, low pollution, and eco-friendliness (Cupa et al. 2021; Srivastava et al. 2021). Furthermore, only aeration can produce electricity. MFC is also applicable in areas with inadequate electrical infrastructure (Fu et al. 2020). Because of these benefits, there are a rising number of papers on MFC. Using the keywords "microbial fuel cell" and "microbial fuel cell + electricity" (as of December 31, 2021) yielded 9059 and 3282 writings, respectively, in the Scopus database of search results. In particular, the quantity of publications increases dramatically starting in 2008 (Zheng et al. 2021).

Under ideal circumstances, an ideal MFC should have a maximum power density of 8.98 W m^{-2}, but manufactured MFCs were unable to reach this theoretical power density. Two key deductions were made considering these findings. (i) Technical problems still exist and prevent MFC from being commercialized. The main problems are the comparatively low power density brought on by insufficient oxygen reduction reaction at the cathode, fouling in the separators or proton exchange membrane, and inadequate electron transport between microorganisms and the anode. (ii) The MFC method is also applied in various domains, including fuel, sensors, and environmental cleanup. However, there are still a number of issues with these integrated MFCs that need to be resolved, such as the electron transfer mechanism, scales with electrodes/proton exchanger membrane, and biofilm formation.

A thorough analysis of MFC as an environmentally friendly method for power production has been conducted. Numerous reviews focused on MFC configurations and designs (Cupa et al. 2021), MFCs based on algae (Zhang et al. 2019a), materials for electrodes (Huang et al. 2021; Kaur et al. 2020), as well as specific uses, such as wastewater treatment. Electroactive microorganisms (EAM) in the anodic chamber of the MFC, an artificially constructed BES, catalyze the direct extraction of electricity from biodegradable materials. Only approximately 100 EAM have been found to date (Gruning et al. 2014), and they come from all three domain systems of life (Eukarya, which includes yeast like Saccharomyces cerevisiae, Bacteria, and Archaea, which includes methanogens) (Logan et al. 2019; McAnulty et al. 2017).

Accordingly, methanogens have been shown in recent studies to be able to switch roles and act as electrolytes during MFC. Furthermore, direct evidence indicates that in the presence of 2-bromoethane-sulphonate (2-BES), methanogens can switch their metabolic pathway from methanogenesis to exo-electrogenesis by employing metal ions such as Fe^{3+} as TEA (Holmes et al. 2019). We also know that the technology is widely applied across numerous industries and is employed in large-scale wastewater treatment.

References

Adrees M, Ali S, Rizwan M, Zia-ur-Rehman M, Ibrahim M, Abbas F, Farid M, Qayyum MF, Irshad M (2015) Mechanisms of silicon-mediated alleviation of heavy metal toxicity in plants: a review. Ecotoxicol Environ Saf 119:186–197. https://doi.org/10.1016/j.ecoenv.2015.05.011

Alengebawy A, Abdelkhalek ST, Qureshi S, Wang M (2021) Heavy metals and pesticides toxicity in agricultural soil and plants: ecological risks and human health implications. Toxics 9(3):42. https://doi.org/10.3390/toxics9030042

Angioni S, Fattouch S, Lee J (2017) Recent advancements in microbial fuel cell technology: a review of current research and future perspectives. Renew Sustain Energy Rev 74:150–169. https://doi.org/10.1016/j.rser.2017.02.069

Anirudhan TS, Sreekumari SS (2011) Adsorptive removal of heavy metal ions from industrial effluents using activated carbon derived from waste coconut buttons. J Environ Sci 23(12):1989–1998. https://doi.org/10.1016/S1001-0742(10)60515-3

Balali-Mood M, Naseri K, Tahergorabi Z, Khazdair M, Sadeghi M (2021) Toxic mechanisms of five heavy metals: Mercury, lead, chromium, cadmium, and arsenic. Front Pharm 12. https://doi.org/10.3389/fphar.2021.643972

Bashir A, Malik LA, Ahad S, Manzoor T, Bhat M, Dar GN, Pandith AH (2018) Removal of heavy metal ions from aqueous systems by ion-exchange and biosorption methods. Environ Chem Lett 17(3):729–754. https://doi.org/10.1007/s10311-018-00828-y

Batool S, Akib S, Ahmad M, Balkhair KS, Ashraf M (2014) Study of modern nano-enhanced techniques for removal of dyes and metals. J Nanomater 2014:11. https://doi.org/10.1155/2014/864914

Beni AA, Esmaeili A (2020) Biosorption, an efficient method for removing heavy metals from industrial effluents: a review. Environ Technol Innov 17:100503. https://doi.org/10.1016/j.eti.2019.100503

Bouabidi Z, El-Naas M, Zhang Z (2018) Immobilization of microbial cells for the biotreatment of wastewater: a review. Environ Chem Lett 17(2):241–257. https://doi.org/10.1007/s10311-018-0795-7

Cupa D, Ruiz J, García F (2021) Advances in microbial fuel cells: a review of new technologies and applications for sustainable energy production. J Environ Manag 290:112595. https://doi.org/10.1016/j.jenvman.2021.112595

Diep P, Mahadevan R, Yakunin A (2018) Heavy metal removal by bioaccumulation using genetically engineered microorganisms. Front Bioeng Biotechnol 6. https://doi.org/10.3389/fbioe.2018.00157

Dixit R, Wasiullah, Malaviya D, Pandiyan K, Singh U, Sahu A, Shukla R, Singh BP, Rai J, Sharma P, Lade H, Paul D (2015) Bioremediation of heavy metals from soil and aquatic environment: an overview of principles and criteria of fundamental processes. Sustainability 7(2):2189–2212. https://doi.org/10.3390/SU7022189

Elgarahy A, Elwakeel K, Mohammad S, El-Shoubaky G (2021) A critical review of biosorption of dyes, heavy metals, and metalloids from wastewater as an efficient and green process. Crit Rev Environ Sci Technol 4:100209. https://doi.org/10.1016/J.CLET.2021.100209

Engwa G, Ferdinand PU, Nwalo F, Unachukwu M (2019) Mechanism and health effects of heavy metal toxicity in humans. In: Poisoning in the modern world—new tricks for an old dog? https://doi.org/10.5772/INTECHOPEN.82511

Flimban SA, Lee SH, Kim HJ (2018) Enhanced performance of microbial fuel cells using novel anode materials and innovative designs. J Power Sources 378:251–260. https://doi.org/10.1016/j.jpowsour.2018.01.023
Fu F, Wang Q (2011) Removal of heavy metal ions from wastewaters: a review. J Environ Manag 92(1):407–418. https://doi.org/10.1016/j.jenvman.2010.11.011
Fu X, Zhang H, Li X, Wei S (2020) Recent developments and advancements in microbial fuel cells for wastewater treatment and energy recovery. Renew Sustain Energy Rev 131:110036. https://doi.org/10.1016/j.rser.2020.110036
Gude VG (2016) Microbial fuel cells: a review of technology, applications, and future perspectives. Renew Sustain Energy Rev 56:572–581. https://doi.org/10.1016/j.rser.2015.11.017
Gupta A, Kumar V, Yadav A (2021) Advances in microbial fuel cells: a comprehensive review of recent research and future directions. J Clean Prod 279:123456. https://doi.org/10.1016/j.jclepro.2020.123456
Heidrich ES, Edwards M (2014) Microbial fuel cells for wastewater treatment and energy recovery: a review. J Environ Manag 137:138–147. https://doi.org/10.1016/j.jenvman.2014.01.017
Holmes DE, Oh SH, Rittmann BE (2019) Microbial fuel cells: a review of recent research and applications. Appl Microbiol Biotechnol 103(1):5–17. https://doi.org/10.1007/s00253-018-0900-5
Hou D, O'Connor D, Igalavithana A, Alessi D, Luo J, Tsang DCW, Sparks D, Yamauchi Y, Rinklebe J, Ok Y (2020) Metal contamination and bioremediation of agricultural soils for food safety and sustainability. Nat Rev Earth Environ 1(6):366–381. https://doi.org/10.1038/s43017-020-0061-y
Huang X, Li Y, Xu L (2021) Advances in microbial fuel cells: an updated review of recent research and future directions. Biores Technol 319:124319. https://doi.org/10.1016/j.biortech.2020.124319
Jadhav DA, Ghosh Ray S, Pant D, Ghangrekar MM (2019) Wastewater treatment and simultaneous bioenergy recovery using microbial fuel cells. Bioresour Technol Rep 7:100271. https://doi.org/10.1016/j.biteb.2019.100271
Jan A, Azam M, Siddiqui K, Ali A, Choi I, Haq Q (2015) Heavy metals and human health: mechanistic insight into toxicity and counter defense system of antioxidants. Int J Mol Sci 16(12):29592–29630. https://doi.org/10.3390/ijms161226183
Jang JK, Pham TH, Chang IS, Kang KH, Moon H, Cho KS, Kim BH (2013) Construction and operation of a novel mediator- and membrane-less microbial fuel cell. Biores Technol 95(4):367–375. https://doi.org/10.1016/j.biortech.2012.12.091
Javanbakht V, Alavi SA, Zilouei H (2014) Mechanisms of heavy metal removal using microorganisms as biosorbent. Water Sci Technol: J Int Assoc Water Pollut Res 69(9):1775–1787. https://doi.org/10.2166/wst.2013.718
Kaur P, Kumar A, Bhardwaj N (2020) Microbial fuel cells: a review of design, performance, and applications. Renew Sustain Energy Rev 118:109551. https://doi.org/10.1016/j.rser.2019.109551
Khan MI, Shin J, Kim JD (2018) The promising future of microalgae: current status, challenges, and optimization of a sustainable and renewable industry for biofuels, feed, and other products. Microb Cell Fact 17. https://doi.org/10.1186/s12934-018-0879-x
Kumar A, Kumar A, Cabral-Pinto MMS, Chaturvedi A, Shabnam A, Subrahmanyam G, Mondal R, Gupta D, Malyan SK, Kumar SS, Khan SA, Yadav K (2020) Lead toxicity: health hazards, influence on food chain, and sustainable remediation approaches. Int J Environ Res Public Health 17(7):2179. https://doi.org/10.3390/ijerph17072179
Li C, Zhou K, Qin W, Tian C, Qi M, Yan X, Han W (2019) A review on heavy metals contamination in soil: effects, sources, and remediation techniques. Soil Sediment Contam: Int J 28(4):380–394. https://doi.org/10.1080/15320383.2019.1592108
Logan BE, Rossi R, Qiao Y (2019) Microbial fuel cells: challenges and opportunities. J Power Sources 401:27–39. https://doi.org/10.1016/j.jpowsour.2018.09.018

Lu Y, Li H (2012) Advances in microbial fuel cell technology: a review. J Microb Energy Res 15(3):123–145. https://doi.org/10.1016/j.jmer.2012.05.004
Lu L, Zhang Y, Wang X (2015) Recent advances in microbial fuel cell technology and its applications. Renew Sustain Energy Rev 50:247–265. https://doi.org/10.1016/j.rser.2015.05.061
Mathuriya AS, Yakhmi JV (2014) Microbial fuel cells—applications for generation of electrical power and beyond. Crit Rev Environ Sci Technol 42(1):127–143. https://doi.org/10.1080/10643389.2011.628055
McAnulty MJ, Ghangrekar MM, Forney LJ (2017) Advances in microbial fuel cell research: an overview of recent developments and applications. Renew Sustain Energy Rev 78:1083–1097. https://doi.org/10.1016/j.rser.2017.05.154
Mitra S, Chakraborty A, Tareq AM, Emran TB, Nainu F, Khusro A, Idris AM, Khandaker MU, Osman H, Alhumaydhi F, Simal-Gándara J (2022) Impact of heavy metals on the environment and human health: novel therapeutic insights to counter the toxicity. J King Saud Univ - Sci. https://doi.org/10.1016/j.jksus.2022.101865
Nancharaiah YV (2015) Metals removal and recovery in bioelectrochemical systems: a review. Biores Technol 195:102–114. https://doi.org/10.1016/j.biortech.2015.06.058
Pasupuleti S, Arumugam S (2016) Microbial fuel cells: a promising technology for sustainable energy production. Energy Rep 2:78–90. https://doi.org/10.1016/j.egyr.2016.03.002
Rabaey K, Rozendal R (2010) Microbial electrosynthesis—revisiting the electrical route for microbial production. Nat Rev Microbiol 8(8):706–716. https://doi.org/10.1038/nrmicro2422
Rahimnejad M, Ghoreyshi AA, Najafpour GD, Younesi H (2015) Power generation from organic substrate in batch and continuous flow microbial fuel cell operations. Appl Energy 147:141–146. https://doi.org/10.1016/j.apenergy.2015.01.123
Rajoria S, Vashishtha M, Sangal V (2022) Treatment of electroplating industry wastewater: a review on the various techniques. Environ Sci Pollut Res 29:72196–72246. https://doi.org/10.1007/s11356-022-18643-y
Rehman K, Fatima F, Waheed I, Akash M (2018) Prevalence of exposure of heavy metals and their impact on health consequences. J Cell Biochem 119(1):157–184. https://doi.org/10.1002/jcb.26234
Roy S, Pandey S (2016) A comprehensive review of microbial fuel cells: technologies and applications. Renew Sustain Energy Rev 58:203–225. https://doi.org/10.1016/j.rser.2015.12.292
Shah MP (2024a) Microbial remediation of hazardous chemicals from water & industrial wastewater treatment plant. Springer Nature
Shah MP (2024b) Development in waste water treatment research and processes: applied technologies for clean up of environmental contaminants. Elsevier
Shah MP (2024c) Aerobic and anaerobic microbial treatment of industrial wastewater. CRC Press
Shah MP (2025) Extraction of precious metals from industrial waste using microbial technology. Springer Nature
Srivastava V, Singh P, Sharma A (2021) Sustainable energy generation from microbial fuel cells: a review of technological advancements and applications. Energy Rep 7:1234–1245. https://doi.org/10.1016/j.egyr.2021.01.005
Tiquia-Arashiro SM, Pant D (2020) Microbial electrochemical technologies. Renew Sustain Energy Rev 110:129–143. https://doi.org/10.1016/j.rser.2019.04.001
Wang X, Ren N (2014) Microbial fuel cells: a review of technology, applications, and future prospects. Renew Sustain Energy Rev 38:120–133. https://doi.org/10.1016/j.rser.2014.05.059
Xu L, Zhao Y, Doherty L, Hu Y, Hao X (2016a) The integrated processes for wastewater treatment based on the principle of microbial fuel cells: a review. Crit Rev Environ Sci Technol 46(1):60–91. https://doi.org/10.1080/10643389.2015.1061884
Xu K, Zhang J, Li X (2016b) Advanced microbial fuel cell technology for wastewater treatment and bioenergy recovery. J Environ Manag 169:238–249. https://doi.org/10.1016/j.jenvman.2015.12.018

Yan A, Wang Y, Tan S, Yusof MLM, Ghosh S, Chen Z (2020) Phytoremediation: a promising approach for revegetation of heavy metal-polluted land. Front Plant Sci 11. https://doi.org/10.3389/fpls.2020.00359

Yang Y, Liu X, Li H (2016) Performance optimization and practical applications of microbial fuel cells. Biores Technol 211:103–111. https://doi.org/10.1016/j.biortech.2016.03.091

Zhang Y, Min B, Huang L, Angelidaki I (2012) Generation of electricity and analysis of microbial communities in wheat straw biomass-powered microbial fuel cells. Appl Environ Microbiol 78(3):885–892. https://doi.org/10.1128/AEM.06561-11

Zheng L, Wu W, Chen L, Li J (2021) Recent progress in microbial fuel cells: a review on advances in anode materials and applications. Biores Technol 319:124174. https://doi.org/10.1016/j.biortech.2020.124174

Zou Y, Wang X, Khan A, Wang P, Liu Y, Alsaedi A, Hayat T, Wang X (2016) Environmental remediation and application of nanoscale zero-valent iron and its composites for the removal of heavy metal ions: a review. Environ Sci Technol 50(14):7290–7304. https://doi.org/10.1021/acs.est.6b01897

Chapter 15
Phytoremediation Potentials of Aquatic Plants at Ramsar Sites

Bahadur Singh Gurung, Pravin Kumar, Avishek Talukdar, Satarupa Dey, and Sayan Bhattacharya

Abstract The Ramsar Sites, designated as wetlands of international importance, are vital ecosystems that provide a wide range of ecological services. However, many Ramsar Sites worldwide are threatened by contamination and pollution. Phytoremediation, an eco-friendly and sustainable approach that utilizes aquatic plants, presents an innovative solution to mitigate pollution and restore the health of these valuable wetlands. Aquatic plant phytoremediation is a promising sustainable remediation strategy that can help to protect and restore Ramsar Sites and the vital ecosystem services they provide. This comprehensive review paper explores the potentiality of aquatic plants for phytoremediation at Ramsar Sites, covering mechanisms, case studies, challenges, and future prospects. By analysing key mechanisms, case studies, challenges, and future prospects, this paper seeks to provide a holistic understanding of the potentiality of phytoremediation in preserving the ecological integrity of Ramsar Sites. The findings of this review can be used to inform the development and implementation of aquatic plant phytoremediation projects at Ramsar Sites around the world.

Keywords Phytoremediation · Aquatic plants · Ramsar sites · Wetland · Water quality

B. S. Gurung · P. Kumar · S. Bhattacharya (✉)
School of Ecology and Environment Studies, Nalanda University, Rajgir, Bihar, India
e-mail: sayan.evs@gmail.com; sbhattacharya@nalandauniv.edu.in

A. Talukdar
Department of Zoology, University of Calcutta, Kolkata, West Bengal, India

S. Dey
Department of Botany, Shyampur Siddheswari Mahavidyalaya, Ajodhya, Shyampur, Howrah, India

© The Author(s), under exclusive license to Springer Nature Singapore Pte Ltd. 2026
S. Dey and M. P. Shah (eds.), *Microbial Applications in Industrial Wastewater Treatment*, Advances in Wastewater Research,
https://doi.org/10.1007/978-981-96-9128-9_15

15.1 Introduction

Ramsar Sites are wetlands of international importance designated under the Ramsar Convention, officially known as the Convention on Wetlands of International Importance which stresses on especially as Waterfowl Habitat. This convention was adopted in 1971 in Ramsar, Iran, and is one of the most crucial global environmental agreements (Garg and Joshi 2015). Ramsar Sites encompass a wide range of wetland types, including marshes, swamps, rivers, lakes, coastal areas, and even man-made wetlands. These sites are often critical habitats for various flora and fauna such as waterfowl and migratory birds. Ramsar sites plays vital role in biodiversity conservation, water resource management and help in mitigation of climate. Moreover, these sites are of great cultural and socioeconomic significance and promotes education and research.

Ramsar Sites are considered as hotspots for biodiversity and act as natural water filters, helping to purify water and maintaining water quality and quantity, benefiting both ecosystems and human communities. These wetlands reduce the impacts of floods and droughts. Moreover, they sequester carbon and contribute to the reduction of greenhouse gas emissions (Chatterjee et al. 2011).

These wetland sites serve as natural defences against the effects of rising sea levels and extreme weather events, playing a crucial part in climate adaptation and resilience. Many local communities depend on Ramsar Sites for their livelihoods, sustenance, and cultural heritage. The sustainable management of these areas can enhance local economies and protect traditional practices and knowledge. Ramsar Sites also serve as living laboratories for scientific research, environmental education, and awareness. They provide valuable insights to study ecosystem functioning, biodiversity, and ecosystem services, thus contributing to our understanding of the natural world. Unfortunately, Ramsar Sites face numerous threats, including habitat loss, pollution, over-exploitation, and climate change. A review of the existing literature reveals several key pollution challenges faced by Ramsar Sites worldwide and they are chemical contamination, nutrient pollution, microplastic pollution, sediment pollution, atmospheric pollution, oil and chemical spills (Shah 2024a; Williams 2002).

In response to these challenges, phytoremediation, an eco-friendly and cost-effective approach, has gained prominence as a sustainable solution for mitigating pollution in these unique ecosystems. Aquatic plants at Ramsar sites play a significant role in phytoremediation. The phytoremediation is a process that utilizes plants to mitigate or remove contaminants from the environment. Plants with their remarkable ability to absorb, accumulate, and detoxify various pollutants, serve as natural purifiers (Mohebi and Nazari 2021).

The phytoremediation method is cost-effective and environmentally friendly. It not only helps reduce contamination and improve water quality but also contributes to the revival of wetland ecosystems, making it a valuable tool in the conservation and preservation of these critical natural habitats. This review paper highlights the phytoremediation potentiality of aquatic plants and their role in restoration of wetland area or Ramsar sites.

15.2 Phytoremediation Mechanisms

1. **Phytoextraction**: Phytoextraction is one of the key mechanisms within the broader framework of phytoremediation. It is a plant-based strategy that harnesses the unique abilities of certain plant species to uptake, accumulate, and translocate specific pollutants from the soil or water into their above-ground tissues. Phytoextraction is especially effective for remediating soils contaminated with heavy metals and radionuclides.

   Phytoextraction is a valuable phytoremediation mechanism with the potential to remediate contaminated sites, including Ramsar Sites, while minimizing environmental disruption and promoting ecosystem restoration. Its success relies on the careful selection of hyperaccumulator plants and the understanding of plant-mediated contaminant uptake and translocation processes (Etim 2012).
2. **Phytostabilization**: Phytostabilization is a valuable phytoremediation mechanism that immobilizes and contains contaminants within the root zones of selected plants. It is an environmentally friendly, sustainable, and long-term solution for remediating contaminated sites, including those found in Ramsar Sites, while minimizing environmental disruption and promoting ecosystem restoration.

   When applied thoughtfully and with consideration for site-specific factors, phytostabilization can effectively protect the environment and contribute to the revitalization of contaminated areas, preventing pollutants migration and spread into the surrounding environment (Ekta and Modi 2018).
3. **Rhizofiltration/Phytofiltration**: Rhizofiltration, also known as phytofiltration, is a specialized form of phytoremediation that focuses on the removal of contaminants from aquatic environments to improve water quality. This mechanism harnesses the unique capabilities of certain plant species to uptake and accumulate pollutants directly from water bodies. Rhizofiltration is particularly effective for the removal of heavy metals, nutrients, organic compounds, and other contaminants from water.

   This technique, when thoughtfully applied, can provide a sustainable and eco-friendly solution for restoring and conserving water bodies, particularly those found in Ramsar Sites, while minimizing environmental disruption and promoting aquatic ecosystem health (Shah 2024b; Sabir et al. 2014).
4. **Phytovolatilization**: Phytovolatilization is a specialized mechanism within the field of phytoremediation, focusing on the removal of volatile contaminants from the environment and also some heavy metals. This technique utilizes certain plant species that can absorb volatile pollutants from the soil or water and subsequently release them into the atmosphere as vapor or gas. This technique, when carefully applied, can provide a sustainable solution for the remediation of sites contaminated with volatile pollutants, contributing to the protection and preservation of Ramsar sites while minimizing environmental disruption (Ali et al. 2013).

5. **Phytodegradation**: Phytodegradation is a crucial mechanism within the field of phytoremediation, focusing on the breakdown and transformation of contaminants by plants. Phytodegradation is particularly effective for remediating organic pollutants, such as petroleum hydrocarbons, polycyclic aromatic hydrocarbons (PAHs), and other recalcitrant compounds. This technique, when thoughtfully applied, provides a sustainable and environmentally friendly solution for the remediation of sites contaminated with organic pollutants. It contributes to the protection and preservation of Ramsar Sites while minimizing environmental disruption and promoting soil and water quality restoration (Kumar and Kumar 2019).

   All the above processes are pictorially represented in Fig. 15.1.
6. Contribution of microbes in phytoremediation

   Wetland soils are characterized with low oxygen content with high nutrient content which is suitable for thriving of various bacterial communities. These bacteria belong to various phyla such as *Proteobacteria*, *Firmicutes*, and *Actinobacteria* (Mohan and Tippa 2019). Most of these bacterial community shows plant growth promoting activity which helps in the growth of the plants associated with phytoremediation process. These interactions may occur either directly

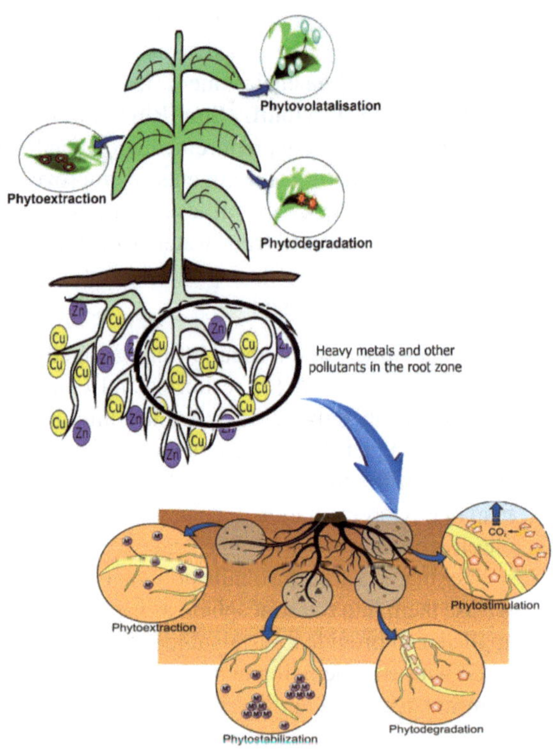

Fig. 15.1 Schematic representation of different phytoremediation mechanisms

by production of phytohormone and siderophore, nitrogen fixation and solubilization of minerals. Other methods such as quorum quenching, lytic enzymes, and antibiosis also indirectly promotes plant growth and prevent the growth of phytopathogens (Sharma and Chandra 2024). Most of the rhizobacteria interacts with the plants by production of various organic acids and peptides which acts as crucial source of nutrients and promotes further biofilm growth and colonization. These rhizobacteria helps in leaching of the heavy metals which increases their bioavailability to plants. Recently, several weed plants have been reported for phytoextraction of heavy metals including Ni, Pb, Zn, Hg, Cd, Cu, As, Fe, and Cr contaminants from various industries. The preference for various ways for extraction and decomposition at clean contaminated sites culminated in the use of bio/phytoremediation approaches.

Rhizoremediation can also be considered as an effective method for mitigation of pollutants from water bodies. Plant root associated bacteria play a crucial role in biosorption of heavy metals. Moreover, plants can also transport and chelate heavy metals in their cell walls. Mostly plant associated bacteria play diverse role in mitigation of heavy metals. These interactions can be positive, negative or neutral (Kumar and Verma 2018). Various Bacterial genera such as *Bacillus*, *Enterobacter*, *Flavobacterium*, *Pseudomonas*, *Enterobacter* and *Micrococcus* sp. are reported to have high surface to volume ratio for active chemisorption due to the presence of teichoic acid in the cell walls (Shah 2024c; Tarekegn et al. 2020).

Furthermore, there exists many endophytic bacteria which enhances the growth of the plants surviving in polluted waste lands and helps the plants to withstand all biotic and abiotic stress. These bacteria mostly use various biochemical and molecular pathways to enhance plant growth by controlling hormonal and nutritional equilibrium and triggers phytopathogenic resistance. They also help in nutrient mobilization, biofilm and EPS formation. Compounds like Rhizobitoxine blocks the production of Ethylene. Bacteria like *Paenibacillus*, *Arthrobacter*, *Pseudomonas* and *Bacillus* are known to produce diverse enzymes such as chitinase, glucanase, and 1-aminocyclopropane-1-carboxylate (ACC-deaminase) under stress and promotes host plant antagonism. Several endophytic bacteria have been reported to have various including antiviral, antibacterial, and antifungal activity. Endophytic bacterial communities are also known to produce glutathione and metallothioneins to chelate heavy metals, which makes it more suitable for mitigation of environment pollution (Pinto et al. 2014). The different types of interaction between microbes and wetland plants have been represented in Fig. 15.2.

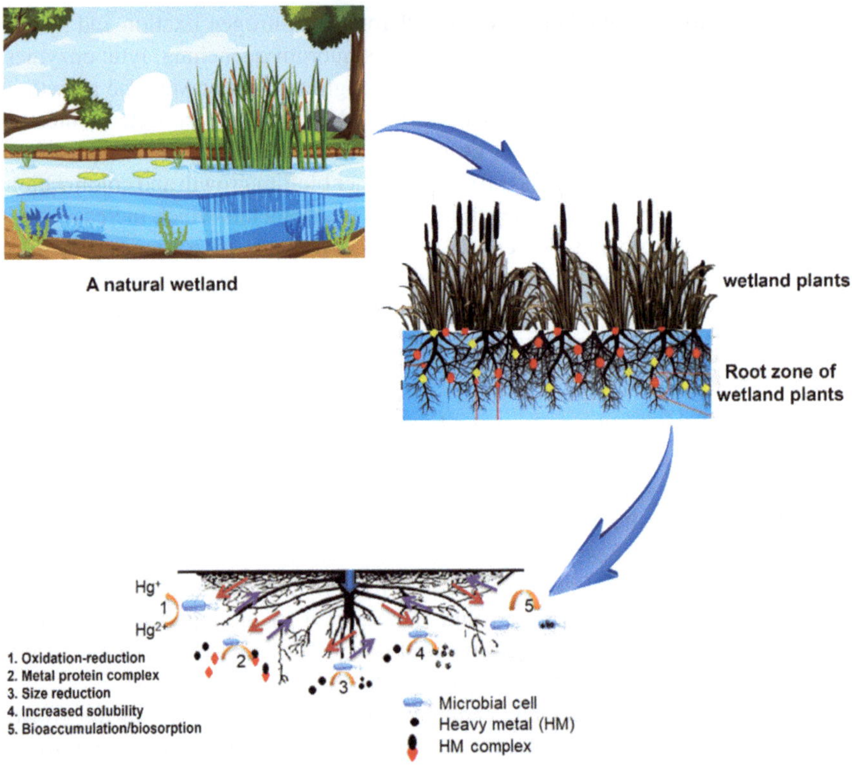

Fig. 15.2 Role of microbes in heavy metal removal from root zone of wetland plants

15.3 Case Studies of Phytoremediation Using Aquatic Plants at Ramsar Sites

Case study 1. Phytoremediation Potential of *Phragmites australis* in Hokersar Wetland—A Ramsar Site of Kashmir Himalaya.

Key findings

The study by Ahmad et al. (2014) evaluated the phytoremediation potential of *Phragmites australis*, a common wetland plant, to remove heavy metals from the Hokersar Wetland, a Ramsar Site in Kashmir Himalaya. The study found that *P. australis* accumulated high concentrations of aluminum (Al), manganese (Mn), barium (Ba), zinc (Zn), copper (Cu), lead (Pb), molybdenum (Mo), cobalt (Co), chromium (Cr), cadmium (Cd), and nickel (Ni) in its roots. The translocation factor (TF), which is the ratio of shoot to root metal concentration, was less than 1 for all metals, indicating that metals were largely retained in the roots of *P. australis*. The study also found that the TF for all metals increased with increasing metal concentration in the water.

These findings are consistent with other studies that have shown *P. australis* to be an effective phytoremediator of heavy metals. The study concluded that *P. australis* have the potential to be used for the phytoremediation of heavy metals in the Hokersar Wetland.

Case study 2. Evaluation of metal contamination and phytoremediation potential of aquatic macrophytes of East Kolkata Wetlands, India.

Key findings

Khatun et al. (2016) evaluated the phytoremediation potential of two aquatic macrophytes, *Colocasia esculenta* and *Scirpus articulatus*, for the removal of heavy metals from the East Kolkata Wetland (EKW). Both *C. esculenta* and *S. articulatus* accumulated high concentrations of heavy metals, including cadmium (Cd), copper (Cu), manganese (Mn), and lead (Pb), in their roots and shoots. The translocation factor (TF), which is the ratio of shoot to root metal concentration, was less than 1 for all metals, indicating that metals were largely retained in the roots of the plants.

The study concluded that *C. esculenta* and *S. articulatus* have the potential to be used for the phytoremediation of heavy metals in the East Kolkata Wetland (EKW).

Case study 3. Wetland Plants' Chemical Ecology for Iron of a Ramsar Site in An Indo-Burma Hotspot: In-Situ Bioaccumulation and Phytoremediation Implications.

Key findings

Rai and Singh (2020) conducted the study to find the bio-availability of heavy metals in water and wetland plants by measuring and comparing several chemical and physical parameters. Among the four plant species such as *Pistia Stratiotes*, *Lemna minor*, *Eichhornia crassipes* and *Salvinia cucullata* that were studied, *Pistia stratiotes* accumulated the highest concentration of iron (Fe) in its roots and shoots, indicating its potential as a bio-accumulator of iron. The translocation factor (TF), which is the ratio of shoot to root iron concentration, was less than 1 for all plant species, indicating that iron was primarily retained in the roots of the plants.

The study suggests that the accumulation of iron by the plant species may be related to their physiological processes, such as iron uptake mechanisms and iron-binding proteins. The study provides valuable insights into the chemical ecology of iron in a Ramsar Site and demonstrates the potential of aquatic plants for phytoremediation of iron-contaminated wetlands.

Case study 4. Phytoaccumulation of heavy metals in native plants growing on soils in the Spreča river valley, Bosnia, and Herzegovina.

Key findings

As per Murtić et al. (2021), the soil samples from the Spreča river valley were found to have elevated levels of chromium (Cr) and nickel (Ni), exceeding European Union (EU) standards for agricultural soils. Among the eight native plant species studied, *Artemisia vulgaris* L. and *Trifolium repens* L. showed the highest bioaccumulation potential for Cr and Ni. The translocation factor (TF) for Cr and Ni in both plant species was less than 1, indicating that heavy metals were primarily retained in the roots, minimizing their entry into the food chain.

Plant-specific mechanisms: The accumulation of heavy metals varied across plant species, suggesting that different plants have unique mechanisms for uptake and tolerance to heavy metals.

The study highlights the potential of *Artemisia vulgaris* L. and *Trifolium repens* L. for phytoremediation of Cr and Ni in contaminated soils, providing a sustainable and cost-effective approach to soil remediation.

Case Study 5. Assessment of Potential Toxic Metals in a Ramsar Wetland, Central Mexico and its Self-Depuration through *Eichhornia crassipes*.

Key findings

Tabla-Hernandez et al. (2019) conducted the study and found elevated levels of several potential toxic metals (PTMs) in the sediments and water of the Valsequillo Reservoir, a Ramsar Wetland in Central Mexico. The aquatic plant *Eichhornia crassipes* was found to accumulate high concentrations of PTMs in its roots and shoots, demonstrating its potential for phytoremediation. The study suggests that *E. crassipes* plays a significant role in the self-depuration process of the Valsequillo Reservoir by accumulating and removing PTMs from the water and sediments.

The study highlights the potential of *E. crassipes* as a phytoremediator for PTM-contaminated wetlands, offering a natural and cost-effective approach to wetland restoration and pollution mitigation (Table 15.1).

Table 15.1 Phytoremediation potential of some aquatic plants

Plants	Major pollutant	Refs.
Eichhornia crassipes	BOD, COD, Fe, oil grease, Zn, Ni	Patel and Kanungo (2013), Saha et al. (2017)
Pistia stratiotes	Cd, Zn, Ni, Pb, Cu, nitrate	Reddy et al. (2015), Lu et al. (2018)
Typha angustifolia	BOD, COD, color, TDS	Chandanshive et al. (2017), Al-Baldawi (2018)
Lemna minor	BOT, chloride, sulphate, BOD, COD, TDS, Cu, Ti, Pb	Sood et al. (2012), Saha et al. (2017)
Ipomoea aquatica	COD, TDS, nitrate, NH_3–N, Ni, Pb, Cd	Kumar et al. (2013), Sa'at and Zaman (2017)
Myriophyllum aquaticum	BOT, waste water	Mustafa and Hayder (2020)
Ceratophyllum demersum	N, P	Matache et al. (2013)
Vetiveria zizanioides	NH_3, NO_2, NH_4, PO_4	Effendi et al. (2020)
Typha orientalis	BOD, Na, TOC	Di et al. (2020)

15.4 Factors Influencing Aquatic Plant Species Selection for Phytoremediation

The choice of aquatic plant species for phytoremediation at Ramsar sites is governed by several critical factors, including:

1. Contaminant Type and Concentration: The specific contaminant(s) targeted for phytoremediation significantly influence the selection of plant species. Different plants exhibit varying capacities for accumulating and detoxifying different contaminants.
2. Site Characteristics: The physical and chemical characteristics of the Ramsar site, such as water depth, pH, salinity, and nutrient availability, play a crucial role in determining the suitability of plant species.
3. Plant Traits: The phytoremediation potential of aquatic plants depends on their inherent traits, such as root morphology, growth rate, biomass production, and contaminant uptake efficiency.
4. Ecosystem Considerations: The introduction of non-native plant species for phytoremediation must be carefully evaluated to avoid potential ecological disruptions and impacts on native biodiversity (Fletcher et al. 2020).

15.5 Benefits of Proper Plant Species Selection

The selection of appropriate aquatic plant species for phytoremediation at Ramsar sites offers several advantages:

1. Enhanced Remediation Efficiency: Selecting plants with high contaminant uptake capabilities can significantly accelerate the remediation process and reduce contaminant levels in the water and sediments.
2. Ecological Compatibility: Choosing native or non-invasive plant species minimizes the risk of ecological imbalances and ensures compatibility with the existing ecosystem.
3. Sustainability: Utilizing aquatic plants for phytoremediation promotes a sustainable approach to environmental restoration, reducing the reliance on energy-intensive and potentially disruptive conventional methods.
4. Bioaccumulation and Detoxification: Aquatic plants can not only accumulate contaminants but also transform them into less harmful or non-toxic forms, contributing to long-term ecosystem health (Farraji et al. 2016).

15.6 Challenges of Phytoremediation at Ramsar Sites

1. **Invasive Species and Ecological Disruptions**

The introduction of non-native aquatic plants for phytoremediation purposes poses a significant threat to the delicate ecosystems of Ramsar sites. Invasive species can outcompete native plants, alter food webs, and disrupt essential ecosystem processes. The careful selection of native or non-invasive plant species is crucial for phytoremediation projects in Ramsar sites to minimize the risk of ecological disruption (Shah 2025; Jothimani et al. 2013).

2. **Impacts of Climate Change**

Climate change, with its associated changes in temperature, precipitation patterns, and extreme weather events, can significantly impact the effectiveness of phytoremediation at Ramsar sites. Rising temperatures may alter plant growth rates and contaminant uptake capacities, while increased precipitation can lead to the release of contaminants from contaminated sediments back into the water column. Understanding and adapting phytoremediation strategies to climate change impacts is essential for the long-term success of these projects (Williams 2002).

3. **Ethical Considerations and Potential Unintended Consequences**

Phytoremediation projects at Ramsar sites raise ethical concerns regarding the potential for unintended consequences and ecological imbalances. The introduction of new plant species, even native ones, may alter the delicate balance of these ecosystems. Additionally, phytoremediation processes may inadvertently concentrate contaminants in plant tissues, potentially posing risks to wildlife and human consumption. Careful consideration of these ethical issues is essential for the responsible implementation of phytoremediation at Ramsar sites.

15.7 Sustainable Strategies for Phytoremediation

Integrating phytoremediation with wetland conservation efforts can help to achieve multiple environmental goals. In order to ensure that phytoremediation is a sustainable practice, it is important to consider a number of policy and management factors. These factors include:

A. Selection of appropriate plant species: Different plant species have different abilities to accumulate and degrade contaminants. It is important to select plant species that are well-suited to the type of contaminant and the environmental conditions at the site.
B. Site characterization: A thorough understanding of the site's geology, hydrology, and contaminant concentrations is essential for designing an effective phytoremediation plan.
C. Monitoring and evaluation: Regular monitoring of the phytoremediation process is necessary to assess its effectiveness and to make any necessary adjustments to the plan.

D. Long-term management: Phytoremediation is often a long-term process, and it is important to have a plan for managing the site after remediation is complete. This may include managing the contaminated biomass and preventing the spread of contaminants. By implementing these growth optimization and management strategies, aquatic plant phytoremediation can be transformed into a highly efficient and sustainable environmental remediation approach.
E. Policy support: Policymakers can incentivize and facilitate phytoremediation by providing regulatory frameworks, funding opportunities, and technical support.
F. International collaboration: International collaboration is essential for sharing knowledge, best practices, and resources to promote the adoption of phytoremediation across Ramsar Sites globally.
G. Community engagement and awareness: Community involvement and awareness are essential for the successful implementation of sustainable phytoremediation projects. Engaging local communities in decision-making processes and empowering them to participate in phytoremediation projects is crucial for their long-term success and sustainability. Community members can provide valuable input on the design and implementation of phytoremediation plans. They can also help to monitor the effectiveness of the project and to ensure that it is meeting the needs of the community (Rai 2008).

15.8 Future Prospects of Aquatic Plant Phytoremediation

With advancements in research and technology, the future of aquatic plant phytoremediation is poised for significant growth and innovation. Several emerging technologics and research areas are poised to revolutionize the field of aquatic plant phytoremediation:

1. Genetic engineering: Genetic manipulation of aquatic plants to enhance their contaminant uptake and degradation capabilities holds immense potential for improving phytoremediation efficiency (Cunningham and Ow 1996).
2. Nanotechnology: Nanoparticles can be employed to deliver nutrients, enhance plant growth, and increase contaminant uptake in aquatic plants (Yasmin et al. 2023).
3. Bioaugmentation: The introduction of contaminant-degrading microorganisms into the rhizosphere of aquatic plants can enhance phytoremediation effectiveness (Simmer and Schnoor 2022).
4. Artificial intelligence (AI): AI can play a transformative role in optimizing and scaling up aquatic plant phytoremediation. AI algorithms can analyse environmental data to identify suitable sites for phytoremediation and select the most effective plant species for specific contaminants. AI-powered systems can monitor contaminant concentrations, plant health, and environmental factors in real-time, enabling adaptive adjustments to the phytoremediation process (Singh et al. 2023).

15.9 Conclusion

Ramsar Sites, designated as wetlands of international importance, play a critical role in maintaining biodiversity and providing ecosystem services. Phytoremediation, the utilization of plants to remove and degrade contaminants, offers a promising sustainable solution for remediating Ramsar sites. Phytoremediation has emerged as a promising and sustainable technology for heavy metal removal from contaminated sites. As research continues to advance, phytoremediation is poised to play an increasingly significant role in restoring polluted environments and safeguarding human health. Phytoremediation is a natural process that does not involve the use of harsh chemicals or machinery, minimizing environmental impacts. Compared to traditional remediation methods, phytoremediation is often more cost-effective, particularly for large-scale remediation projects. Phytoremediation can improve water quality, reduce nutrient loading, and enhance habitat for diverse flora and fauna, contributing to the overall health of Ramsar Sites.

In conclusion, aquatic plant phytoremediation offers a promising pathway towards a cleaner and healthier environment. Stakeholders and policymakers play a pivotal role in wetland preservation and the promotion of phytoremediation as a sustainable remediation strategy for the benefit of present and future generations.

References

Ahmad SS, Reshi ZA, Shah MA, Rashid I, Ara R, Andrabi SM (2014) Phytoremediation potential of Phragmites australis in Hokersar wetland—a Ramsar site of Kashmir Himalaya. Int J Phytorem 16(12):1183–1191

Al-Baldawi IA (2018) Removal of 1,2-dichloroethane from real industrial wastewater using a subsurface batch system with Typha angustifolia L. Ecotoxicol Environ Saf 147:260–265

Ali H, Khan E, Sajad MA (2013) Phytoremediation of heavy metals—concepts and applications. Chemosphere 91(7):869–881

Chandanshive VV, Rane NR, Tamboli AS, Gholave AR, Khandare RV, Govindwar SP (2017) Co-plantation of aquatic macrophytes Typha angustifolia and Paspalum scrobiculatum for effective treatment of textile industry effluent. J Hazard Mater 338:47–56

Chatterjee S, Chetia M, Singh L, Chattopadhyay B, Datta S, Mukhopadhyay SK (2011) A study on the phytoaccumulation of waste elements in wetland plants of a Ramsar site in India. Environ Monit Assess 178:361–371

Cunningham SD, Ow DW (1996) Promises and prospects of phytoremediation. Plant Physiol 110(3):715

Di L, Li Y, Nie L, Wang S, Kong F (2020) Influence of plant radial oxygen loss in constructed wetland combined with microbial fuel cell on nitrobenzene removal from aqueous solution. J Hazard Mater 394:122542

Effendi H, Utomo BA, Pratiwi NT (2020) Ammonia and orthophosphate removal of tilapia cultivation wastewater with Vetiveria zizanioides. J King Saud Univ-Sci 32:207–212

Ekta P, Modi NR (2018) A review of phytoremediation. J Pharmacogn Phytochem 7(4):1485–1489

Etim EE (2012) Phytoremediation and its mechanisms: a review. Int J Environ Bioenergy 2(3):120–136

Farraji H, Zaman NQ, Tajuddin R, Faraji H (2016) Advantages and disadvantages of phytoremediation: a concise review. Int J Env Tech Sci 2:69–75
Fletcher J, Willby N, Oliver DM, Quilliam RS (2020) Phytoremediation using aquatic plants. In: Phytoremediation: in-situ applications, pp 205–260
Garg A, Joshi B (2015) Ecosystem sustenance of Upper Ganga Ramsar site through phytoremediation. Geophytology 45(2):175–180
Jothimani P, Ponmani S, Sangeetha R (2013) Phytoremediation of heavy metals—a review. Int J Res: Stud Biosci 1:17–23
Khan S, Masoodi TH, Pala NA, Murtaza S, Mugloo JA, Sofi PA, Kumar A (2023) Phytoremediation prospects for restoration of contamination in the natural ecosystems. Water 15(8):1498
Khatun A, Pal S, Mukherjee AK, Samanta P, Mondal S, Kole D et al (2016) Evaluation of metal contamination and phytoremediation potential of aquatic macrophytes of East Kolkata Wetlands, India. Environ Health Toxicol 31
Kumar V, Kumar P (2019) A review on feasibility of phytoremediation technology for heavy metals removal. Arch Agric Environ Sci 4(3):326–341
Kumar A, Verma JP (2018) Does plant-microbe interaction confer stress tolerance in plants: a review? Microbiol Res 207:41–52
Kumar SR, Arumugam T, Anandakumar C, Balakrishnan S, Rajavel D (2013) Use of plant species in controlling environmental pollution. Bull Environ, Pharmacol Life Sci 2:52–63
Lu B, Xu Z, Li J, Chai X (2018) Removal of water nutrients by different aquatic plant species: an alternative way to remediate polluted rural rivers. Ecol Eng 110:18–26
Matache ML, Marin C, Rozylowicz L, Tudorache A (2013) Plants accumulating heavy metals in the Danube River wetlands. J Environ Health Sci Eng 11:1–7
Mohan S, Tippa A (2019, May) Role of wetland soil bacteria in enhancing the phytoremediation process through bioavailability phenomenon. In: World environmental and water resources congress 2019. American Society of Civil Engineers, Reston, VA, pp 1–10
Mohebi Z, Nazari M (2021) Phytoremediation of wastewater using aquatic plants, a review. J Appl Res Water Wastewater 8(1):50–58
Murtić S, Zahirović Ć, Čivić H, Sijahović E, Jurković J, Avdić J, Podrug A (2021) Phytoaccumulation of heavy metals in native plants growing on soils in the Spreča river valley, Bosnia and Herzegovina. Plant Soil Environ 67(9):533–540
Mustafa HM, Hayder G (2020) Recent studies on applications of aquatic weed plants in phytoremediation of wastewater: a review article. Ain Shams Eng J 1–11
Patel DK, Kanungo VK (2013) Treatment of domestic wastewater by potential application of a submerged aquatic plant Hydrilla verticillate Casp. Recent Res Sci Technol 4:56–61
Pinto E, Aguiar A, Ferreira I (2014) Influence of soil chemistry and plant physiology in the phytoremediation of Cu, Mn, and Zn. Crit Rev Plant Sci 33:351–373
Rai PK (2008) Heavy metal pollution in aquatic ecosystems and its phytoremediation using wetland plants: an ecosustainable approach. Int J Phytorem 10(2):133–160
Rai PK, Singh MM (2020) Wetland plants' chemical ecology for iron of a Ramsar site in an Indo-Burma Hotspot: in-situ bioaccumulation and phytoremediation implications. Nat Environ Pollut Technol 19(4):1607–1615
Reddy SSG, Raju AS, Kumar BM (2015) Phytoremediation of sugar industrial water effluent using various hydrophytes. Int J Environ Sci 5:1147
Sa'at SKM, Zaman NQ (2017) Phytoremediation potential of palm oil mill effluent by constructed wetland treatment, pp 49–54
Sabir M, Waraich EA, Hakeem KR, Öztürk M, Ahmad HR, Shahid M (2014) Phytoremediation: mechanisms and adaptations. Soil remediation and plants: prospects and challenges, vol 85, pp 85–105
Saha P, Shinde O, Sarkar S (2017) Phytoremediation of industrial mines wastewater using water hyacinth. Int J Phytorem 19:87–96
Shah MP (2024a) Microbial remediation of hazardous chemicals from water & industrial wastewater treatment plant. Springer Nature

Shah MP (2024b) Development in waste water treatment research and processes: applied technologies for clean up of environmental contaminants. Elsevier

Shah MP (2024c) Aerobic and anaerobic microbial treatment of industrial wastewater. CRC Press

Shah MP (2025) Extraction of precious metals from industrial waste using microbial technology. Springer Nature

Sharma P, Chandra R (2024) Phytoremediation mechanism and role of plant growth promoting rhizobacteria in weed plants for eco-restoration of hazardous industrial waste polluted site: a review. Environ Sci Pollut Res 31:42495–42520. https://doi.org/10.1007/s11356-024-33910-w

Simmer RA, Schnoor JL (2022) Phytoremediation, bioaugmentation, and the plant microbiome. Environ Sci Technol 56(23):16602–16610

Singh P, Pani A, Mujumdar AS, Shirkole SS (2023) New strategies on the application of artificial intelligence in the field of phytoremediation. Int J Phytorem 25(4):505–523

Singh S, Kumar V, Dhanjal DS, Parihar P, Ramamurthy PC, Singh J (2022) Phytoremediation of heavy metals, metalloids, and radionuclides: prospects and challenges. In: Phytoremediation technology for the removal of heavy metals and other contaminants from soil and water, pp 253–276

Sood A, Uniyal PL, Prasanna R, Ahluwalia AS (2012) Phytoremediation potential of aquatic macrophyte, Azolla. Ambio 41:122–137

Tabla-Hernandez J, Rodriguez-Espinosa PF, Mendoza-Pérez JA, Sánchez-Ortíz E, Martinez-Tavera E, Hernandez-Ramirez AG (2019) Assessment of potential toxic metals in a ramsar wetland, Central Mexico and its self-depuration through Eichhornia crassipes. Water 11(6):1248

Tarekegn MM, Salilih FZ, Ishetu AI (2020) Microbes used as a tool for bioremediation of heavy metal from the environment. Cogent Food Agric 6(1):1783174

Williams JB (2002) Phytoremediation in wetland ecosystems: progress, problems, and potential. Crit Rev Plant Sci 21(6):607–635

Yasmin T, Imadi SR, Gul A (2023) Nanotechnology in phytoremediation: application and future. In: Phytoremediation: management of environmental contaminants, vol 7, pp 427–442

Chapter 16
Genomics and Proteomics Associated with Bioremediation of Microplastics

Siddhartha Das and R. P. N. Sravya Sree

Abstract Application of genomics and proteomics in the bioremediation of microplastics represents a cutting-edge approach to addressing environmental pollution. Conjoint integration enables the identification and characterization of specific genes, enzymes, and metabolic pathways involved in the breaking down of these persistent pollutants. Through genomic sequencing, researchers can pinpoint microorganisms with natural abilities to degrade microplastics, uncovering the genetic blueprints that govern these processes. Metagenomics, which involves the collective genome analysis of microbial communities in contaminated environments, allows for the discovery of novel plastic-degrading organisms and the elucidation of their functional roles in situ. Proteomics complements genomics by revealing the active proteins and enzymes that facilitate microplastic degradation. By profiling the proteome of these microorganisms, scientists can identify key enzymes such as hydrolases, oxygenases, and reductases that catalyze the degradation of plastic polymers into smaller, less harmful molecules. This integrative approach thus holds promise for significantly reducing the ecological footprint of microplastics.

Keywords Microplastics · Metagenomics · Proteomics · Bioaugmentation

16.1 Introduction

Microplastics, defined as plastic particles less than 5 mm in diameter, have emerged as a significant environmental concern due to their pervasive presence in ecosystems and their potential impacts on both ecological and human health. Originating from a variety of sources, including the breakdown of larger plastic debris, microbeads from personal care products, and synthetic fibers from clothing, these tiny pollutants are now found in oceans, freshwater systems (Wang et al. 2019b), soil, and

S. Das (✉) · R. P. N. Sravya Sree
Department of Plant Pathology, M.S. Swaminathan School of Agriculture, Centurion University of Technology and Management, Paralakhemundi, Odisha, India
e-mail: siddhartha.das10@gmail.com

© The Author(s), under exclusive license to Springer Nature Singapore Pte Ltd. 2026

S. Dey and M. P. Shah (eds.), *Microbial Applications in Industrial Wastewater Treatment*, Advances in Wastewater Research,
https://doi.org/10.1007/978-981-96-9128-9_16

even the air. The persistence of microplastics in the environment is primarily due to their resistance to natural degradation processes, leading to accumulation and biomagnification in food chains. As they can adsorb toxic chemicals and be ingested by a wide range of organisms, including zooplankton, fish, (Amaral-Zettler et al. 2020) and ultimately humans, microplastics pose a complex challenge that intersects with issues of pollution, public health, and environmental sustainability. Addressing the microplastics problem requires innovative solutions, among which bioremediation has shown promising potential. The bioremediation process can be enhanced through genetic engineering to improve the efficiency and specificity of these microorganisms (Bacha et al. 2023). Despite the promise of bioremediation, several challenges need to be addressed for its practical application on a large scale. These include understanding the ecological impacts of introducing new microbial species into the environment, optimizing conditions for microbial activity, and ensuring that the by-products of degradation are non-toxic and environmentally benign (Shen et al. 2019). Moreover, integrating bioremediation with other waste management practices, such as mechanical recycling and chemical depolymerization (Shah 2025; Thiounn and Smith 2020), could provide a more comprehensive solution to plastic pollution. Public awareness and policy measures also play crucial roles in managing microplastic pollution. Regulatory frameworks that limit the use of microplastics in consumer products, promote sustainable plastic alternatives, and encourage proper waste management practices are essential for reducing the source of microplastics (Sanniyasi et al. 2021).

16.2 Diverse Sources of Microplastics

Microplastics originate from two different sources. Microplastics are primarily derived from a variety of sources, including polyethylene, polystyrene, polyvinyl chloride, polyamide nylon 6, polypropylene, and household and medical items. It is known that many personal care products, including toothpaste, scrubs, cleaning supplies, and cosmetics, contain irregularly shaped microplastics with a diameter of 0.5 to less than 0.1 mm. These microplastics are commonly advertised as "microbeads" or "micro-exfoliates" and are a source of primary microplastics (Fendall and Sewell 2009). However, as microplastic pollutants are effectively eliminated by the skimming and settling treatment processes, the wastewater treatment plants now in operation demonstrate that tertiary water treatment is not a source of microplastic contamination (Carr et al. 2016; Kay et al. 2018). Secondary microplastics are created when large plastic objects or particles are widely broken up in the presence of environmental factors such high temperatures, UV light, stress, reactive ozone, oxidation, and air pressure (John et al. 2021; Tiwari et al. 2020). The only way polymeric materials may resist oxidative-thermal breakdown is if stabilizers and antioxidants are introduced. Secondary microplastics are also produced through physical abrasion. Furthermore, it is recognized that biological agents including bacteria, fungi, and algae create a wide variety of enzymes that are essential for the breakdown of microplastics (Chia et al. 2020; John et al. 2021; Othman et al. 2021;

Manzi et al. 2022; Miri et al. 2022; Shah 2024a; Zhu et al. 2022). The majority of secondary microplastics originate from large plastic objects made of the same polymers that are improperly disposed of in land and water systems. Physical abrasion weakens the chemical bonds in these objects, which then causes oxidative-thermal breakdown (Gerritse et al. 2020). The breakdown of synthetic fibers while washing clothing and industrial processes like thermal cutting of polystyrene are two other secondary sources of microplastics. The overproduction of items made of polyethylene, polypropylene, and as well as polyethylene terephthalate has been attributed to the growing use of single-use plastics.

16.3 Role of Microplastics in Aquatic-Terrestrial Ecosystem and Modern Civilization

Microplastics serve as a vehicle for the spread of various contaminants, including ingested heavy metals, bacterial fish pathogens, multidrug-resistant *E. coli* (Wang et al. 2019a), persistent organic pollutants, etc. (Enders et al. 2015; Viršek et al. 2017; Caruso 2019; Song et al. 2020). Additionally, they may leach chemicals added during their manufacturing process (Groh et al. 2019; Bacha et al. 2023). Microplastics could have a significant effect on the aquatic fora and fauna (Sathicq et al. 2021). Furthermore, microplastics offer a fresh environment for the development of microbial films made of bacteria, fungi, and algae, and they may also disseminate microbial infections and resistance to antibiotics (Zettler et al. 2013; Guo et al. 2020). Because most microplastics fall into the same size range as the items that zooplankton typically eats, they may unintentionally make their way up the food chain (Gregory 2009). The choice of microplastics over food particles may result in the loss of energy resources and have sublethal effects on the reproductive cycle of the species (Enders et al. 2015).

Microplastics' effects on the diversity and expansion of the microalgal population often vary. Sjollema et al. (2016) reported that *Dunaliella tertiolecta* microalgae development was adversely affected by uncharged polystyrene particles at high concentrations (250 mg/l) and decreasing particle size. High microplastic concentrations hampered the growth of *Spirulina* sp., especially due to shading effects and reduced light intensity, which harmed photosynthesis and microalgal cell walls (Khoironi and Anggoro 2019). Capolupo et al. (2020) found that in the case of *Raphidocelis subcapitata* and *Skeletonema costatum* the cell growth is stunted because of leaching of additives. Conversely, *Raphidocelis subcapitata* demonstrated a greater rate of growth when exposed to plastic microbeads (Cannif and Hoang 2018). *Dunaliella salina* also demonstrated a comparable increase in cell growth and photosynthetic activity when exposed to bigger microplastics (Chae et al. 2019). The trace concentration of added chemicals that leach out of microplastics, such as stabilizers, phthalates, and endocrine disruptors, appears to be connected to this acceleration of growth (Chae et al. 2019). It is yet unclear how microplastics may affect people, and further research is needed (Vethaak and Legler 2021).

16.4 Steps for Microplastic Degradation

Most of the conventional methods for degrading microplastics that have been discussed involve feeding plastic scrap back into the heating cycle of the processing unit as the main technique; waste can also be made into new plastic products by mixing it with a virgin polymer, which can drastically reduce production costs; and in some situations, plastic wastes can be changed chemically or thermochemically to be recycled in the industrial loop. The majority of them combine with the organic materials in a landfill, which is utilized for anaerobic digestion or composting, resulting in excessive pollution and the creation of hazardous substances like lead and cadmium as well as toxic metals like phthalates, tetrabromobisphenol A, polybrominated diphenyl ethers, and dioxins. Microplastic particles are currently disposed of using different types of physical and chemical techniques, such as recycling, landfilling, and burning. Chemical recycling techniques like pyrolysis are quite commonly used in the business world (Thiounn and Smith 2020). The plastic trash is transformed into a mixture of charred and tarry materials using the slow pyrolysis processes, which are carried out at three different kinds of temperatures like: 300, 425, and 550 °C (Dussud et al. 2018). The recovery of thermal energy from the pyrolysis of polypropylene, polystyrene, and polypropylene has been the subject of numerous studies. The unsuitable for recycling contaminated, mixed, or degraded leftovers can be utilized as feedstocks for waste-to-energy techniques including pyrolysis, which is an endothermic cracking process without oxidation, and incineration, which oxidizes polymers. Microplastics can be broken down physically, chemically, or biologically. The biological breakdown process involves a wide range of enzymes (Padervand et al. 2020; Bacha et al. 2023). Phases like the disintegration of polymers into smaller particles and the subsequent disintegration of the smaller particles into oligomers, dimers, and monomers comprise the basic process. Microbes help with the subsequent phases of mineralization that come after this degradation. Following the completion of mineralization, a variety of intermediate molecules are produced, which provide energy for the growth of microbes and release carbon dioxide (Gironi and Piemonte 2011). Extracellular enzymes such as lipases, esterases, lignin peroxides, laccases, and manganese peroxides are among the many that are necessary for the degradation of microplastics. According to Taniguchi et al. (2019), these enzymes convert microplastics into carbonyl or alcohol residues and increase their hydrophilicity. Hydrolases—a class of enzymes that break down plastic surfaces by accelerating chain cleavage reactions—include lipases, esterases, and cutinases. Although they do not penetrate the polymer, their action creates surface cracks, and the resulting monomers are taken up by microbes' cytoplasm and eventually go through various metabolic pathways. Even though a lot of study has been done utilizing extracellular enzymes in the biodegradation of microplastics, the involvement of intracellular enzymes in the degradation of microplastics is still poorly understood, and the pathways involved in the uptake of monomers remain unclear. Following the fragmentation of the microplastics, the tricarboxylic acid cycle and β-oxidation pathway are usually utilized by the cell to metabolize the metabolic intermediates containing carbonyl and hydroxyl groups (Taniguchi et al. 2019). The full mineralization of plastic waste into H_2O, CO_2, N_2, and CH_4 occurs after this procedure (Zettler et al. 2013). Researchers have thoroughly examined how microplastics colonize surfaces by breaking down a consortium and creating a biofilm on the particles (Fig. 16.1).

Numerous mechanisms, including as biofouling and plasticizer degradation, lead to the attachment process of the bacteria. This is followed by an attack on the polymer's backbone, which is then linked to hydration and microbe penetration into the polymer structure. Furthermore, a number of elements must be present for effective biodegradation, including the presence of prospective microbial degrading organisms with the appropriate enzymes and metabolic pathways as well as other environmental factors consisting temperature, pH, salt, and moisture content (Raddadi and Fava 2019; Syranidou et al. 2017; Miri et al. 2022). The polymer's surface and structure, as well as its crystalline and amorphous areas, crystal size, and lamellar thickness, all affect how biodegrades microplastics. According to Shabbir et al., polyhydroxyalkanoates depolymerase enzymes hydrolyze the amorphous chain structures

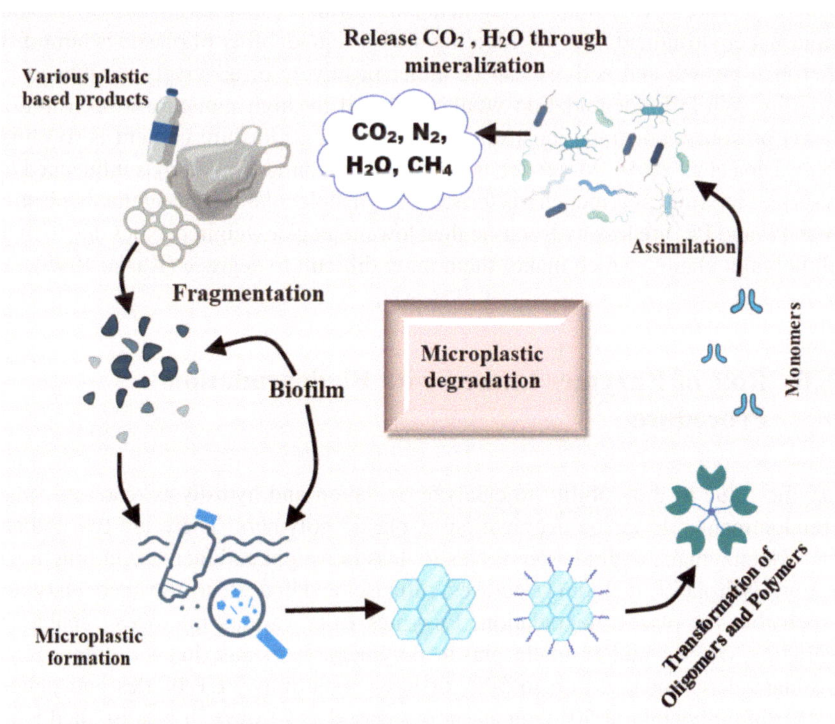

Fig. 16.1 Schematic representation of microplastic degradation

on the fragmentation film surface, causing the crystalline chain structures to erode (Shah 2024b; Shabbir et al. 2020).

16.5 Factors Involved in Plastic Biodegradation Procedure

The biodegradability of plastics is influenced by both biotic and abiotic influences. Temperature, pH, light, and humidity are abiotic variables that have a big impact on biodegradation (Gewert et al. 2018; Oluwasina et al. 2019; Singh et al. 2019; Arisa-Tarazona et al. 2020). These elements promote the plastic polymers' hydrolysis, which results in chain scission. Because of the scission, biotic agents, or microorganisms, can further break down polymers. Temperature affects both the variety and activity of microbes (Zoungranan et al. 2020). Over 30 °C increases the quantity of microbial species but slows down the breakdown of plastic, improving the biodegradation rate by 20% (Zoungranan et al. 2020). In addition, pH encourages the growth of microorganisms and enzymatic activity that influences biodegradation. MPs displayed fractures and brittleness at 0 °C and pH 3 and 11 (Arisa-Tarazona et al. 2020). Moreover, using UV light for photolysis can be a pre-treatment method and improve plastic degradation. When synthetic plastics are kept in presence of UV light for a whole year, smaller fragments are formed (Song et al. 2017). Another important environmental factor affecting the biodegradability of plastic is humidity. Microbial growth and activity can be either positively or adversely stimulated by humidity. While biodegradation would be aided by high moisture level, dilution effects prevent biodegradation from occurring when moisture content is too high (Oluwasina et al. 2019). Moreover, the plastic's total biodegradation is influenced by its surface area and polymer characteristics. High molecular weight synthetic plastics (such PE and PP) are less hydrophilic than low molecular weight plastics due to their full polymer chains, which makes them more difficult to degrade (Kawai 1995).

16.6 Role of Enzymes Linked with Biodegradation Procedure

Enzymes, due to their ability to catalyze oxidation and hydrolysis reactions, play a predominant role in the degradation of plastic polymers. There are two indications for enzymatic degradation of plastic: loss in weight and increase in functional groups. Hydrolases, including esterases, cutinases and lipases, are the major enzymes responsible for plastic degradation. Enzymes from various microbes capable of degrading various human-made plastic polymers. It means that its cutinase-like hydrolase depolymerized up to 50% of low-crystallinity PET bottles and pellets of their initial weight at 50 °C in as short a period as 21 days. It was isolated from a leaf-branch compost metagenomic library by LC-cutinase, which could hydrolyze low-crystallinity PET films, causing 50% weight loss at 50 °C in 7 days. Yoshida et al. found *Ideonella sakaiensis* 201-F6, which forms MHETase and PETase, a hydrolase, upon its adhesion to PET substrate. These enzymes can degrade PET into the general constituents, including terephthalic acid, ethylene glycol, bis(2-hydroxyethyl) terephthalate, and mono (2-hydroxyethyl) terephthalate. Terephthalic acid is reduced to protocatechuic acid, PCA, which is then subjected to PCA 3, 4 dioxygenase cleavages to yield 4-carboxy-2-hydroxymuconic acid. This compound was further dehydrogenated into 2-pyrone-4, 6 dicarboxylic acids, which was then routed into the TCA cycle. Ethylene glycol, on the other hand, was converted into glycoaldehyde by the action of the enzyme glycoaldehyde reductase, which further oxidized it into glycolate. Glycolate is converted to glyoxylate by the action of the enzyme glycolate oxidase in the TCA cycle. The glyoxylate condenses with acetyl CoA and forms malate. Santo et al. isolated Rhodococcus ruber C208, which degraded polyethylene through the secretion of extracellular laccase. Induction and activation of the laccase enzyme require copper ions. Polyethylene films' amorphous parts oxidized with laccase resulted in the biodegradation process. This has been verified by FTIR analysis and has resulted in the formation of readily accessible carbonyl groups and a drop in molecular weight of the polyethylene film. Comparison of the transcription levels of alkB1 and alkB2, two alkane monooxygenases expressed in Pseudomonas aeruginosa E7, using qRT-PCR for their capacity to degrade low molecular weight polyethylene was performed by Jeon et al. When *P. aeruginosa* E7 was incubated on low molecular weight polyethylene medium as the sole carbon source, the transcription level of alkB2 was higher than that of alkB1. The surface-binding domain and catalytic domain of polyurethane esterase isolated from *Comamonas acidovorans* TB-35 were both required for polyurethane degradation. PU esterase hydrolyzed polyurethane to yield ethylene glycol and polyisocyanate. The metabolite ethylene glycol is further converted to glycolic acid and glycoaldehyde before entering the TCA cycle. Schimdt et al. examined the degradability of polyester PU Elastollan B85A-10 and C85A-10 by four polyester hydrolases: LC Cutinase, TfCut2, Tcurl278, and Tcur0390. The results showed that LCC caused Elastollan B85A-10 and C85A-10 to lose 4.9% and 4.1% of their weight after incubation at 70 °C for 200 h, respectively. At 70 °C, it was observed that LCC was most effective, which might be related to its higher thermostability. Moreover, FTIR results showed that ester bond breakage was responsible for the weight losses.

16.7 Role of Microorganisms in Microplastic Degradation

Plastics are categorized into seven major types based on how frequently they are used: (1) Polyvinyl Chloride (PVC or Vinyl), (2) High-Density Polyethylene (HDPE), (3) Low-Density Polyethylene (LDPE), (4) Polypropylene (PP), (5) Polystyrene (PS or Styrofoam), (6) Polyethylene Terephthalate (PET or PETE), and (7) Other. Many of them can be degraded. The specific biodegradable plastics are polyamide, polyacrylonitrile, butadiene styrene, acrylic, acetyl cellulose, cellulose triacetate, alkyd, cellophane, epoxy resin, and poly (butylene adipate-co-teraphthalate). Thus, a variety of industries use the aforementioned plastics. Several researches conducted in the last few years have documented the potential of different microbes to degrade various synthetic polymers. An estimated 50% or more of plastic garbage is disposed of in landfills, with the remaining 19% being burned. 22% of poorly managed plastic trash ends up in terrestrial and marine habitats, with only around 9% of plastic garbage being recycled.

16.7.1 Microplastic Degrading Algal Microbes

Microalgae and their toxins and enzymes can be used to efficiently aid in the biological decomposition of polymeric material (Moog et al. 2019; Chia et al. 2020; Manzi et al. 2022). The main advantage is that they can develop in a variety of conditions where most microplastics are prevalent, unlike the bacterial system, which requires a rich carbon supply for growth (Yan et al. 2016). Microalgae have been observed to colonize plastic surfaces in wastewater streams (Shah 2024c; Hirooka et al. 2005). Plastic is broken down by ligninolytic and exopolysaccharide enzymes, which are produced in response to this adhesion. These polymers mainly serve as a source of carbon, increasing the quantity of proteins and carbohydrates within cells and accelerating their growth. Scanning electron microscopy has just been used to identify the surface degradation or disintegration of low-density polyethylene sheet due to algae infestation (Sanniyasi et al. 2021). Algal biodegradation mostly happens through a variety of mechanisms, including fouling, hydrolysis, penetration, and corrosion (Chia et al. 2020). It was also demonstrated that *Oscillatoria subbrevis* and *Phormidium lucidum* could colonize and degrade low-density polyethylene without pro-oxidative compounds or pretreatment. Many species of bacteria and algae, like *Stephanodiscus hantzschii*, *Chlorella vulgaris*, *Chlamydomonas mexicana*, and *Chlorella fusca* var. *vacuolated*, degraded an estrogenic component, bisphenol A, commonly present in polymers. Most of the time, the decomposition of microplastics is associated with the formation of biofilms on the polymer surface. Numerous cyanobacterial strains, including those from the genus *Microcystis*, species *Rivularia*, *Pleurocapsa*, *Synechococcus*, *Prochlorothrix*, *Leptolyngbya Calothrix*, and *Scytonema*, were also able to form films on the microplastic polymers (Muthukrishnan et al. 2019; Dussud et al. 2018; Debroas et al. 2017). Besides cyanobacterial species, biofilms also comprise diatoms, thus contributing to photosynthesis (Sarmah and Rout 2018). The progress on different biotechnological approaches nowadays allows developing a huge number of transgenic microalgal cell factories that are capable of producing and secreting enzymes needed for plastic degradation. Genetic modification allowed the green microalgae *Chlamydomonas reinhardtii* to produce polyethylene terephthalate hydrolase, an enzyme with activity in degrading polyethylene terephthalate films and terephthalic acid (Ji et al. 2014; Kim et al. 2020). *P. tricornutum* produced polyethylene terephthalate hydrolase and catalyzed the polymer polyethylene terephthalate glycol as well as polyethylene terephthalate, was also successfully modified in a similar manner (Moog et al. 2019).

16.7.2 Microplastic Degrading Fungal Microbes

Most of the diverse range of organisms that comprise fungi are called saprotrophs; they are also referred to as obligatory or opportunistic parasites. They are incredibly adaptable and may thrive in a wide range of aquatic and terrestrial eco systems in a variety of environmental conditions. Their resistance to a wide range of toxic substances and metals is matched by their capacity to produce a diverse array of extracellular enzymes and naturally occurring biosurfactants, including hydrophobins, which can degrade complex polymers into simpler monomers and provide microorganisms with carbon and electrons. *Aspergillus niger* and *Penicillium pinophilum*, which mediated the degradation in low-density polyethylene, required physical pretreatment procedures, such as thermooxidization at 80 °C for 15 days. After incubation over 30 months, the percentages of degradation were 0.57 and 0.37%. In a similar vein, *Aspergillus* and *Lysinibacillus* species demonstrated 15.8% and 29.5%, respectively, of the biodegradation of UV-irradiated and non-UV-irradiated polymer films, respectively (Esmaeili et al. 2013). Numerous intracellular and extracellular enzymes that are capable of catalyzing a wide range of reactions and breaking down petroleum-based polymers are produced by fungi. An important part of the fungal adaptation and detoxification processes is played by the intracellular enzymes (Schwartz et al. 2018). The metabolism of aliphatic, alicyclic, and aromatic substances is aided by the oxidation and conjugation reactions associated with the enzyme systems of the cytochrome P450 family, epoxidases and transferases. Deamination, sulfoxidation, desulfuration, epoxidation, and dehalogenation are only a few of the activities they perform (Shin et al. 2018). The cytochrome P450 family of enzymes use cofactors such heme, NADPH+ H+, and FAD to aid in the development of the spore wall and maintenance of the integrity of the hyphal wall.

16.7.3 Microplastic Degrading Bacterial Microbes

Using microbes to degrade microplastics has been the subject of numerous studies. Microplastics can be broken down by bacteria in a wide range of habitats, including compost, wastewater, sludge, polluted sediments, and municipal landfills (Mehmood et al. 2016). They can also break down microplastics in harsh weather conditions, like Antarctic soils, mangroves, and marine sediments. Moreover, the gut microbiota of earthworms has been found to harbor bacteria that degrade microplastic. It is well known that microorganisms commonly pick up the ability to activate the enzyme machinery responsible for breaking down microplastics when they live in contaminated regions. Both pure cultures and bacterial consortiums can be utilized to degrade microplastics. In contrast to that, pure cultures bear some advantages over the degradation process itself and offer a meaningful tool for investigating metabolic processes of relevance. It is also much easier to control the influences of environmental factors like pH, temperature, surfactants, or substrate properties on the degradation process itself (Janssen et al. 2002). The very slow disintegration rate is the major drawback.

Pseudomonas stutzeri, *Vibrio* sp. and *Pseudomonas chlororaphis*, were used in a consortium by Cacciari et al. (1993) to break down polypropylene in the first investigation on microplastics biodegrading microorganisms. It was also noted in the same study that the starch addition improved the biodegradation capacity. *Pseudomonas chlororaphis*, *Pseudomonas stutzeri*, and *Vibrio* sp. were used in a consortium by Cacciari et al. (1993) to break down polypropylene in the first investigation on microplastics biodegrading microorganisms. Starch was also found to improve the biodegradation capacity in the same study. Later on, *Bacillus subtilis*, *B. fexus*, *Pseudomonas stutzeri*, and *Rhodococcus rhodochrous* were used in a consortium to biodegrade polypropylene, as described by Arkatkar et al. (2010) and Fontanella et al. (2013). According to Kowalczyk et al. (2016), these microbial isolates were discovered to form a biofilm through the isolation of *Achromobacter xylosoxidans*. After 40 days of incubation, two bacterial strains from the genera *Bacillus* and *Rhodococcus* that were isolated from mangrove sediments showed polypropylene degradation efficiencies of 4.0% and 6.4%, respectively, according to a study by Auta et al. (2017). Furthermore, they mentioned that *Bacillus cereus* and *Bacillus gottheilii* may be able to degrade microplastics (Auta et al. 2017). *B. gottheilii*-induced microplastics were responsible for the weight loss percentages of 6.2%, 3.0%, 3.6%, and 5.8% for polyethylene, polyethylene terephthalate, polypropylene, and polystyrene, respectively.

16.7.4 Synergy Between Man-Made Synthetic Polymers and Insect Gut Microbes

Nowadays, scientists in their quest to find the best plastic biodegradation method have way more choices at their disposal than before, thanks to the ability of insect larvae to eat and digest manmade polymers, mainly plastic. Reportedly, the larvae of Coleopteran beetles consume and break down plastics. Members of the coleopteran order—such as *Plesiophthalmus davidis*, *Zophobas asatratus*, *Tribolium castaneum*, *Tenebrio molitor* and *Tenebrio obscurus*, the super-worm, and others—have this special skill. In addition to the coleopteran fauna, *Achroia grisella*, the lesser wax moth, *Galleria mellonella*, the Indian meal moth, and other lepidopteran caterpillars are natural decomposers of synthetic polymers such as polystyrene PS, polyethylene PE, polyvinyl chloride PVC, and polypropylene PP (Table 16.1).

Table 16.1 Insect-based bioremediation of microplastics

Sl. No	Type of insects	Scientific name	Types of consumable plastics	Insect gut microbial dynamics	Active enzymes
1	Yellow mealworm	*Tenebrio molitor*	Polystyrene, Polypropylene (PP), Polyethylene (PE)	*Exiguobacterium* sp. strain YT2	Alkaline proteases Alkali-tolerant esterase Hydrolase
2	Rice weevil	*Sitophilus oryzae*	Nylon, Polypropylene (PP), Polyethylene (PE)	*Bacillus subtilis* strain TLO3 *Staphylococcus* sp.	Lipase and Hydrolase
3	Leaser grain borer	*Rhyzopertha dominica*	Polypropylene (PP), Polyethylene (PE)	*Aeromonas liquifaciens*	Lipase Chitinase Protease
4	Indian meal moth	*Plodia interpunctella*	Polyethylene	*Enterobacter asburiae* YT1 *Bacillus* sp. YP1	Esterase
5	Rice moth	*Corcyra cephalonica*	Polyethylene	*Staphylococcus saprophyticus*	Information is not available as of now
6	Red flour beetle	*Tribolium castaneum*	Polystyrene (PS)	*Acinetobacter* sp. AnTc-1	Lipase Chitinase Protease
7	Super worms	*Zophobas atratus*	Polystyrene (PS)	*Pseudomonas* sp. strain DSM 50071 *Klebsiella pneumoniae* *Alcaligenes* sp. *Acinetobacter* sp.	Monooxygenase Lipase Cutinase Esterase Polyurethanase
8	Greater wax moth	*Galleria mellonella*	Polyethylene, Polystyrene	*Massilia* sp. FS1903 *Acinetobacter* sp. Bacillus sp. *Serratia* sp., *Enterobacter* sp. D1, *Aspergillus flavus*	Manganese Peroxidase Hydrogen peroxide Lac and Lignin Peroxidase (LiP) Lipase Protease Polyurethanase

16.8 Role of Enzymes Linked with Biodegradation Procedure

Enzymes, the biological catalysts, play a pivotal role in the biodegradation process, breaking down complex organic matter into simpler substances. These microscopic workhorses possess unique active sites that bind to specific substrates, accelerating chemical reactions without being consumed. In the realm of biodegradation, enzymes produced by microorganisms are particularly crucial. For instance, bacteria and fungi secrete enzymes capable of degrading cellulose, lignin, and other recalcitrant compounds present in plant matter. These enzymes, such as cellulase and lignin peroxidase, hydrolyze complex polymers into smaller units like glucose and organic acids, respectively. Moreover, enzymes facilitate the breakdown of pollutants, such as hydrocarbons in oil spills, through a process called bioremediation (Zadjelovic et al. 2020). Microorganisms equipped with enzyme arsenals can degrade these harmful substances into harmless byproducts, restoring the environment. Thus, enzymes emerge as indispensable tools in nature's recycling system, contributing significantly to the maintenance of ecological balance and the remediation of environmental contamination.

16.8.1 Role of Genomics in Microplastic Bioremediation

The role of genomics in microplastic bioremediation is an emergent and revolutionary field, holding the promise to address One of the most serious environmental issues of our time. Microplastics, minute plastic particles less than 5 mm in size, pervade oceans, rivers, and soils, posing a significant hazard to human health and ecosystems because of their tenacity and propensity to accumulate hazardous chemicals (De-la-Torre 2020). Traditional methods of plastic degradation are inadequate given the scale and resilience of microplastics, necessitating innovative solutions like bioremediation. Genomics, the study of the complete set of DNAs within an organism, offers a profound tool in this regard, providing insights into the genetic makeup of organisms that can degrade or transform plastics. By leveraging genomics, scientists can identify, understand, and enhance the capabilities of microorganisms with the natural propensity to break down plastics. The emergence of high-throughput sequencing technologies has catalyzed this field, enabling the comprehensive analysis of microbial communities from diverse environments. Through metagenomics, the collective genome of these communities can be sequenced and analyzed, revealing enzymes and pathways involved in plastic degradation. This approach has already led to the discovery of various bacteria and fungi capable of degrading polyethylene, polyethylene terephthalate (PET), and other plastics. For instance, *Ideonella sakaiensis*, a bacterium discovered in a PET recycling plant, harbors PETase and MHETase enzymes that can hydrolyze PET into its monomers. Genomic studies of such organisms elucidate the genetic basis of their plastic-degrading abilities, paving the way for genetic engineering to enhance these capabilities. Synthetic biology, an offshoot of genomics, allows for the design and construction of microbial strains with optimized plastic-degrading functions. By inserting or modifying genes responsible for plastic degradation, scientists can create robust strains capable of efficiently breaking down microplastics in various environmental conditions. Moreover, genomics facilitates the understanding of microbial interactions within biofilms, complex communities of microorganisms that adhere to surfaces, including plastics. Biofilms play a crucial role in microplastic degradation, as they provide a stable environment for microorganisms to colonize and act on plastic surfaces. Genomic insights into biofilm formation and maintenance can lead to strategies that enhance the colonization and activity of plastic-degrading microbes. The application of genomics extends to the monitoring and assessment of bioremediation efforts as well. Through techniques like quantitative PCR and metatranscriptomics, the abundance and activity of plastic-degrading genes in environmental samples can be tracked, providing real-time feedback on the effectiveness of bioremediation strategies. This allows for adaptive management practices, where interventions can be adjusted based on genomic data to optimize outcomes.

16.8.2 Model Key Microorganisms Involved in Microplastic Degradation

16.8.2.1 Bacterial Species (e.g., *Ideonella sakaiensis*)

The issue of microplastic pollution has become a major environmental concern with adverse effects on marine and terrestrial ecosystems, human health, and biodiversity. Among the various approaches to mitigate this issue, the role of microbial degradation has gained attention, particularly with the discovery of the bacterial species *Ideonella sakaiensis*. This species, which was discovered in 2016 by a team of Japanese scientists, has demonstrated an amazing capacity to degrade polyethylene terephthalate (PET), a typical plastic found in a variety of goods like bottles, clothes, and packaging. The degradation process involves a two-step enzymatic reaction facilitated by two key enzymes: PETase and MHETase. PETase, the first enzyme is responsible for catalyzing the hydrolysis of PET into the intermediate molecule mono (2-hydroxyethyl) terephthalic acid (MHET). The bacteria can then proceed with their further metabolism of ethylene glycol and terephthalic acid, which are the monomers that are left over after MHETase breaks down MHET.

16.8.2.2 Fungal Species (e.g., *Aspergillus* sp.)

Microplastic pollution is an escalating environmental concern, with profound implications for ecosystems and human health. Among the innovative solutions to this problem, the role of fungal species, particularly Aspergillus sp., in microplastic degradation has garnered significant attention. Fungi, renowned for their robust enzymatic systems and versatile metabolic pathways, are proving to be potent agents in breaking down complex polymers. *Aspergillus* species, a diverse genus of fungi, have demonstrated considerable efficacy in this domain because they can generate a large variety of extracellular enzymes, such as hydrolases and oxidases, which can cleave the chemical bonds in synthetic polymers (Sánchez 2009). These enzymes degrade microplastics into smaller molecules, which can then be further assimilated and mineralized by microbial communities, thereby reducing the persistence of these pollutants in the environment. The biodegradation process generally consists of colonization of microplastic surfaces by fungal hyphae, followed by the secretion of enzymes that initiate the breakdown of the plastic material. This process not only mitigates the accumulation of microplastics but also facilitates the transformation of these pollutants into less toxic substances. Research indicates that *Aspergillus* species can degrade plastics of various types, including polyethylene, polystyrene, and polypropylene, which are among the most prevalent pollutants. The metabolic pathways of *Aspergillus* sp. involve initial adhesion to the plastic surface, followed by biofilm formation and the secretion of plastic-degrading enzymes. These enzymes, such as laccases, peroxidases, and esterases, act synergistically to break down the polymer chains into oligomers and monomers. Subsequent assimilation of these smaller molecules by the fungal cells completes the degradation process.

16.8.3 Genetic Engineering Tools and Application for the Enhancement of Microplastic Biodegradation

Molecular scissors, or designed nucleases, are used in gene editing to change DNA sequences. The three primary gene editing technologies are CRISPR-Cas, ZFN, and TALENs. These tools operate by creating double-strand breaks in the target gene sequence, which are then repaired through the nonhomologous end joining (NHEJ) or homology-directed repair (HRD) pathways. Zinc finger proteins (ZFPs) are artificial restriction enzymes that contain Folk1 cleavage domain, which creates double strand breaks (DSB) at the target site, and alpha-helices, which are 30 amino acid long and function as DNA binding domains. Together, these domains form zinc finger nucleases (ZFN). Target-specific gene editing is made possible by ZFPs' ability to identify 18 bp unique DNA sequences. Depending on the target site, four to six ZFPs can interact with the cleavage domain. The TAL proteins that are used by transcription activator-like effector nucleases (TALENs) are composed of two domains: one for recognizing certain sequences and another for cleaving sequences that cause double strand breaks (DSBs) in the target DNA. By introducing brief segments of invasive genomes (spacers) into the CRISPR locus, proteins known as clustered regularly interspaced short palindromic repeats (CRISPR) and their accompanying Cas function as an adaptive immune system for microorganisms. CRISPR systems come in three main varictics, with several variants identified. The type II system is the most well-characterized of these. It consists of the transactivating crRNA, the mature guide crRNA, and the Cas9 nuclease. Transactivating crRNA creates a CRISPR complex and uses the mature guide crRNA to interact with the target DNA. Cas9 endonuclease causes double strand breaks that are subsequently fixed by the machinery of the host cell. This causes either the insertion or deletion of genes, which disrupts the open reading frames of the genes.

16.8.4 Genetic Pathways for Microplastic Degradation

Microplastics, ubiquitous pollutants resulting from the breakdown of larger plastic debris, pose significant environmental and health risks. Addressing this, scientists have increasingly focused on genetic pathways for microplastic degradation. These pathways, primarily found in bacteria and fungi, involve enzymes such as hydrolases, oxidases, and depolymerases. Research has identified bacteria like *Ideonella sakaiensis*, which uses PETase and MHETase enzymes to degrade polyethylene terephthalate (PET) into its monomers. Similarly, fungi such as *Aspergillus niger* produce enzymes capable of breaking down polyurethane. The genetic mechanisms underlying these processes involve complex regulatory networks that activate these enzymes in response to the presence of microplastics. Advances in genomics and bioinformatics have allowed for the identification and cloning of genes responsible for plastic degradation, enabling their expression in more efficient microbial hosts. Synthetic biology further enhances these efforts by creating genetically engineered microorganisms with optimized pathways for faster and more complete degradation. The effectiveness and range of biodegradation can be increased by using these modified bacteria to specifically target particular kinds of plastics. However, challenges remain in scaling these solutions for environmental application, including ensuring the survival and activity of these organisms in diverse and sometimes harsh natural conditions.

16.8.5 Key Genes and Enzymes Involved in Plastic Degradation (e.g., PETase, MHETase)

Plastic pollution, particularly from polyethylene terephthalate (PET), has become a pressing environmental issue, prompting research into biodegradation pathways facilitated by specific genes and enzymes. Among the most significant enzymes are PETase and MHETase, which are produced by certain bacteria such as *Ideonella sakaiensis*. PETase initiates the degradation process by hydrolyzing PET into mono(2-hydroxyethyl) terephthalic acid (MHET), a simpler compound. The PETase gene, *IsPETase*, encodes this enzyme, which has shown a remarkable ability to break down PET at ambient temperatures, a property that distinguishes it from other hydrolases. Subsequently, MHETase, encoded by the *IsMHETase* gene, further degrades MHET into terephthalic acid (TPA) and ethylene glycol, compounds that are easier for microbial communities to assimilate and metabolize. These enzymes work synergistically to convert PET, a highly durable and recalcitrant polymer, into environmentally benign substances. Research has also identified other genes and enzymes in different microbial species that contribute to plastic degradation, such as esterase enzymes in *Thermobifida fusca* and cutinases in *Fusarium solani pisi*. Genetic engineering and protein engineering efforts have aimed to enhance the activity and stability of these enzymes under various environmental conditions, thereby improving their efficiency in plastic waste management applications. Moreover, understanding the regulatory networks and metabolic pathways associated with these genes offers insights into optimizing microbial consortia for large-scale bioremediation. The discovery and enhancement of these key genes and enzymes present a promising solution to mitigating the environmental impact of plastic waste, paving the way for more sustainable waste management practices and the development of eco-friendly plastics that are more amenable to biological degradation.

16.9 Horizontal Gene Transfer and Its Role in Spreading Biodegradation Capabilities

Horizontal gene transfer (HGT) is a crucial mechanism in the microbial world, facilitating the spread of genetic material across different species, thereby contributing significantly to the diversification and evolution of microorganisms. Unlike vertical gene transfer, which involves the transmission of genetic material from parent to offspring, HGT allows for the direct acquisition of new genes from the environment or from other organisms. This process is particularly pivotal in spreading biodegradation capabilities among microbial communities. Microorganisms that possess genes encoding for specific biodegradation pathways can transfer these genes to other microbes, enabling them to metabolize and break down various pollutants and complex organic compounds. This exchange of genetic material can occur through mechanisms such as transformation, transduction, and conjugation. For instance, bacteria exposed to pollutants like hydrocarbons or heavy metals can acquire resistance genes and metabolic pathways from neighboring bacteria that have already adapted to such conditions. This transfer not only enhances the recipient's survival prospects in contaminated environments but also accelerates the microbial community's overall capacity to degrade and detoxify harmful substances. Consequently, HGT plays a fundamental role in environmental bioremediation processes, where microbial consortia are employed to clean up contaminated sites. By facilitating the rapid dissemination of biodegradation genes, HGT contributes to the dynamic adaptability and resilience of microbial ecosystems. This capability is particularly beneficial in responding to anthropogenic environmental challenges, such as oil spills and industrial waste contamination.

16.10 System Biology Methods for the Microplastic Bioremediation

The systems biology method sheds light on how microbial systems react differently in various environmental settings. The interdisciplinary field of inquiry that combines multiomics and computational biology makes this possible. Microbial consortia are crucial for the effective biodegradation of xenobiotic substances through cooperative metabolic activities. As a result, in silico analysis can be used to forecast the drug's ultimate fate, which may involve partial or complete breakdown into a harmless component. The several databases that offer information on the metabolic degradation pathway, the genes and enzymes involved in chemical biodegradation, and chemical biodegradation. Molecular docking analysis can be performed using a variety of software programs, including Autodock, Schrödinger, Vina, YASARA, Glide, GEMDOCK, PatchDOCK web server, GOLD, and DARWIN. This procedure makes it possible to understand an enzyme's function in the biodegradation process and to show how an enzyme can cling to a pollutant. Bioinformatic analysis can be utilized to speculate about the mechanisms of degradation. Molecular dynamic simulations can help overcome the protein's static position in molecular docking studies. The stability of a ligand-receptor complex proposed by molecular docking studies can be ascertained by molecular dynamic simulations because they allow for conformational changes in both the ligand and the enzyme in complexes during biological activity. Abalone, Amsterdam Density Functional (ADF), and Desmond are the programs that can be used to explore molecular simulations. The multiomics approach allows the discovery of uncharacterized biosynthetic gene clusters in the genomes of sequenced organisms. It combines metagenomics, proteomics, transcriptomics, genomics, and metabolomics. Because metagenomics can break down oil, petroleum, polymers, and other hydrocarbons by directly analyzing the genomes of nonculturable microbes, it is gaining a lot of attention. Fang et al. investigated the biodegradation genes responsible for the breakdown of atrazine, HCH (hexachlorocyclohexane), and DDT (dichlorodiphenyltrichloroethane) using metagenomics. After performing genome annotation, the genes involved in biodegradation can be identified. Studies utilizing transcriptome analysis were carried out on the marine Bacillus species AIIW2, yielding valuable information about the hydrolytic enzymes implicated in the degrading process of PET. To verify the PET degradation process, many techniques were employed such as the weight loss assay, Fourier transform infrared spectroscopy (FTIR), scanning electron microscopy (SEM) analysis, and detection of the hydrolytic products.

16.11 Proteomics in Understanding Microplastic Bioremediation

Proteomics, the large-scale study of proteins, plays a crucial role in understanding microplastic bioremediation, offering insights into the complex biological processes that microorganisms utilize to degrade microplastics. Microplastics, pervasive environmental pollutants, pose significant ecological and health risks due to their persistence and potential to accumulate toxins. By employing proteomic techniques, researchers can identify and characterize the proteins expressed by microorganisms during the degradation of microplastics. This involves analyzing the proteome, or the entire set of proteins produced by these organisms, under various conditions to determine which proteins are involved in breaking down microplastics. These proteins can include enzymes that specifically target plastic polymers, facilitating their breakdown into smaller, less harmful molecules. Through proteomics, it is possible to map out the metabolic pathways activated in response to microplastic exposure, shedding light on the enzymatic mechanisms and regulatory networks that govern biodegradation.

16.11.1 Analysis of Protein Expression in Response to Microplastic Exposure

Microplastics are ubiquitous in the environment and have the potential to affect biological systems, making the investigation of protein expression in response to microplastic exposure an important field of study. Microplastics, are plastic particles less than 5 mm in size, can enter ecosystems and organisms through various pathways, leading to potential toxicological effects. Recent studies have utilized advanced proteomic techniques to investigate how exposure to microplastics affects protein expression profiles in different organisms, including marine species and terrestrial wildlife.

Researchers can recognize and measure changes in protein expression that arise in response to environmental stressors through the large-scale study of proteins known as proteomics. It has been demonstrated that exposure to microplastics causes oxidative stress, inflammatory reactions, and changes in metabolic pathways. These effects are mirrored in the altered expression of the proteins involved in these processes.

For instance, proteins related to stress response, detoxification, and immune function are often upregulated, while those associated with normal cellular functions may be downregulated. These protein expression changes can provide insights into the mechanisms of toxicity and the biological pathways affected by microplastic exposure.

16.11.2 Post-translational Modifications and Enzyme Activity

An essential function of post-translational modifications (PTMs) is to control the activity of enzymes, influencing a variety of cellular processes and signaling pathways. These modifications occur after protein synthesis, altering the enzyme's structure, stability, localization, and interaction with other molecules. Common types of PTMs include phosphorylation, glycosylation, ubiquitination, acetylation, and methylation. Phosphorylation, the process by which kinases add a phosphate group; may either activate or inhibit the activity of an enzyme. serving as a molecular switch in signal transduction pathways. Glycosylation, the attachment of carbohydrate moieties, can affect enzyme folding, stability, and cell surface recognition. Ubiquitination often marks enzymes for degradation via the proteasome, thus regulating protein levels and activity. Acetylation, particularly on lysine residues, can influence enzyme activity by altering charge interactions and protein conformation. Methylation, commonly associated with histones, affects chromatin structure and gene expression, indirectly influencing enzyme production and activity. These modifications are reversible, allowing dynamic regulation in response to cellular conditions and external stimuli. PTMs can also induce conformational changes in enzymes, exposing or obscuring active sites, thus modulating catalytic efficiency. The interplay between different PTMs adds a layer of complexity, as multiple modifications can occur on a single enzyme, leading to diverse regulatory outcomes. This crosstalk between PTMs ensures precise control over enzyme function, essential for maintaining cellular homeostasis and responding to environmental changes. Disruptions in PTMs are linked to several illnesses, such as cancer, neurological conditions, and metabolic syndromes, emphasizing their significance in health and disease.

16.11.3 Mechanisms of Plastic Degradation at the Molecular Level

Plastic degradation at molecular level involves several intricate mechanisms that break down complex polymer chains into smaller, less harmful compounds (Singh and Sharma 2008). The primary processes include photodegradation, thermal degradation, biodegradation, and chemical degradation. Photodegradation is driven by ultraviolet (UV) light, which provides the energy needed to break the links between the polymers in plastic, which leads to chain scission and the formation of free radicals. Thermal degradation occurs when heat induces molecular motion that leads to the breaking of covalent bonds, often resulting in the formation of smaller, volatile molecules. Biodegradation is facilitated by microbes that secrete enzymes that can break the polymer chains, such bacteria and fungus. These enzymes, like lipases, esterases, and proteases, act on specific bonds within the polymer structure, converting plastics into monomers and oligomers that microbes can further metabolize. Chemical degradation encompasses a variety of reactions, including hydrolysis, oxidation, and solvolysis. Hydrolysis involves the reaction of plastic polymers with water, leading to the cleavage of ester bonds in polyesters. Oxidation, often initiated by free radicals or metal catalysts, can cause significant molecular changes, including chain scission and cross-linking. Solvolysis, the reaction with solvents, can break down polymers by incorporating solvent molecules into the polymer backbone, thereby reducing molecular weight.

16.11.4 Enhancing Bioremediation Through Genomic and Proteomic Insights

Enhancing bioremediation through genomic and proteomic insights represents a cutting-edge approach to addressing environmental contamination. By leveraging the detailed genetic and protein expression profiles of microorganisms involved in bioremediation, scientists can gain a deeper understanding of the metabolic pathways and regulatory networks that drive the breakdown of pollutants. Genomic studies allow the identification of key genes responsible for the degradation of specific contaminants, facilitating the engineering of more efficient microbial strains. Proteomics complements this by revealing the actual proteins expressed under various environmental conditions, thus providing real-time data on the functional state of the bioremediation agents. This dual approach can lead to the development of optimized consortia of microorganisms with enhanced capabilities to detoxify environments contaminated with complex mixtures of pollutants.

16.11.5 Genetic Modification to Improve Microbial Degradation Capabilities

Genetic modification to improve microbial degradation capabilities holds significant promise in addressing environmental pollution and waste management challenges. By engineering microorganisms through techniques like CRISPR-Cas9, scientists can enhance their ability to break down complex pollutants, including plastics, petroleum by-products, and toxic chemicals, which are otherwise resistant to natural degradation processes. This biotechnological advancement involves manipulating microbial DNA to express enzymes that specifically target and decompose these contaminants more efficiently and at a faster rate. For instance, genetically modified bacteria have been developed to degrade polyethylene terephthalate (PET), a common plastic, into harmless by-products, showcasing a potential solution to plastic waste accumulation. Moreover, improving microbial degradation capabilities can also aid in bioremediation of contaminated soils and water bodies, reducing the ecological footprint of industrial activities (Chia et al. 2021). These engineered microbes can be tailored to thrive in various environmental conditions, ensuring their effectiveness across diverse habitats. However, the application of genetically modified organisms (GMOs) in environmental contexts raises concerns regarding ecological balance and potential unintended consequences.

16.12 Synthetic Biology Approaches to Develop New Bioremediation Agents

Synthetic biology offers innovative approaches to developing new bioremediation agents by engineering organisms to address environmental contaminants more efficiently and specifically. By manipulating genetic material, scientists can create microorganisms with enhanced abilities to degrade pollutants such as heavy metals, petroleum hydrocarbons, and plastics. These engineered organisms can be tailored to thrive in specific environments, ensuring optimal performance in diverse conditions. Techniques like CRISPR-Cas9 allow precise genetic modifications, enabling the insertion of metabolic pathways that can break down complex pollutants into harmless byproducts. Furthermore, synthetic biology facilitates the construction of biosensors that detect pollutants and activate bioremediation pathways only when necessary, reducing the risk of unintended environmental impacts. Advanced modeling and computational tools aid in designing these organisms, predicting their behavior, and ensuring their safety and efficacy in real-world applications. The integration of synthetic biology with traditional bioremediation strategies holds promise for creating robust, adaptable, and sustainable solutions to environmental pollution, addressing challenges that conventional methods cannot effectively tackle. As research progresses, these synthetic biology-derived bioremediation agents are expected to become vital tools in the global effort to restore and protect ecosystems from anthropogenic harm.

16.13 Protein Engineering to Enhance Enzyme Stability and Efficiency

Protein engineering has emerged as a pivotal tool in enhancing enzyme stability and efficiency, revolutionizing various industrial and biomedical applications. By leveraging techniques such as directed evolution and rational design, scientists can introduce specific mutations to enzymes, optimizing their performance under diverse conditions. These modifications often lead to increased thermal stability, resistance to harsh chemical environments, and improved catalytic activity. Directed evolution mimics natural selection, allowing the creation of enzyme variants with desirable traits through iterative rounds of mutation and selection. In contrast, rational design relies on detailed knowledge of enzyme structure and function, enabling targeted changes at the molecular level. Advances in computational modeling and high-throughput screening have significantly accelerated the protein engineering process, facilitating the identification of beneficial mutations with greater precision. Enhanced enzymes are crucial in the production of pharmaceuticals, biofuels, and food processing, where stability and efficiency directly impact cost-effectiveness and sustainability.

16.14 Challenges and Future Directions

Bioremediation of microplastics presents a promising yet challenging frontier in environmental science. The ubiquitous presence of microplastics in marine and terrestrial ecosystems poses significant threats to biodiversity and human health, necessitating innovative solutions. Bioremediation, leveraging microorganisms or plants to degrade pollutants, offers a sustainable approach to addressing this issue. However, the complex and resistant nature of microplastics makes biodegradation difficult. Microplastics' diverse chemical compositions, sizes, and environmental interactions hinder the effectiveness of microbial degradation. Additionally, the potential toxicity of plastic additives and the release of harmful byproducts during degradation pose risks to ecosystems and human health. The development of efficient bioremediation strategies requires a deep understanding of the microbial degradation pathways and the environmental factors influencing these processes. Research has identified several promising microbial strains capable of degrading microplastics, yet their efficacy in natural settings remains uncertain. Enhancing microbial activity through genetic engineering or optimizing environmental conditions could improve degradation rates, but these approaches raise ethical and ecological concerns. Moreover, the integration of bioremediation with other remediation technologies, such as chemical or physical methods, could enhance overall effectiveness but requires careful consideration of potential synergies and trade-offs. Future research should focus on identifying robust microbial consortia, understanding the mechanisms of microplastic degradation at a molecular level, and developing scalable applications. Collaboration across disciplines, including microbiology, materials science, and environmental engineering, is essential to address the multifaceted challenges. Additionally, public awareness and policy support are crucial to foster the implementation of bioremediation technologies. Addressing the challenges of microplastic bioremediation will not only mitigate pollution but also advance our understanding of microbial capabilities and ecosystem resilience, paving the way for a cleaner and more sustainable future.

16.15 Conclusion

The integration of genomics and proteomics in the bioremediation of microplastics heralds a transformative approach to addressing one of the most pervasive environmental pollutants. Genomics provides comprehensive insights into the genetic blueprints of microorganisms capable of degrading microplastics, enabling the identification and manipulation of genes responsible for the synthesis of enzymes that break down these pollutants. Through advanced genomic techniques, such as metagenomics and CRISPR-Cas9 gene editing, scientists can enhance the metabolic pathways of microbes, optimizing their plastic-degrading capabilities. Concurrently, proteomics complements this genetic understanding by elucidating the structure, function, and interactions of proteins involved in the biodegradation process. By analyzing the proteome, researchers can identify key enzymes and regulatory proteins that facilitate the breakdown of microplastics into less harmful byproducts, thereby improving the efficiency of bioremediation strategies. The synergy between genomics and proteomics allows for a more precise and holistic approach to engineering microbial communities that are not only effective at degrading a wide range of plastic polymers but also resilient in diverse environmental conditions. This interdisciplinary approach also enables the development of bioinformatics tools and computational models that predict microbial behavior and enzyme activity in various ecosystems, ensuring the scalability and sustainability of bioremediation efforts. Moreover, understanding the genetic and protein expression responses of these microorganisms to environmental stressors can inform the design of robust microbial consortia capable of thriving in polluted habitats. As microplastic pollution continues to threaten marine and terrestrial ecosystems, the application of genomics and proteomics in bioremediation represents a promising avenue for mitigating this crisis. By harnessing the natural degradative potential of microorganisms and enhancing it through biotechnological innovations, we can develop effective, eco-friendly solutions to restore and protect our planet's health. Ultimately, the integration of these 'omics' technologies in bioremediation not only offers a powerful tool for combating plastic pollution but also exemplifies the broader potential of microbial biotechnology in solving complex environmental challenges.

References

Amaral-Zettler LA, Zettler ER, Mincer TJ (2020) Ecology of the plastisphere. Nat Rev Microbiol 18(3):139–151. https://doi.org/10.1038/s41579-019-0308-0

Arisa-Tarazona MC, Villareal-Chiu JF, Hernández-López JM, De la Rosa JR, Barbieri V, Siligardi C et al (2020) Microplastic pollution reduction by a carbon and nitrogen-doped TiO2: effect of pH and temperature in the photocatalytic degradation process. J Hazard Mater 395(122632):122632. https://doi.org/10.1016/j.jhazmat.2020.122632

Arkatkar A, Juwarkar AA, Bhaduri S, Uppara PV, Doble M (2010) Growth of *Pseudomonas* and *Bacillus biofilms* on pretreated polypropylene surface. Int Biodeterior Biodegrad 64(6):530–536. https://doi.org/10.1016/j.ibiod.2010.06.002

Auta HS, Emenike CU, Fauziah SH (2017) Screening of Bacillus strains isolated from mangrove ecosystems in Peninsular Malaysia for microplastic degradation. Environ Pollut 231:1552–1559. https://doi.org/10.1016/j.envpol.2017.09.043

Bacha AUR, Nabi I, Zaheer M, Jin W, Yang L (2023) Biodegradation of macro- and micro-plastics in environment: a review on mechanism toxicity and future perspectives. Sci Total Environ 858:160108. https://doi.org/10.1016/j.scitotenv.2022.160108

Cacciari I, Quatrini P, Zirletta G, Mincione E, Vinciguerra V, Lupattelli P, Giovannozzi Sermanni G (1993) Isotactic polypropylene biodegradation by a microbial community: physicochemical characterization of metabolites produced. Appl Environ Microbiol 59(11):3695–3700. https://doi.org/10.1128/aem.59.11.3695-3700.1993

Cannif PM, Hoang TC (2018) Microplastic ingestion by Daphnia magna and its enhancement on algal growth. Sci Total Environ 633:500–507. https://doi.org/10.1016/j.scitotenv.2018.03.176

Capolupo M, Sørensen L, Jayasena KDR, Booth AM, Fabbri E (2020) Chemical composition and ecotoxicity of plastic and car tire rubber leachates to aquatic organisms. Water Res 169:115270. https://doi.org/10.1016/j.watres.2019.115270

Carr SA, Liu J, Tesoro AG (2016) Transport and fate of microplastic particles in wastewater treatment plants. Water Res 91:174–182. https://doi.org/10.1016/j.watres.2016.01.002

Caruso G (2019) Microplastics as vectors of contaminants. Mar Pollut Bull 146:921–924. https://doi.org/10.1016/j.marpolbul.2019.07.052

Chae Y, Kim D, An YJ (2019) Effects of micro-sized polyethylene spheres on the marine microalga Dunaliella salina: focusing on the algal cell to plastic particle size ratio. Aquat Toxicol 216:105296. https://doi.org/10.1016/j.aquatox.2019.105296

Chia WY, Tang DYY, Khoo KS, Lup ANK, Chew KW (2020) Nature's fight against plastic pollution: algae for plastic biodegradation and bioplastics production. Environ Sci Ecotechnol 4:100065. https://doi.org/10.1016/j.ese.2020.100065

Chia RW, Lee JY, Kim H, Jang J (2021) Microplastic pollution in soil and groundwater: a review. Environ Chem Lett 19(6):4211–4224. https://doi.org/10.1007/s10311-021-01297-6

Debroas D, Mone A, Ter Halle A (2017) Plastics in the North Atlantic garbage patch: a boat-microbe for hitchhikers and plastic degraders. Sci Total Environ 599–600:1222–1232. https://doi.org/10.1016/j.scitotenv.2017.05.059

De-la-Torre GE (2020) Microplastics: an emerging threat to food security and human health. J Food Sci Technol 57:1601–1608. https://doi.org/10.1007/s13197-019-04138-1

Dussud C, Meistertzheim AL, Conan P, Pujo-Pay M, George M, Fabre P, Coudane J, Higgs P, Elineau A, Pedrotti ML, Gorsky G, Ghiglione JF (2018) Evidence of niche partitioning among bacteria living on plastics, organic particles and surrounding seawaters. Environ Pollut 236:807–816. https://doi.org/10.1016/j.envpol.2017.12.027

Enders K, Lenz R, Stedmon CA, Nielsen TG (2015) Abundance, size and polymer composition of marine microplastics ≥10 μm in the Atlantic Ocean and their modelled vertical distribution. Mar Pollut Bull 100(1):70–81. https://doi.org/10.1016/j.marpolbul.2015.09.027

Esmaeili A, Pourbabaee AA, Alikhani HA, Shabani F, Esmaeili E (2013) Biodegradation of low-density polyethylene (LDPE) by mixed culture of *Lysinibacillus xylanilyticus* and *Aspergillus niger* in soil. PLoS One 8(9):e71720. https://doi.org/10.1371/journal.pone.0071720

Fendall LS, Sewell MA (2009) Contributing to marine pollution by washing your face: microplastics in facial cleansers. Mar Pollut Bull 58(8):1225–1228. https://doi.org/10.1016/j.marpolbul.2009.04.025

Fontanella S, Bonhomme S, Brusson J, Pitteri S, Samuel G, Pichon G et al (2013) Comparison of biodegradability of various polypropylene films containing pro-oxidant additives based on Mn, Mn/Fe or Co. Polym Degrad Stab 98:875–884. https://doi.org/10.1016/j.polymdegradstab.2013.01.002

Gerritse J, Leslie HA, de Tender CA, Devriese LI, Vethaak AD (2020) Fragmentation of plastic objects in a laboratory seawater microcosm. Sci Rep 10(1):1–16. https://doi.org/10.1038/s41598-020-67927-1

Gewert B, Plassman M, Sandblom O, MacLeod M (2018) Identification of chain scission products released to water by plastic exposed to ultraviolet light. Environ Sci Technol Lett 5:272–276. https://doi.org/10.1021/acs.estlett.8b00119

Gironi F, Piemonte V (2011) Bioplastics and petroleum-based plastics: strengths and weaknesses. Energy Sources Part A Recov Util Environ Eff 33(21):1949–1959. https://doi.org/10.1080/15567030903436830

Gregory MR (2009) Environmental implications of plastic debris in marine settings—entanglement, ingestion, smothering, hangers-on, hitch-hiking and alien invasions. Philos Trans R Soc Lond B Biol Sci 364(1526):2013–2025. https://doi.org/10.1098/rstb.2008.0265

Groh KJ, Backhaus T, Carney-Almroth B, Geueke B, Inostroza PA, Lennquist A, Leslie HA, Mafni M, Slunge D, Trasande L, Warhurst AM (2019) Overview of known plastic packaging associated chemicals and their hazards. Sci Total Environ 651:3253–3268. https://doi.org/10.1016/j.scitotenv.2018.10.015

Guo XP, Sun XL, Chen YR, Hou L, Liu M, Yang Y (2020) Antibiotic resistance genes in biofilms on plastic wastes in an estuarine environment. Sci Total Environ 745:140916. https://doi.org/10.1016/j.scitotenv.2020.140916

Hirooka T, Nagase H, Uchida K, Hiroshige Y, Ehara Y, Nishikawa JI, Nishihara T, Miyamoto K, Hirata Z (2005) Biodegradation of bisphenol A and disappearance of its estrogenic activity by the green alga *Chlorella fusca* var. *vacuolata*. Environ Toxicol Chem: Int J 24(8):1896–1901. https://doi.org/10.1897/04-259R.1

Janssen PH, Yates PS, Grinton BE, Taylor PM, Sait M (2002) Acidobacteria actinobacteria proteobacteria verrucomicrobia. Appl Environ Microbiol 68(5):2391–2396. https://doi.org/10.1128/AEM.68.5.2391-2396.2002

Ji MK, Kabra AN, Choi J, Hwang JH, Kim JR, Abou-Shanab RA, Oh YK, Jeon BH (2014) Biodegradation of bisphenol A by the freshwater microalgae *Chlamydomonas mexicana* and *Chlorella vulgaris*. Ecol Eng 73:260–269. https://doi.org/10.1016/j.ecoleng.2014.09.070

John J, Nandhini AR, Velayudhaperumal Chellam P, Sillanpää M (2021) Microplastics in mangroves and coral reef ecosystems: a review. Environ Chem Lett. https://doi.org/10.1007/s10311-021-01326-4

Kawai F (1995) Breakdown of plastics and polymers by microorganisms. Adv Biochem Eng Biotechnol 52:151–194. https://doi.org/10.1007/BFb0102319

Kay P, Hiscoe R, Moberley I, Bajic L, McKenna N (2018) Wastewater treatment plants as a source of microplastics in river catchments. Environ Sci Pollut Res 25(20):20264–20267. https://doi.org/10.1007/s11356-018-2070-7

Khoironi A, Anggoro S (2019) Evaluation of the interaction among microalgae *Spirulina* sp, plastics polyethylene terephthalate and polypropylene in freshwater environment. J Ecol Eng 20:161–173. https://doi.org/10.12911/22998993/108637

Kim JW, Park SB, Tran QG, Cho DH, Choi DY, Lee YJ, Kim HS (2020) Functional expression of polyethylene terephthalate degrading enzyme (PETase) in green microalgae. Microb Cell Fact 19(1):1–9. https://doi.org/10.1186/s12934-020-01355-8

Kowalczyk A, Chyc M, Ryszka P, Latowski D (2016) *Achromobacter xylosoxidans* as a new microorganism strain colonizing high density polyethylene as a key step to its biodegradation. Environ Sci Pollut Res 23(11):11349–11356. https://doi.org/10.1007/s11356-016-6563-y

Manzi HP, Abou-Shanab RA, Jeon BH, Wang J, Salama ES (2022) Algae: a frontline photosynthetic organism in the microplastic catastrophe. Trends Plant Sci S1360–1385(22):00157–00161. https://doi.org/10.1016/j.tplants.2022.06.005

Mehmood CT, Qazi IA, Hashmi I, Bhargava S, Deepa S (2016) Biodegradation of low-density polyethylene (LDPE) modified with dye sensitized titania and starch blend using *Stenotrophomonas pavanii*. Int Biodeterior Biodegrad 113:276–286. https://doi.org/10.1016/j.ibiod.2016.01.025

Miri S, Saini R, Davoodi SM, Pulicharla R, Brar SK, Magdouli S (2022) Biodegradation of microplastics: better late than never. Chemosphere 286:131670. https://doi.org/10.1016/j.chemosphere.2021.131670

Moog D, Schmitt J, Senger J, Zarzycki J, Rexer KH, Linne U, Erb T, Maier UG (2019) Using a marine microalga as a chassis for polyethylene terephthalate (PET) degradation. Microb Cell Fact 18(1):1–15. https://doi.org/10.1186/s12934-019-1220-z

Muthukrishnan T, Al Khaburi M, Abed RMM (2019) Fouling microbial communities on plastics compared with wood and steel: are they substrate- or location-specific? Microb Ecol 78:361–374. https://doi.org/10.1007/s00248-018-1303-0

Oluwasina O, Olaleyem F, Olusegun S, Oluwasina O, Mohallen N (2019) Influence of oxidized starch on physicomechanical, thermal properties, and atomic force micrographs of cassava starch bioplastic film. Int J Biol Macromol 135:282–293. https://doi.org/10.1016/j.ijbiomac.2019.05.150

Othman AR, Hasan HA, Muhamad MH, Ismail NI, Abdullah SRS (2021) Microbial degradation of microplastics by enzymatic processes: a review. Environ Chem Lett 19(4):3057–3073. https://doi.org/10.1007/s10311-021-01197-9

Padervand M, Lichtfouse E, Robert D, Wang C (2020) Removal of microplastics from the environment. A review. Environ Chem Lett 18(3):807–828. https://doi.org/10.1007/s10311-020-00983-1

Raddadi N, Fava F (2019) Biodegradation of oil-based plastics in the environment: existing knowledge and needs of research and innovation. Sci Total Environ 679:148–158. https://doi.org/10.1016/j.scitotenv.2019.04.419

Sánchez C (2009) Lignocellulosic residues: biodegradation and bioconversion by fungi. Biotechnol Adv 27(2):185–194. https://doi.org/10.1016/j.biotechadv.2008.11.001

Sanniyasi E, Gopal RK, Gunasekar DK, Raj PP (2021) Biodegradation of low-density polyethylene (LDPE) sheet by microalga, Uronema Africanum Borge. Sci Rep 11(1):1–33. https://doi.org/10.1038/s41598-021-96315-6

Sarmah P, Rout J (2018) Efficient biodegradation of low-density polyethylene by cyanobacteria isolated from submerged polyethylene surface in domestic sewage water. Environ Sci Pollut Res 25(33):33508–33520. https://doi.org/10.1007/s11356-018-3079-7

Sathicq MB, Sabatino R, Corno G, Di Cesare A (2021) Are microplastic particles a hotspot for the spread and the persistence of antibiotic resistance in aquatic systems? Environ Pollut 279:116896. https://doi.org/10.1016/j.envpol.2021.116896

Schwartz M, Perrot T, Aubert E, Dumarçay S, Favier F, Gérardin P, Gelhaye E (2018) Molecular recognition of wood polyphenols by phase II detoxification enzymes of the white rot Trametes versicolor. Sci Rep 8(1):1–11. https://doi.org/10.1038/s41598-018-26601-3

Shabbir S, Faheem M, Ali N, Kerr PG, Wang LF, Kuppusamy S, Li Y (2020) Periphytic biofilm: an innovative approach for biodegradation of microplastics. Sci Total Environ 717:137064. https://doi.org/10.1016/j.scitotenv.2020.137064

Shah MP (2024a) Microbial remediation of hazardous chemicals from water & industrial wastewater treatment plant. Springer Nature

Shah MP (2024b) Development in waste water treatment research and processes: applied technologies for clean up of environmental contaminants. Elsevier

Shah MP (2024c) Aerobic and anaerobic microbial treatment of industrial wastewater. CRC Press

Shah MP (2025) Extraction of precious metals from industrial waste using microbial technology. Springer Nature

Shen M, Zeng G, Zhang Y, Wen X, Song B, Tang W (2019) Can biotechnology strategies effectively manage environmental (micro) plastics? Sci Total Environ 697:134200. https://doi.org/10.1016/j.scitotenv.2019.1342001596-6

Shin J, Kim JE, Lee YW, Son H (2018) Fungal cytochrome P450s and the P450 complement (CYPome) of *Fusarium graminearum*. Toxins 10(3):112. https://doi.org/10.3390/toxins10030112

Singh B, Sharma N (2008) Mechanistic implications of plastic degradation. Polym Degrad Stab 93:561–584. https://doi.org/10.1016/j.polymdegradstab.2007.11.008

Singh RK, Ruj B, Sadhukhan A, Gupta P (2019) Thermal degradation of waste plastics under non-sweeping atmosphere: part 1: effect of temperature, product optimization, and degradation mechanism. J Environ Manag 239:395–406. https://doi.org/10.1016/j.jenvman.2019.03.067

Sjollema SB, Redondo-Hasselerharm P, Leslie HA, Kraak MH, Vethaak AD (2016) Do plastic particles affect microalgal photosynthesis and growth? Aquat Toxicol 170:259–261. https://doi.org/10.1016/j.aquatox.2015.12.002

Song YK, Hong SH, Jang M, Han GM, Jung SW, Shim WJ (2017) Combined effects of UV exposure duration and mechanical abrasion on microplastic fragmentation by polymer type. Environ Sci Technol 51:4368–4376. https://doi.org/10.1021/acs.est.6b06155

Song J, Jongmans-Hochschulz E, Mauder N, Imirzalioglu C, Wichels A, Gerdts G (2020) The travelling particles: investigating microplastics as possible transport vectors for multidrug resistant E. coli in the Weser estuary (Germany). Sci Total Environ 720:137603. https://doi.org/10.1016/j.scitotenv.2020.137603

Syranidou E, Karkanorachaki K, Amorotti F, Franchini M, Repouskou E, Kaliva M, Vamvakaki M, Kolvenbach B, Fava F, Corvini PFX, Kalogerakis N (2017) Biodegradation of weathered polystyrene films in seawater microcosms. Sci Rep 7(1):1–12. https://doi.org/10.1038/s41598-017-18366-y

Taniguchi I, Yoshida S, Hiraga K, Miyamoto K, Kimura Y, Oda K (2019) Biodegradation of PET: current status and application aspects. ACS Catal 9(5):4089–4105. https://doi.org/10.1021/acscatal.8b05171

Thiounn T, Smith RC (2020) Advances and approaches for chemical recycling of plastic waste. J Polym Sci 58(10):1347–1364. https://doi.org/10.1002/pol.20190261

Tiwari N, Santhiya D, Sharma JG (2020) Microbial remediation of micro-nano plastics: current knowledge and future trends. Environ Pollut 265:115044. https://doi.org/10.1016/j.envpol.2020.115044

Vethaak AD, Legler J (2021) Microplastics and human health. Science 371(6530):672–674. https://doi.org/10.1126/science.abe5041

Viršek MK, Lovšin MN, Koren Š, Kržan A, Peterlin M (2017) Microplastics as a vector for the transport of the bacterial fish pathogen species *Aeromonas salmonicida*. Mar Pollut Bull 125(1–2):301–309. https://doi.org/10.1016/j.marpolbul.2017.08.024

Wang B, Xu J, Gao J, Fu X, Han H, Li Z, Yao Q (2019a) Construction of an *Escherichia coli* strain to degrade phenol completely with two modified metabolic modules. J Hazard Mater 373:29–38. https://doi.org/10.1016/j.jhazmat.2019.03.055

Wang Z, Qin Y, Li W, Yang W, Meng Q, Yang J (2019b) Microplastic contamination in freshwater: first observation in lake ulansuhai, yellow river basin, China. Environ Chem Lett 17(4):1821–1830. https://doi.org/10.1007/s10311-019-00888-8

Yan N, Fan C, Chen Y, Hu Z (2016) The potential for microalgae as bioreactors to produce pharmaceuticals. Int J Mol Sci 17(6):962. https://doi.org/10.3390/ijms17060962

Zadjelovic V, Chhun A, Quareshy M, Silvano E, Hernandez-Fernaud JR, Aguilo-Ferretjans MM, Bosch R, Dorador C, Gibson MI, Christie-Oleza JA (2020) Beyond oil degradation: enzymatic potential of Alcanivorax to degrade natural and synthetic polyesters. Environ Microbiol 22:1356–1369. https://doi.org/10.1111/1462-2920.14947

Zettler ER, Mincer TJ, Amaral-Zettler LA (2013) Life in the "plastisphere": microbial communities on plastic marine debris. Environ Sci Technol 47(13):7137–7146. https://doi.org/10.1021/es401288x

Zhu B, Wang D, Wei N (2022) Enzyme discovery and engineering for sustainable plastic recycling. Trends Biotechnol 40(1):22–37. https://doi.org/10.1016/j.tibtech.2021.02.008

Zoungranan Y, Lynda E, Dobi-Brice KK, Tchirioua E, Bakary C, Yannick DD (2020) Influence of natural factors on the biodegradation of simple and composite bioplastics based on cassava starch and corn starch. J Environ Chem Eng 8:104396. https://doi.org/10.1016/j.jece.2020.10439

Correction to: Paradigm Shift of Food Industry Waste into High Valued Resources Through Valorisation as an Integrated Sustainable Approach

Manisha Mohapatra and Sanjana Singh

Correction to:
Chapter 12 in: S. Dey and M. P. Shah (eds.), *Microbial Applications in Industrial Wastewater Treatment*, Advances in Wastewater Research, https://doi.org/10.1007/978-981-96-9128-9_12

The original version of this chapter was revised: the chapter title has been corrected from "Resource Recovery and Valorization of Food Industry Waste" to "Paradigm Shift of Food Industry Waste into High Valued Resources Through Valorisation as an Integrated Sustainable Approach".

The updated version of this chapter can be found at
https://doi.org/10.1007/978-981-96-9128-9_12

© The Author(s), under exclusive license to Springer Nature Singapore Pte Ltd. 2026

S. Dey and M. P. Shah (eds.), *Microbial Applications in Industrial Wastewater Treatment*, Advances in Wastewater Research,
https://doi.org/10.1007/978-981-96-9128-9_17

Index

© The Editor(s) (if applicable) and The Author(s), under exclusive license to Springer Nature Singapore Pte Ltd. 2026

S. Dey and M. P. Shah (eds.), *Microbial Applications in Industrial Wastewater Treatment*, Advances in Wastewater Research,
https://doi.org/10.1007/978-981-96-9128-9

MIX
Papier aus verantwortungsvollen Quellen
Paper from responsible sources
FSC® C105338
FSC
www.fsc.org

If you have any concerns about our products,
you can contact us on
ProductSafety@springernature.com

In case Publisher is established outside the EU,
the EU authorized representative is:
Springer Nature Customer Service Center GmbH
Europaplatz 3, 69115 Heidelberg, Germany

Printed by Libri Plureos GmbH
in Hamburg, Germany